AF540803

First Published-2003

ISBN 81-7141-733-7

Published by
DISCOVERY PUBLISHING HOUSE
4831/24, Ansari Road, Prahlad Street,
Darya Ganj, New Delhi-110002 (India)
Phone: 23279245 • Fax: 91-11-23253475
E-mail:dphtemp@indiatimes.com

Printed at: **Tarun Offset Printers, Delhi-53**

Movement and Locomotion in Animals

By

Ashok Kumar
Homeopathic Pharmacopolea Laboratory (HPL),
CGO-I, Hapur,
Ghaziabad (U.P.)
India

DISCOVERY PUBLISHING HOUSE
NEW DELHI-110002

Movement and Locomotion in Animals

Preface

Locomotion is the mechanical act belonging to a new branch of science, the biomechanisms. It is one of the liveliest and most fascinating branch dealing with the animal locomotion: how birds fly and fishes swim, how slugs crawl, how legged animals (and people) run, and much else besides. The bewildering variety of styles of movement used by creatures ranging in size from amoebas to whales presents innumerable challenges to biomechanicists. The present title describes the many ingenious experiments and illuminating ideas that have given in our present understanding of animal movement, and it points to some of the problems that remain to be solved.

The title is not intended to be comprehensive nor could it be at length, but it concentrates as putting across the basic principles of the subject as briefly and lucidly as possible. It does this with the aid of careful selected examples–some recent and other classic of the field and with numerous illustrations. The aim is to enthuse the reader with this active and exciting area of research and to lay a solid foundation on which further study of its various facets may be based.

It has been constant endeavour of the author to furnish maximum substance, keeping in view the limitations of the size of the book. Efforts have been made to condense the matter as far as practicable. It is hoped that this book will not only meet the requirement of students but will also be useful as a guidelines to the teachers in their teaching.

There can be no claim of originality except in the manner of treatment and presentation and much of the information has been gathered from the books and scientific journals available in different libraries.

The author expresses his thanks to his friends and colleagues whose constant inspiration have initiated him in bringing out this book.

Special thanks are given to Mr. Vasan and staff of M/s Discovery publishers for their whole hearted co-operation in the publication of this book.

Author

CONTENTS

CHAPTER 1

THE MOTOR FOR ANIMAL MOVEMENT

Some animal movements are intensely beautiful: think, for example, of the soaring of eagles and the running of gazelles. Other are very puzzling. How does an insect walk upside down on the eeiling or the underside of a leaf? How do amoebas crawl? Yet other movements may seem easy to explain if you know only a little but become profoundly puzzling when you learn more.

It the present title all forms of animals locomotion: walking, running, and jumping; crawling and climbing; swimming and flight have been dealt. In addition to animal movement, it discusses human movement and the movements of single–celled organisms like amoebas. It explains (as far as our current knowledge allows) how animals move and how their bodily structure and their patterns of movement are adapted to save energy and maximize performance.

One aspect of this book may surprise you. You will expect to find biological concepts in a book on animals movement, but you may be less prepared for ideas from engineering and physical science. Yet you will find physical ideas cropping up on almost every page, for a very good reason. The primary goal of the present title is to make you to understand how animals move and how evolution has adjusted their structure and movements to maximize performance. If you wanted to understand how a motor vehicle moved and how it was designed for high performance, you would apply your knowledge of engineering and physical science. The same physical laws apply to the movement of living things as to the working of machinery, so we need physical ideas to understand the movements of animals, and many other biological problems. Engineering, physics, and biology are far less separate from each other in modern science than they often seem to be in the school curriculum.

A good way to start, if we were studying motor vehicles, would be to learn about the engines that drive them. Similarly, in this book about animal movement, it seems sensible to start with the muscles that power the movements of most animals. We will soon find ourselves using such physical concepts as force, work, and energy.

Exerting Force with Muscles

There is more of muscular tissue than of any other in the bodies of many animals: for example, muscle accounts for 63 percent of the body weight of trout and 46 percent of the body weight of the Uganda kob antelope. In human being 40% of the body weight is only due to muscles. Muscle is responsible for driving the body's movements. It powers the galloping of horses, the flight of bees, and even the swimming of jellyfishes. Only in very small animals such as rotifers and other tiny plankton are body movements driven principally by motors other than muscle.

To power movement, muscles must exert forces: forces to support the animal's weight, forces to accelerate or decelerate the animal, and forces to overcome the resistance of the air or water through which the animal moves. These forces are transmitted to the environment through structures such as limbs and fins. The muscles of our legs, for example, exert forces on the skeleton, making our feet push on the ground, and the muscles of a fish's tail exert forces that make the tail fin push on the water.

Much of what we know about how muscles work and about the forces they can exert comes from experiments on muscles or parts of muscles from the legs of small animals such as frogs and rats. If these are removed from the animal's body very soon after death and placed in solutions that resemble its body fluids, they can be kept alive for many hours. During this time they can be made to contract by electrical stimulation, and their properties can be investigated.

These experiments have confirmed that stout muscles are stronger than slender ones: the force that a muscle can exert depends on its cross-sectional area. To take account of this, physiologists often divide the force by the cross-sectional area. The result of this simple calculation is the stress that a muscle can exert. Experiments on muscles from frogs, rats, and other vertebrates show that however stout or slender these muscles are, all can exert up to about 0.3 newton (that is, 30 grams force) per square millimetre of muscle section—44 pounds force per square inch. This is not a very large stress (it is approximately the same as the stress in a rubber band stretched by 30 percent of its original length), but muscles with large cross-sectional areas can exert very large forces. For example, the cross-sectional areas of muscles in the legs of living people can be measured by computer-aided tomography, an X-ray technique often known as a CAT scan that produces pictures representing sections through the body. Measurements of the hamstring muscles at the back of the thighs of healthy young men show that they have a total cross-sectional area of about 15,000 square millimetres, enough to exert 4500 newtons, or 1000 pounds force. More slender muscles exert proportionately smaller forces, but even these can be quite substantial; for example, the small muscle that bends the human thumb can exert a force of 130 newtons (30 pounds force).

Like other living tissues, muscle consists of cells, but its cells have an unusual shape: because they are very long and slender, they are called *muscle fibres*. A cat muscle fibre only 40 micrometers (0.4 millimeter) in. diameter may be as much as 40 millimetres (1.6 inches) long. The relatively larger size of these fibers has consequences for their internal structure. Whereas most other cells have just one nucleus each, the genes in a single nucleus could not exert effective control over the metabolism of a cell with the length of a muscle fiber. This problem is solved by spacing out several nuclei along the length of the fiber. Though muscle fibers are much longer than most other cells, they are seldom if ever more than a few centimeters long. To create the long muscles we see pictured in anatomical drawings, fibers are arranged in bundles called fascicles, which may be much longer than the individual fibers. Similarly, woollen threads are very much longer than the fibers from which they are spun. The muscles at the back of an adult human's thigh have fascicles that can be as long as 250 millimeters, but it seems unlikely that any individual fiber is more than a small fraction of that length.

The fascicles in a muscle are able to develop tension and shorten, pulling on the bones and so moving the joints. Muscles can pull but they cannot push, so at least two muscles are needed to work a joint. For example, we bend our knees by shortening the hamstring muscles at the back of the thigh and straighten them by shortening the quadriceps muscles in front of the thigh. Shortening either of these groups of muscles stretches the other.

Muscles obviously cannot pull on a joint unless they are somehow connected to it. Some muscle fascicles connect directly to bone at their ends, but in many cases the attachment is made through tendons. In mammalian muscle (for example, beef) the muscle fascicles are red or pink and the tendons are white. The muscle fibers are the motors that power the body, and the tendons are simply the ropelike connections between the fibers and the bones.

Muscle fascicles run straight along almost the whole length of the hamstring muscles at the back of the thigh; only a little tendon is attached at each end. In contrast, the fascicles in some other muscles run obliquely between two or more long tendons. This "pennate" arrangement increases the force that the muscle can exert by packing in a larger number of fascicles. For example, the pennate muscles of the human calf (the *gastrocnemius* and *soleus* muscles), which connect by the Achilles tendon to the heel, can probably exert more than twice the force of the hamstrings even though they have only 0.9 times the volume. The inevitable penalty for packing in the extra fascicles is that they have to be short and so cannot lengthen and shorten very much. This restricts the range of angles through which the muscles can move their joints. As a rough general rule, muscle fascicles work well over a range of lengths in which the minimum is 70 percent of the maximum. A hamstring muscle fascicle, for example, might extend to a maximum length of 140 millimeters and contract to a minimum of 100, a range of 40 millimeters. A much shorter fascicle from the soleus muscle might extend only to 35 millimeters and contract to 25, a range of 10 millimeters. These ranges are not the maximum possible, but at longer and shorter lengths the forces that the muscles could exert would be severely reduced. The penalty for having a strong soleus muscle is that its short fascicles are effective only over a small range of length, allowing only a small range of ankle angles. A change in length of 10 millimeters in the soleus is only enough to move the ankle through an angle of about 12 degrees.

To understand how muscles shorten and why they can shorten only over a limited range of lengths, we need to look at the fine detail of their structure.

Muscle Structure and Arrangement

Vertebrates have three major types of muscle: cardiac muscle, of the heart, smooth muscle, of the viscera, and striated muscle, of the skeletal system.

Cardiac Muscle

Cardiac muscle cells are unusual in that they are all laterally fused to, or anastomosed with, each other. Such fusion is thought to provide more efficient conduction of the action potentials generated by the heart's pacemaker than if there were no fusion of cells. Also unique to the heart muscle, or *myocardium,* are numerous *intercalated discs* of tough connective tissue that join the muscle cells to each other at their ends. These discs give great tensile strength to the entire myocardium.

Smooth Muscle

The cells of smooth muscle, also known as involuntary or nonstriated muscle,are spindle-shaped, being tapered at both ends. Each fiber,as a muscle cell is also called, has a single, centrally located nucleus and faint longitudinal striations called *myofibrils,* the contractile units of the muscle cell. Smooth muscle is the major muscle type in the internal organs of the vertebrate body; it makes up the muscular component of artery walls, gland-ducts, and digestive tract and is controlled by the autonomic nervous system. Smooth muscle constitutes but a small fraction of the vertebrate body's musculature. In many invertebrates, such as the common earthworm, smooth muscle is the only muscle tissue in their bodies.

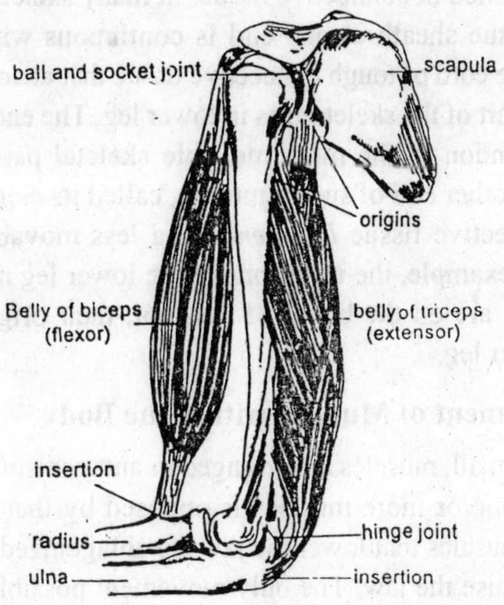

Fig 1.1: ***Muscles and bones of the upper arm, showing origin, insection and belly of a muscle, and the antagonistic arrangement of the biceps and triceps.***

Striated Muscle

By far the greatest mass of the vertebrate body is composed of striated muscle. Striated muscle is so named because its fibers are prominently striped with alternate dark and light bands. These bands are readily visible under the light microscope, even without staining. Like smoth muscle, striated mucle is made up of many individual muscle fibers. Striated muscle is under conscious control and therefore is also known as skeletal or voluntary muscle.

Striated muscle fibers are multinucleate, with the nuclei lying along the cell periphery, just inside the plasma membrane, which is known as the *sarcolemma.* Mitochondria in striated muscle cells are large and numerous, as is appropriate for cells that expend great amounts of energy. Each muscle fiber contains many myofibrils. Striated muscle cells are found throughout the animal kingdom, from sponges to insects to mammals. Even some protozoans have striated contractile fibrils. Striated muscles are most common, however, in arthropods and vertebrates.

Groups of striated muscle fibers are organized into muscle bundles, or *fascicull.* A muscle is composed of several to many faciculi. The bundles,as well as the entire muscle are enclosed in a sheath of tough connective tissue called *perimysium* or *fascia.* Individual muscle fibers are also ensheathed in connective tissue. In many skeletal muscles, the connective tissue sheath at one end is continuous with a *tendon,* a glistening white cord of tough connective tissue that attaches the muscle to a movable part of the skeleton, as in lower leg. The end of the muscle attached by tendon to the more movable skeletal part is termed its *insertion*. The other end of such a muscle, called its *origin*, is attached by short connective tissue *ligaments* to a less movable part of the skeketon. For example, the insertions of the lower leg muscles, which move the foot, are on the bones of the foot; their origins are on the upper and lower leg.

The Arrangement of Muscles within the Body

Most, but not all, muscles are arranged in antagonistic sets, in which the action of one or more muscles is opposed by that of others. For example, the muscles that lower the jaw are antagonized in their action by those that raise the jaw. The only movement possible for a muscle is *contraction*, that is, shortening. For a structure to be moved in any direction, it must be pulled in that direction by one or more muscles. Thus, all moving body parts have at least two sets of muscles. One of the clearest examples of the antagonism of two striated muscles is provided by the actions of the *biceps femoris* and the *triceps femoris* of the upper arm. The circular and longitudinal smooth muscles of the intestine and other tubular organs are also arranged in antagonistic sets. Contraction of the longitudinal fibers of the intestine shortens and thickens it at the same time, stretching the circular muscles. When the circulars muscles contract, the intestinal diameter is reduced, forcing it to lengthen—thereby stretching the longitudinal muscles.

Physiology of Striated Muscle

Muslcle function has long been investigated at two levels, one being the classical physiology and mechanics of entire muscles, the other being the biochemistry and microscopic structure of the muscle cell. Classical muscle physiology studies the bases for muscular fatigue, physical factors in contraction, and the varying responses of muslces to stimuli of different strengths and frequencies. Although the principles discussed in this and the following section on mictoscopy and biochemistry of muscle contraction refer to striated muscle, most of them apply to cardiac and smooth muscle as well.

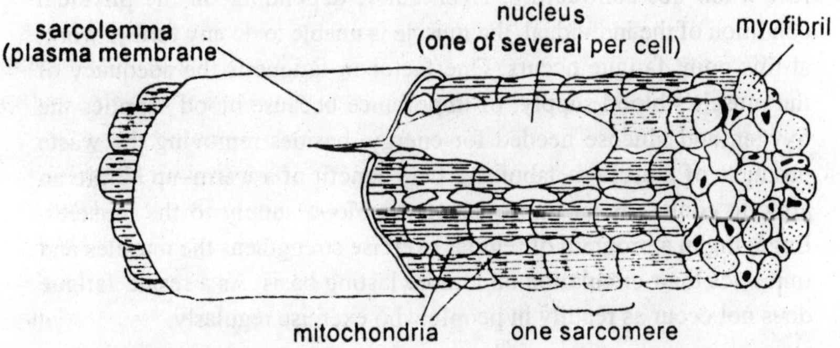

Fig. 1.2 : Sketch of a portion of a single strlated muscle cell, or fiber showing details visible in electron microscopy.

The Motor Unit

When a particular muscle, such as the calf muscle of the leg, contracts, seldom do all of its fibers respond at the same time. Instead, the fibers work in groups, and, as fatigue occurs in one group, another group takes over; the various groups usually take turns in the execution of a given task. It was explained that each motor neuron, with its cell body in the spinal cord, sends out an axon to a skeletal muscle. At the muscle, the axon gives off many b: anches, each branch innervating a separate muscle fiber through its *motor end plate.*

The muscle fibers served by the branches of a single motor neuron are all stimulated whenever that neuron generates a train of impulses. When an action potential reaches a branch point of an axon—always at a node of Ranvier—impulses normally go down both of the branches. As a result, all the muscle fibers served by the branches of one axon are stimulated and contract together. A motor neuron, its branches, and all the muscle fibers they serve from a *motor unit.* For example, a motor unit of a leg muscle may contain as many as 150 muscle fibers. The muscles that control such delicate actions as the movements of the eye, on the other hand, have an average ratio of only three muscle fibers to each motor neuron serving them, providing a much finer degree of control.

Fatigue

When a muscle performs a task for a long time, such as lifting and holding a heavy weight, a large number of motor units is involved. As time goes on, motor units continually take turns, gaining a period of rest when not contracting. Eventually, depending on the physical condition of the individual, the muscle is unable to do any further work; at this point fatigue occurs. One factor in fatigue is the adequacy of the muscle's blood supply, of importance because blood supplies the oxygen and glucose needed for energy, besides removing the waste products of muscle metabolism. One benefit of a warm–up before an athletic contest is an improvement in blood supply to the muscles. Engaging in a program of regular exercise strengthens the muscles and improves their circulation on a more lasting basis. As a result, fatigue does not occur as readily in people who exercise regularly.

The physiological basis of fatigue at the level of the motor unit is of two types. In one type, continual direct stimulation of a nerve attached to a particular muscle results in the depletion of the acetylcholine–containing synaptic vesicles at its motor end plates. The muscle then fails to respond until more acetylcholine is synthesized. That the muscle itself is not fatigued and is still capable of contracting can be shown by stimulating it directly. It contracts, showing that in such a case, fatigue has occurred at the synapse. Another cause of fatigue occurs in an excised muscle deprived of a blood supply. Continued stimulation will deplete the glycogen it contains and will result in an ultimate failure to respond to any stimulation, direct or not. Although both these types of fatigue are readily demonstrated in the laboratory, they are not thought to occur under normal circumstances.

Elastic Components and the Smoothness of Muscular Action

Although isolated muscle fibers contract in a rapid, jerky fashion when stimulated, intact muscles in the body contract very smoothly. One major reason for this is the *elasticity* of muscles, that is, their ability to be stretched. When a given muscle contracts, neighbouring muscles that are not contracting, especially ones that are antagonistic to the contracting muscle, resist and slow its contraction. Another important contribution to muscle elasticity is made by the connective tissues associated with a muscle. Their natural elesticity also slows and thereby imparts smoothness of muscular contraction. The combined elasticity of muscles and associated connective tissues causes the tension, or stretching force, of contraction to be shared by these other tissues. Such internal friction in other tissues slows not only the rate of contraction but of relaxation as well. Although additional factors are involved, the sharing of tension by other tissues is a major reason for the fact that muscles remain contracted for a time after the stimulus that caused their contraction is discontinued. Even an isolated muscle fiber undergoing a single contraction lasting but a few milliseconds may take a full second—1,000 times as long —to relax again. Without the elasticity of muscle and connective tissue, our movements would be quite jerky. This tissue elasticity also keeps contracting muscles from literally tearing themselves apart.

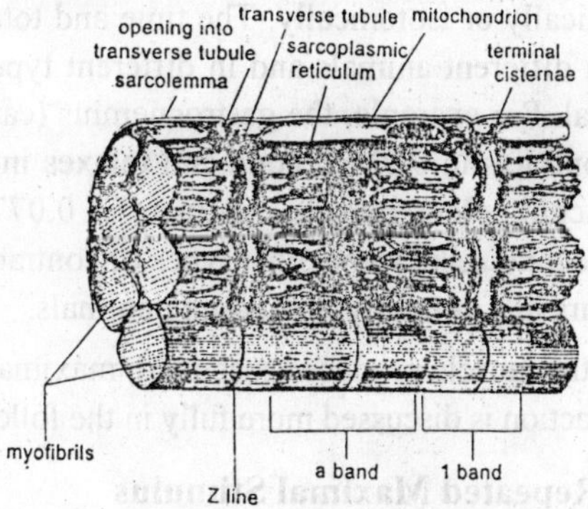

Fig. 1.3 : Control of filament sliding in muscle fibers. A drawing showing the relationships of myofibrils, sarcoplasmic reticulum, and transverse tubules (T tubules) in a muscle fiber.

Types of Muscular Contraction

The contraction of muscles can be classified in various ways. Descriptions of some types of muscular contraction follow.

1. In *isometric contraction,* the muscle is not permitted to shorten. Examples include the muscular activity involved in maintaining posture or that involved in attempting to pick up an object that is bolted to the floor. In such cases, the muscle actually does shorten somewhat but is not perceived to do so because of the elasticity of the attached tendons and other noncontractile connective tissue.

2. In *isotonic contraction,* the degree of tension within the muscle remains the same while the muscle contracts. Theoretically, a steady pull of the arm muscles during a slow swimming stroke could constitute an isotonic contraction. In reality, during ordinary movements of the body's muscles, the load of tension on a muscle continually changes. True isotonic contractions therefore rarely occur normally. Because such contractions can be induced experimentally, they have been a favorite subject of study by physiologists.

3. In *tonic contraction,* a muscle engages in sustained, partial contraction. This is actually a type of isometric contraction maintained over a long period of time. The term is applied to the normal but slight amount of contraction or tension exhibited by all healthy muscles, also referred to as muscles tone. An individual exhibits less muscles tone when deeply anesthetized than when awake.

4. In a *twitch,* a skeletal muscle responds to a single volley of nerve impulses by briefly contracting all of its fibers. A twitch can be produced isometrically or isotonically. The time and total duration of a twitch varies in different animals and in different types of muscles in the same animal. For example, the gastrocnemius (calf) muscle, of a domestic cat contracts in 0.039 second and relaxes in 0.04 second, whereas the adjacent soleus muscle contracts in 0.077 second and relaxes in 0.079 second. Like some other types of contraction, a single twitch rarely occurs in a normally functioning animals.

5. In *tetanus,* the muscle produces sustained, maximal contraction. This type of contraction is discussed more fully in the following section.

The Effects of Repeated Maximal Stimulus

A maximal stimulus is one that achieves the fullest possible contraction; even if the strength of the maximal stimulus is doubled, no greater shortening of the muscle occurs. In a twitch, the fibers of a muscle contract following a single maximal stumulus. However, if a second maximal stimulus is applied before the first contraction cycle is completed, the second contraction is added to the residue of the first and a greater shortening of the muscle results. The two contractions are said to have summed, and the process is referred to as *summation.* The additional shortening results from the recoil of the noncontracting but elastic connective tissue components of the muscle. Even as the muscle fibers begin to relax, these elastic elements contribute to the twitch by "pulling back," much as a stretched rubber band resists being stretched. When a second stimulus is applied while these elastic components are still recoiling, the second contraction is added to this shortening.

If a third maximal stimulus is applied before the end of the second cycle, it is added to the first two, and an even greater shortening of the muscle results. Because the effect, as recorded by a measuring device, appears as a tracing resembling a series of steps on a staircase, it was referred to by early muscle physiologists as treppe (German for "stairs"), or the staircase effect.

What happens if a long series of maximal stimuli is applied to a muscle at short intervals? At first, the muscle continues to shorten with each new stimulus. However, the muscle shortens less and less with each new stimulus until it remains contracted at a constant length, even if stimuli are continued at the same rate. The muscle is then in complete tetanus, or fusion. For most muscles, the tension generated by a tentanic contraction is about four times that of a twitch. In normal functioning, most muscular contractions reach incomplete tetanus.

The Structure and Biochemistry of Striated Muscle Contraction

Muscle fiber structure and function have been intensively investigated by light and electron microscopy and at the molecular level. Our present understanding of muscular contraction reflects the combined results of observations made at all three of these levels of investigation.

Microscopic Appearance of Striated Muscle

Birefringent, or doubly refractive, materials split a light beam into two beams as they refract, or bend, it. In contrast to this reaction, a beam of light entering an isotropic substance such as glass, water, or air at an oblique angle from another medium is refracted to a new direction but still appears as a single beam. When examined by light microscopy in polarized light—light vibrating in only one plane—the dark cross striations of skeletal muscle fibers are found to be

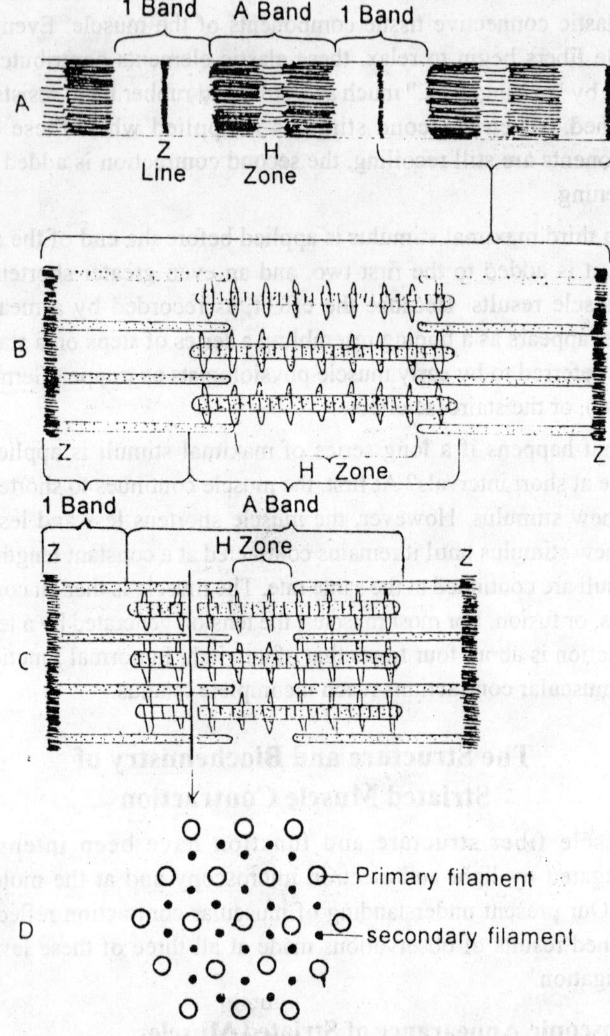

Fig. 1.4 : ***Diagram illustrating the sliding filament hypothesis of the mechanism of muscle contraction.***

anisotropic, or birefringent, while the lighter bands are isotropic. This observation led to the designation of the dark striations as *A bands* and the lighter striations as *I bands.* Close examination under the light microscope or by electron microscopy reveals that the band appearance of each fiber is due to the A and I bandings of the many fine, longitudinal myofibrils of which the fiber is principally composed. Because the myofibrils are closely aligned with each other, the entire fiber appears striped. Each *sarcomere,* or unit of a myofibril, extends from the dark Z *line* in the middle of one I band to the Z line in the middle of the next I band. The lighter band often visible under higher magnification in the middle of the A band is the *H zone.* The dark line at its center is the *M line.*

Viewed by electron microscopy, the endoplasmic reticulum of a striated muscle cell, called its *sarcoplasmic reticulum* (SR), is seen to be an extensive system of smooth tubules, closely investing groups of myofibrils. Cross connections called *terminal cisternae* overlie most of the I band along the myofibrils.

As can be seen in Figure, a slender transverse *T tubule* runs between each pair of terminal cisternae at each Z line. The membranes of the T tubules are continuous with the sarcolemma, and their interior is open to the tissue fluid surrounding the fiber. These deep, tubular incursions of the plasma membrane into the muscle cell deep, tubular incursions of the plasma membrane into the muscle cell provide efficient communication between the plasma membrane and the underlying sarcomeres. The exact location of terminal cisternae and the number of T tubules per sarcomere varies in different species of vertebrates.

Early Biochemical Studies

In the early 1930s, it was found that a moderately large protein molecule, *myosin* (molecular weight 500,000), could be extracted from muscle. Myosin accounts for about half of a muscle's protein content. Another fraction, containing the enzymes of glycolysis, was first removed by washing minced muscle in distilled water. The myosin fraction was then removed by prolonged washing in cold 0.6 molar potassium chloride solution. In time, it was found that myosin has enzymatic activity: it can hydrolyze ATP to ADP and inorganic phosphate, that is, it functions as an ATPase.

Some years later, another, smaller protein (molecular weight 42,000) was extracted and given the name *actin.* In the presence of ATP the small, globular molecules of actin, or G actin, could be polymerized to form a long fibrous protein, F actin. Under the right conditions, myosin could also be made to form a long filament in the presence of actin. This substances was called *actomyosin.*

Actomyosin fibers—prepared by squirting the substance through a fine pipet—contracted when placed in a solution of Ca^{2+} and ATP. This landmark achievement of the 1940s excited the scientific world. For the first time, the principle biochemical components of the contractile mechanism were clearly identified; they had been separately removed from the muscle cell, reassembled, and then made to perform their essential role of contraction.

In the years since the identification of actin and myosin as the principal molecules of vertebrate muscle, contractle elements composed of actin and myosin molecules have been discovered not only in all animals but in many protists and some plants as well. Actin and myosin appear to have evolved very early in the history of life on earth.

Changes in Sarcomere Structure During Stretching and Contraction

Several important observations were made possible when comparisons were made of muscles prepared for examination by light and electron microscopy while (1) being stretched (2) undergoing contraction, and (3) relaxed. Both light and electron microscopy of these preparations showed that A bands remained the same length under all three of these conditions. I bands, on the other hand, lengthened when the fiber was stretched and shortened during its contraction. Similar findings had been reported since the late 1800s but were not generally accepted until the 1950s.

The Fine Structure and Chemistry of Myofibrils: Two Type of Filaments

Myofibrils that were sectioned transversely and viewed by electron microscopy were found to be composed of filaments of two thicknesses. Extracting myosin from the preparations resulted in the disappearance of the thicker of the two filaments; removed of actin caused the disappearance of most of the thin filament structure.

In many cross sections of myofibrils, each thick filament (myosin) was surrounded by several fine filaments (actin). Exceptions were cross sections that passed through the center of the A band or through most regions of the I band. Sections through the center region of the A band usually revealed only the thick myosin filaments. Sections through the I band contained only in the filaments of actin.

The Constant Length of the A Band

Whether relaxed or contracted, the A band remains the same length. When a myofibril in a relaxed (resting) condition is viewed by electron microscopy, the myosin and actin filaments are arranged. In the darkest regions of the A band, filaments of myosin and actin overlap. The light (H zone) area within A band in rest–length myofibrils lacks actin filaments. The dark (M) line at the center of the H zone is due to cross connections between myosin filaments.

The Changing Length of the I Band

Only actin filaments are found in the I band regions. If a fibril contracts, the I band shortens—as does the H zone—because of the further insertion of the actin filaments among the myosin filaments. Note that the Z lines,denoting the ends of the sarcomere, are much farther apart at rest than when the myofibril is contracted.

In the fully contracted state—usually achieved only if the myofibril has been freed of its connective tissue—the H band completely disappears. In full contraction, the actin filaments meet and may even overlap at the center of the sarcomere. The I band almost completely disappears as well, and the Z line approach the edge of the A band.

The Sliding Filament Hypothesis

From these observations of the changes in sarcomers structure during stretching and contraction, largely the work of British biologists Andrew F. and Hugh E. Huxely, it become clear that the contraction of muscles is not due to a shortening of any of its components molecules—a view once widely held. Instead, myosin and actin filaments shorten the myofibril by actively sliding past each other.The myosin filaments can be regarded as retaining their position while the actin filaments move in reference to them. These discoveries changed existing theories of contractility and provided the molecular basis for what is known as the sliding filament theory or muscle contraction.

How can chemical energy be transduced into the mechaincal energy needed to move the filaments past each other? What forces could cause such movenment? What connections, if any, do the parallel thick and thin filaments have with each other? Some of these questions will be considered next.

Cross Bridges Between Myosin and Actin

High–resolution electron micrographs of myofibrils show the existence of fine cross bridges between the myosin and actin filaments. The bridges project at right angles from the surface of the myosin filaments at regular intervals except in a sizable central region in the

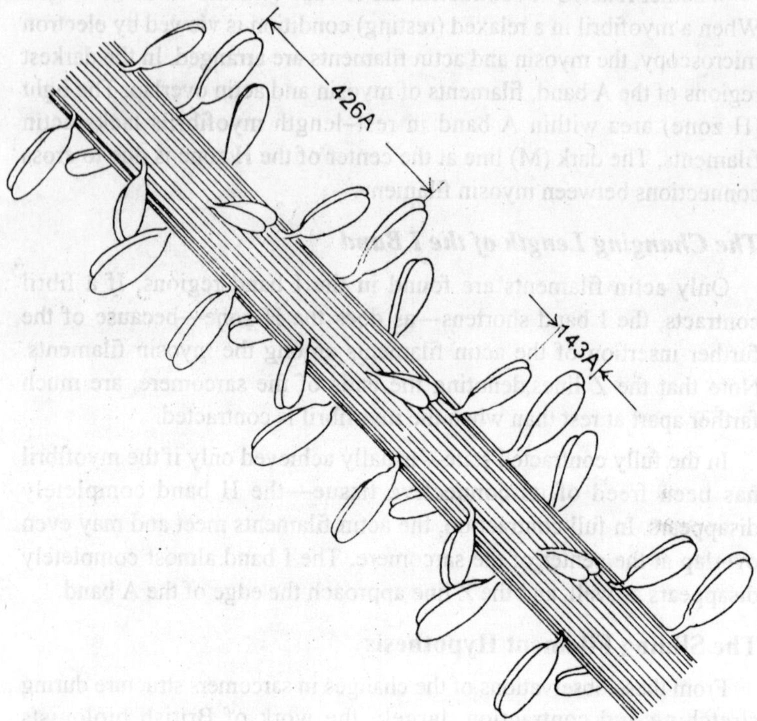

Fig. 1.5 : ***Diagram of cross bridges on a thick (myosin) filament of a striated myofibril.***

H zone from which they are absent. They do not project in only one plane but form a helical (coiled) arrangement, with six bridges for each full turn of the helix. Each of the six thin (actin) filaments surrounding any one thick (myosin) filament is touched by that thick filament as well as by the thick filament on its other sides. Each of the six cross bridges in one full turn of the helix is bounded to the nearest of the six actin filaments surrounding it.

The Proteins of Muscle

The myosin molecule as seen in an electron micrograph is rodlike, with a globular head at one end. Each complex, made up of a globular head and a small part of the shaft, termed *heavy meromyosin,* has ATPase activity. Most of the rodlike part of the myosin molecule is called *light meromyosin.* Unit of several myosin molecules become assembled in solution with the rodlike portions at the center and the globular heads projecting at both ends of the unit in the helical arrangement referred to above. This suggests that filaments of myosin in the intact muscle are formed in a similar way. Actin filaments are made up of a double helical chain of globular actin molecules, resembling two strings of pearls twisted about each other.

Biochemical analysis of actin filaments has revealed the presence of two additional proteins: tropomyosin and troponin. Both are present in small amounts in the intact muscle. *Tropomyosin* is a long, fibrous protein, one molecule of which lies in each of the two grooves formed by the two twisted filaments of actin. The presence of tropomyosin apparently gives rigidity to the actin molecules. The *troponin* molecules are complexes of three polypetides incorporated into the actin–tropomyosin filament at regular intervals. In the absence of Ca^{2+}, actin–myosin cross–bridge formation is blocked by tropomyosin. But when Ca^{2+} concentration is high, Ca^{2+} binds to troponin, causing tropomyosin to move aside and expose binding sites at which the actin–myosin cross bridges can form.

Summary of the Sliding Filament Theory

From the known facts about muscle proteins, the sliding filament theory of muscle contraction has been developed. According to this theory, muscle contraction is believed to occur in the following way:

1. A series of nerve impulses arrives at a motor end plate, releasing acetylcholine into its specialized synaptic cleft.

2. A local depolarizing electrotonic potential, the end plate potential, is generated, followed, in every case, by an action potential that sweeps across the entire surface of the muscle fiber, including its invaginations, the T tubules.

3. The depolarization of the T tubule membrance affects the adjacent terminal cisterna of the sarcoplasmic reticulum by increasing its permeability to Ca^{2+}. Because Ca^{2+} is 500 times more concentrated within the SR than in the sarcoplasm, the increase in the membrane's permeability to Ca^{2+} results in a rapid diffusion of Ca^{2+} into the sarcomere.

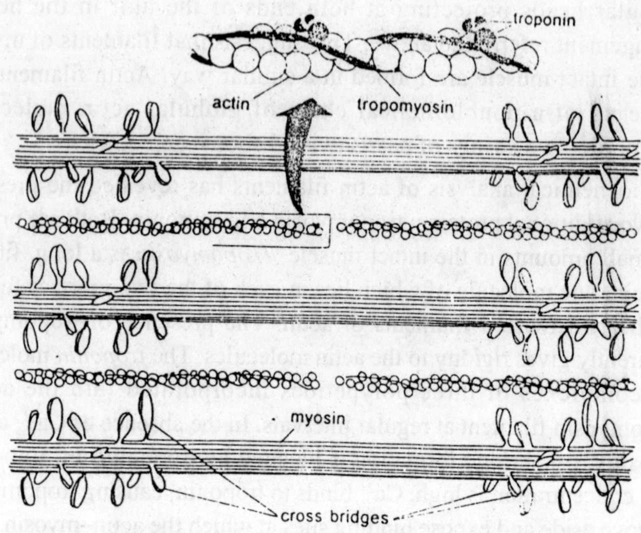

Fig. 1.6 : Probable relationship of actin and myosin filaments to each other during contraction. Note the two connecting cross bridges in the lowest figure.

4. The flood of Ca^{2+} binds calcium to troponin, which displaces tropomyosin, permitting the globular heads of the myosin molecules to extend outward to form temporary bridges with adjacent actin filaments.

5. When the globular head of a myosin molecule comes in contact with an actin filament, ADP and inorganic phosphate molecules associated with the myosin head are released. This changes the shape of the myosin molecule, causing the myosin head to bend backwards. This in turn pulls the actin molecule to which it is attached toward the center of the A band.

6. As all the cross bridges exert this pull at essentially the same time

throughout the myofibril—and, indeed, throughout the entire muscle fiber—the actin filaments are forced to slide past the myosin filaments. The result is that each sarcomere contracts very slightly, no more than the length of a cross bridge. This interaction between actin and myosin is immediately repeated, however, dozens of times per second.

7. In a healthy, rested muscle, ATP is regenerated as rapidly as it is split. As soon as an ATP molecule binds to the myosin head, the cross bridge formed by the head detaches from the actin and immediately snaps back to its original shape. The myofibril, however, at least for a moment, remains slightly contracted. If Ca^{2+} is still present, as it normally is, the process is immediately repeated, occurring at an estimated rate of 50 to 60 times per second. The result is that the actin filaments continue to slide past the myosin filaments and each sarcomere continues to shorten. This shortens the myofibril, thereby producing contraction of the muscle.

Releaxation of Muscle

A few milliseconds after a muscle has received a stimulus adequate to generate a contraction, the flood of Ca^{2+} that is released from the SR beings to be reabsorbed if no further stimulation is applied. As a result of this withdrawal of Ca^{2+} from the sarcoplasm, troponin no longer prevents tropomyosin from shielding the binding sites on actin. The actin and myosin filaments stop interacting, and the muscle slowly relaxes. The mechanism by which Ca^{2+} is reabsorbed involves an ATP–using active process similar to the Na^+/K^+ pump. Thus, both contraction and relaxation require energy.

The maximum shortening speeds is usually stated in lengths per second. For example, a muscle that shortened by a quarter of its length in a quarter of a second would be described as shortening at a rate of one length per second. Different muscles have very different maximum shortening speeds: some muscles in the legs of mice can shorten at 20 lengths per second, whereas others in the legs of tortoises can probably shorten no more than 0.1 length per second. Muscles obviously must be fast enough to do the jobs required of them,but if muscles are unnecessarily fast energy is wasted, as we will see. If the muscles in the legs of mice were not able to shorten so fast, the mice could not run so fast and would be in more danger of capture from predators. Tortoises have no need to run to escape because they can retire into the safety of their shells, so very slow muscles that are very economical of energy are best for them.

Like all bodily processes, muscular movement is powered by the energy that comes from food. For many animals, food is hard to obtain, and there is a strong advantage in using muscles and patterns of movement that are as economical of energy as possible. For this reason, we will want to know about the energy costs of different styles of locomotion and how energy is used by the muscles that power them.

The Energy Costs of Movement

Muscles are fueled by the by carbohydrates and fats obtained from food. Usually, these foodstuffs are oxidized by chemical reactions equivalent to burning:

carbohydrate or fat + oxygen ⟶ carbon dioxide + water

Such reactions release energy, some of which can be used to power muscular movement and the other processes of life. The amount of energy released can be measured in units called joules: for every liter of oxygen that reacts with carbohydrate or fat, about 20,000 joules of energy is released, from amost any foodstuff.

The link between oxygen use and energy is the basis of a very convenient method for finding out how fast animals use energy: if you measure their oxygen consumption, you can calculate how fast energy is being used. We will see later in this book how the method has been used to discover the energy costs of running, swimming, and even flight for a wide variety of animals. However, it does not always work because there are processes (as we shall soon see) that make food energy available to power muscles without any immediate use of oxygen.

The oxygen consumption method has also been used to find out how much energy different muscles need to do different things. Norm Heglund of Harvard University and Giovanni Cavagna of the University of Milan removed muscles from the dead bodies of frogs, toads, and rate and stimulated them electrically to activate them—that is, to set their crossbridges working, exerting forces. They stretched the muscles or allowed them to shorten at controlled rates and measured how fast they removed dissolved oxygen from the surrounding solution. The muscles used oxygen at a low rate even when they were not stimulated and were exerting no force: some energy is needed to sustain life, even in inactivity. Whenever the muscles were stimulated and developed force, they used oxygen much faster. If they were held at constant length while being stimulated, they used oxygen many times faster than when at rest. If they were allowed to shorten, they exerted less force but nevertheless used oxygen as fast as or faster than when at constant length. These results tell us that the faster a muscle is shortening, the faster it must use energy to maintain a particular force. Experiments by other people have shown that if a muscle is being stretched it can exert the same force for less energy cost. Why does a shortening muscle use energy faster than a lengthening or still muscle? Muscles can pull but they cannot push, so to power an animal's movement they must shorten. Only shortening muscles are doing work.

Work is energy delivered by a force that moves the object to which it is attached: no work is done unless there is movement, and this movement must be in the direction of force. The work done by a force equals the force multiplied by the distance moved in the direction of the force:

$$\text{work} = \text{force} \times \text{distance}$$

Work is a form of energy, so it is measured in joules, the standard unit for *all* forms of energy. When a force of I newton moves an object a distance of 1 meter (in its own direction), it does 1 joule of work.

The notion that you are not doing work when you are straining to carry a heavy suitcase may seem odd. But we must realize that the scientific definition of work includes a particular relationship with energy. Whenever work is done, energy is given to the moved object or to its surroundings. When we stand holding a suitcase, no work is done and the suitcase's energy remains unchanged, just as if it were supported by an inanimate structure such as a table. When we carry the case across a level floor, its energy is again constant (provided we carry it at constant speed) and again (if we ignore air resistance) no work is needed. To understand why not, imagine a suitcase on a frictionless trolley on a level floor.If we gave the trolly a push to start it moving, it would continue to move at constant speed, with no need for any additional input of energy, until it hit an obstacle. Real trolleys slow down only because their energy is gradually dissipated by friction in their moving parts and by air resistance.

We still do not fully understand the relationship between the rate of energy consumption of a muscle and the job it is doing, but it seems helpful as a rough guide to think of the energy used by a muscle as the sum of two energy costs:

total energy cost = cost of force + cost of work

The equation shows why muscles use food energy faster when shortening than when exerting the same force and holding constant length. The extra energy is the cost of doing the work. The Principle of Conservation of Energy, one of the firmly established principles of physics, says that energy cannot be created from nothing (nor can it be destroyed). That tells us that for every joule of work that muscles do, they must use at least a joule of food energy. They actually use much more, commonly about 5 joules, because a lot of the food energy is lost as heat. (When we shiver we are using muscular movement to produce heat, to warm ourselves up.) The ratio of work output to food energy input is called efficiency, so if 5 joules of food energy is needed to do 1 joule of work, the efficiency is 1/5, or 0.2. Obviously, high efficiency is desurable, since it will lower the cost of work—the amount of food energy that has to be used in order to do the work:

$$\text{cost of work} = \frac{\text{work}}{\text{efficiency}}$$

The efficiency of a muscle is not constant but is different at different rates of shortening. As a general rule, the least metabolic energy will be used while doing a given amount of work if the muscle shortens at about one third of its maximum shortening speed.

Muscles use food energy when they are exerting force, even if they are holding constant length and so doing no work. Food energy is needed to support the activity of the crossbridges, which do not simply hold tight but continually detach and reattach. Crossbridges go through their cycles of detachment and rettachment whenever a muscle applies force. It is in this activity of the crossbridges that we seek the factors that must affect the energy cost of exerting force. while holding constant length. These factors determine the second energy cost that, together with the cost of work, contributes to the total energy cost.

The larger the force, the greater the cost, because more fascicles have to be activated. Also, the longer the fascicles, the greater the energy cost: it seems obvious that it must cost twice as much energy to exert a force in a fascicle 10,000 sarcomers long as to exert the same force in a fascicle only 5000 sarcomeres long. Finally, it costs twice as much energy to maintain a force for 2 seconds as to hold it for only I.

These factors (force, fascicle length, and time) are not only ones that affect the cost of force. There is also another, the "economy" of the particular muscle:

$$\text{cost of force} = \frac{\text{force} \times \text{fascicle length} \times \text{time}}{\text{economy}}$$

The economy is different for different muscles. It is larger for muscles that can shorten slowly (such as tortoise leg muscles) than for ones that can shorten faster (such as mouse leg muscles). As a rough general rule, doubling the maximum shortening speed halves the economy. The reason is that the crossbridges of faster muscles detach and reattach more repidly, even when the muscle is not shortening.

Muscles cannot do work without exerting force, so the distinction that we have made between the cost of work and the cost of force is rather artificial. However, we think it is a useful step towards understanding how muscles use energy. Unfortunately, our understanding of this fundamental aspect of muscle physiology is still imperfect.

Aerobic and Anaerobic Muscles

Some muscles can only work aerobically, in the presence of oxygen. While aerobic muscles are active, oxygen must be supplied fast enough to oxidize the foodstuffs that they are using as fuel. Consequently, the rate of energy use is limited by the rate at which the lungs or gills can take up oxygen or the bloodstream can transport it. If an animal needs to make a burst of speed to, say, escape a predator or capture prey, oxygen may not reach its muscles fast enough. For–tunately, many animals have other muscles they can call upon, which work without oxygen.

Anaerobic muscles get energy from foodstuffs by processes that do not require oxygen—for example, by converting glucose to lactic acid. Anaerobic processes yield much less energy than does oxidation of the same quantity of food, so these processes would be very wasteful if the products were simple discarded. The more economical alternative is to oxidize some of the products afterward and use the energy so obtained to convert the rest back to carbohydrate. Thus an oxygen debt that has built up rapidly during a burst of violent activity may be repaid slowly while the animal recovers afterward. The advantage is that energy can be used faster to power a burst of activity. The disadvantage is that anaerobic processes cannot continue for long. When our muscles and those of other vertebrate animals work anaerobically, lactic acid accumulates, and we cannot tolerate more than a certain concentration of lactic acid in our bodies, Anaerobic processes enable us to make more intense bursts of activities than would otherwise be possible, but prolonged activities depend on aerobic muscle activity. For that reason, we cannot run marathons as fast as we can sprint. An athlete sprinting 100 meters uses mainly anaerobic metabolism, but one running a 42–kilometer (26–mile) marathon depends almost entirely on aerobic metabolism.

Some muscle fibers are specialized for aerobic work and some for anaerobic. The two types differ in biochemical composition, in microscopic structure, and often in colour. In vertebrate animals, aerobic muscle fibers are generally reddish in colour and anaerobic ones are generally whiter, but in cephalopod mollouses like octopuses and squids the colour are yellowish and white.

The colour differences can be seen easily in some familiar foods. We are accustomed to seeing the swimming muscles of fish at our dinner tables in the form of fillets; these muscles consist mainly of white (anaerobic) muscle but generally have a band of red (aerobic) muscle running along the side, close under the skin. The red muscle is used for sustained swimming and the white for short bursts of speeds. The white breasts of chickens are the muscles that serve to beat the wings. Appropriately, they are anaerobic, since chickens make only very short flights. The darker leg muscles are largely aerobic to enable chickens to run around continually, using their leg muscles. Birds such as pigeons that make prolonged flights have dark muscles on their breasts, composed largely of aerobic fibers.

There are other distinct kinds of muscle with special properties that are important for their functions in locomotion—for example, there are the fibrillar flight muscles that enable small insects to beat their wings at extraordinarily high frequencies.

Many—perhaps most—animals use more energy for locomotion than for any other purpose. Locomotion is generally necessary for obtaining food, finding mates, and escaping predators. Because locomotion is such a large part of the energy budget, there is a big advantage to the animals in keeping energy costs low. So important are these costs that they will be considered repeatedly in the chapters that follow. Measurements of oxygen consumption will enable us to compare the costs of different forms of locomotion.

CHAPTER 2
WALKING AND RUNNING

Successful racehorses gallop at 35 to 40 miles per hour. In this gait the two fore feet are set down in rapid succession, then the two hind feet, followed by the two fore feet again. The spring action of a sheet of tendon allows the back to bend and extend at appropriate stages, lengthening the stride.

Running Adaptations

The animals, which pass their life time exclusively oı, the earth are known as cursorial forms. In these types, speed has developed in a very wonderful manner. The body of cursorial animals is moulded externally in such a way, as to offer minimum resistance to the air, through which they walk or run. The limbs are the main propelling organs of the terrestrial animals. As the hind limbs are the more efficient drivers, they undergo more modifications than the fore-limbs in an organism, whose cursorial adaptation has not gone very far. The limbs have become larger, the foot posture has undergone changes from plantigrade to digitigrade in a large prop-ortion of modernized mamm-als (specially in ungulates), which has occurred for speed adaptation.

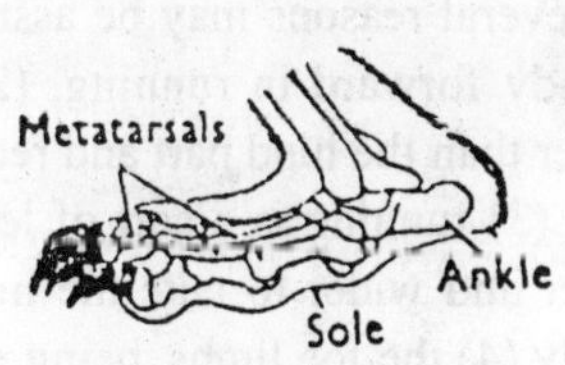

Fig. 2.1. Plantigrade foot of bear.

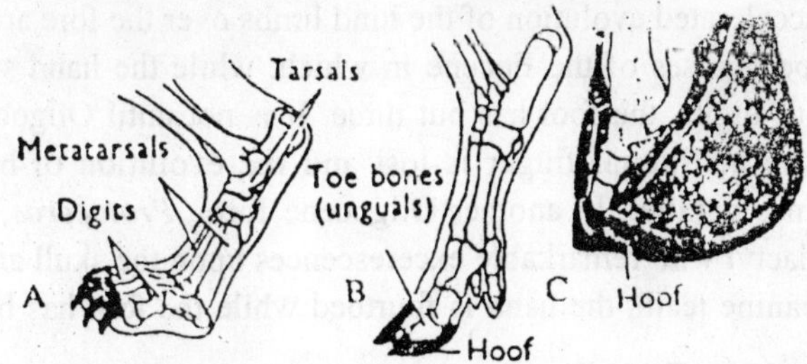

Fig. 2.2. Foot postures of mammals. A-Digitigradefoot of Hyaena B-Unguligrade foot of pig C-Unguligrade foot of Rhinoceros showing supporting pad.

Body Contour

All speedy animals, whether terrestrial, aquatic, or aerial, have the body moulded externally in such a way as to offer the least resistance to the medium through which they pass. Owing to the greater resistance of water, this is especially true of the aquatic, nevertheless a speedy cursorial type also shows it, though not always as well when at rest as it does when in action. A race-horse with head and neck extended, ears thrown back, and every tense muscle of its wonderful body working with machine-like precision, shows the beautiful contour of a perfectly adapted mechanism. The body spindle-shape, the stream-line contour, the lines extended into the neck and head without a break in their curves—all are calculated for swift passage through the air with a minimum of resistance.

Mechanism

Loss of General Utility : The propelling organs in cursorial forms are the limbs exclusively, so that, aside from the resistance-lessening contour, this adaptation concerns itself chiefly with their modification, of which the first is the loss of general utility. This is especially true ot the hind limbs, because they are the more efficient drivers and therefore, especially in forms whose adaptation has not gone very far, are likely to be somewhat is advance of the fore limbs in the degree of their revolution. For this several reasons may be assigned: (1) The extended hands pull the body forward in running, (2) the fore part of the body is usually heavier than the hind part and requires large limbs to support and propel it; (3) running is a sort of leaping on all fours, and the hands are larger and wider to take the implact when the animals falls forward; finally (4) the foe limbs, being nearer the mouth and hence perhaps somewhat concerned in food-getting, are the last to lose their generally utility. Two notable instances of this accelerated evolution of the hind limbs over the fore are the early fourtoed horses of the Eocene in which, while the hand still retains its four digits, the foot has but three. It is not until Oligocence time that the additional finger is lost and the evolution of both limbs becomes parallel. In another Oligocene form, *Protoceras*, a curious artiodactyl with remarkable excerescences upon the skull and dagger-like canine teeth, the hand is fourtoed while the foot has but two.

Change in Foot Posture

The primitive terrestrial foot is plantigrade (Lat, *planta*, sole, and *gradi*, to walk), which means that the entire palm or sole rests on the ground, neither wrist not ankle being raised. Almost the first step in speed adaptation is the lengthening of the limb and this may be accomplished without the actual elongation of a bone, merely by rising upon the toes. While the bear, raccoon, and the primates such as the baboons and man are plantigrade, probably secondarily so, a large proportion of moderinzed mammals have become digitigrade (Let. *digitus*, fingure, toe), walking or running upon the digits themselves, with the bones of the wrist (carpal) and ankle (tarsal), the upper ends of the palm (metacarpal) and the sole bones (metatarsal) clear of the ground. Some of the speediest of animals—dinosaurs, birds, dogs, all mammals in fact but the ungulates—have merely perfected the digitigrade gait, developing special sole-pads for the absorption of the shock of impact, and have never gone beyong it. The ungulates or hoofed animals, on the other hands, walked on the modified nail or hoof (unguligrade, Lat. *ungula*, hoof), the distal toe-bones (unguals) being depressed or flattened and not, with rare exceptions, compressed or claw- like. The hoof has reached the highest degree of perfection in the horses; in other related but non-cursorial types like the rhinoceros the hoof bear little of the weight, as a broad cushion-like pad serves instead.

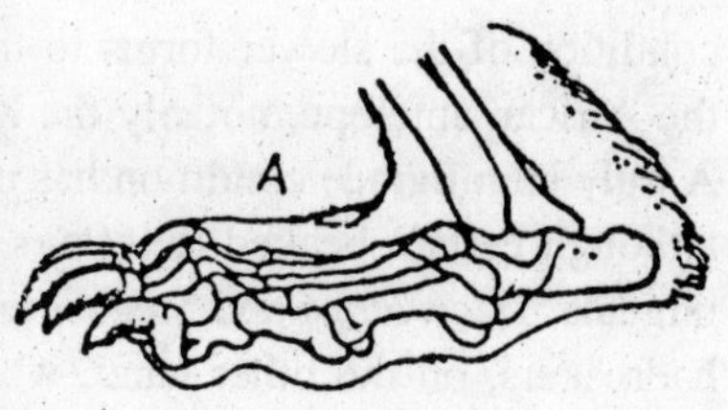

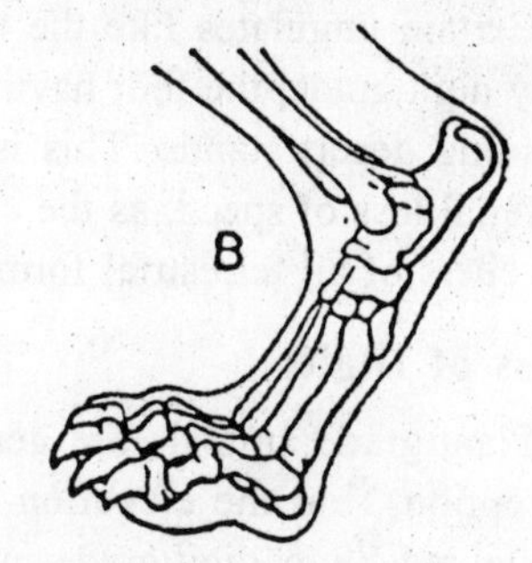

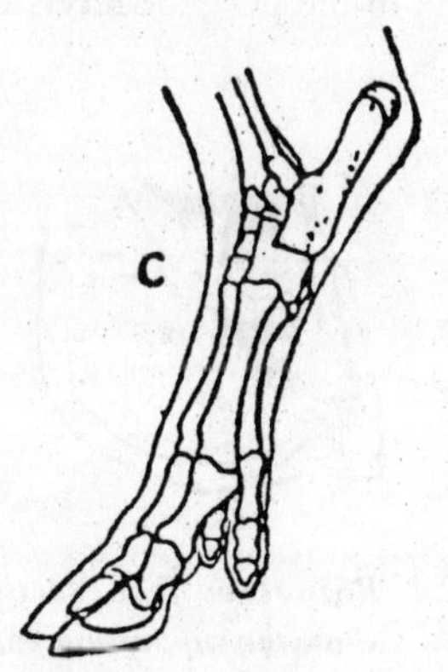

Fig.2.3. Foot postures. A, Plantigrade, bear; B, digitigrade, hyaena; C, unguligrade, pig.

Ungulate animals show all gradations from the semiplantigrade condition of the slower forms to the high, stilted hoofs of certain of the African antelope, notably the Klipspringer (*Oreotragus saltator*). A truly unguligrade condition has never been attained among reptiles, although certain beaked dinosaurs has depressed instead of clawlike unguals. *Triceratops* and *Stegosaurus* were certainly far from speedy; hadrosaurs, on the other hand, which also bore this type of ungual, were bipedal and doubtless possessed a fair measure of speed when well under way, though littel celerity of movement is indicated.

Certain ungulates like the modern camels have become secondarily digitigrade, the foot having retrogressed as an adaptation to the yielding desert stands. This is not accompanied, however, by any material loss of speed, as the camels are among the most remarkable travellers of all terrestrial forms.

Loss of Digits

Plantigrade animals are generally five-toed; there are of course exceptions, but the elevation of the wrist generally carries with it digital reduction, digitigrade animals becoming four-toed, unguligrade four-, three-, two-, or even one-toed, two toes in the artiodactyles and one in the perissodactyls being the irreducible minimum.

Fig. 2.4. Rhinocros foot, sectioned to show the supporting pad.

The frilled of Australia, *Chlamydosaurus*, is five-toed but the lateral toes are shorter than the median ones, which is almost universally ture except in aquatic types such as the seal and otter. Hence when *Chlamydosaurus* runs on its hind feet, as it does when startled, the outer and inner toes are raised off the ground and the animals makes a three-toed track. If this were the habitual gait of the creature, the lateral digits would be rendered practically useless and would follow the course of all useless organs and become reduced, whatever the philosophical explanation of the mean whereby this is accomplished.

Environment as well as speed adaptation has its influence in determining digital reduction, for it will be accelerated if the ground is hard, as in the prairie-evolved horse with but one remaining digit, or the prong-horn antelope of similar environment with two. On the other hand, the Miocene forest horse, *Hypohippus*, retained and lateral toes as functional organs just as the reindeer and caribou (*Rangifer*) have today, as an adaptation to a yielding footing, while contemporary relatives had in each instance evolved much farther along the line of digital reduction.

Concurrent with the loss of digits, especially if the foot be lengthening after the manner to be described below, comes a compacting of the bones of the palm and sole (metapodials) and often this is carried so far as to give rise to actual fusion of these elements into a "cannon-bone". The dinosaurs, with one doubtful exception, never attained an actual fusion, although in many respects, especially in *Ornithomimus* of the Cretaceous, the foot is very bird-like. The birds, on the other hand, always show a fusion of the metatarsals. Among mammals the carnivores do not torm a cannon-bone nor does the marsupial wolf; but all the speed-adapted ungulates do. Ancient ungulates, however, had the metatarsals separate, and we can often witness the fusion in fossil series (camels, etc.) when the proper degree of speed adaptation has been reached.

Among rodents, the jerboa, a three-toed bipedal form, has a foot and metatarsus so wonderfully bird-like that one almost has to count the phalanges of the digits to be sure he has a mammal before him.

Fig.2.5. Hind limb of horse,* Equus caballus, *to show pulley-like joints for the restriction of movement of one plane.

Reduction of Fibula and Ulna

The fore arm and shin, that is, the second segment from the body, have each typically two bones : in the arm, the radius and ulna, and in the leg, the tibia and fibula. These are both developed in slow-moving forms or where the fore limb still has considerable general utility, especially if the rotation of the hand on the arm is retained. On the other hand, cursorial forms, especially if the limbs are exclusively locomotor, tend to reduce the ulna of the arm, the proximal end onlyl being present in extreme cases to form the elbow joint. They also lose the fibula of the leg, which may be reduced to the merest vestige.

Loss of Universal Movement

The entire motion of the limbs becomes pendulum-like, that is, restricted to movement in but one plane, the exception being at the hip and shoulder, where universal movement is still retained through the development of a ball-and-socket articulation. The necessity for this is apparent, first to avoid interference between fore and hind feet when running, since a dog, for instance, at top speed brings his hind feet well in advance and outside of the fore. A second need is that of lying down and rising again, which would be practically impossible were the movement at hip and shoulder restricted to the fore-and-aft plane. With the other articulations, those of ankle and wrist, knee and elbow, and between the digits, the tendency is toward rigid limitation of movement in unguligrade, less so in digitigrade forms. This is accomplished by the development of tongue and groove joints such as were discussed under Kinetogenesis. These are very perfectly shown in the hind limb of a modern horse, as well as at the elbow joint, forming in each instance an articulation permitting movement through a wide arc in one plane of space and none whatever in any other. These joints, while they may be broken, cannot be dislocated.

The limbs are compound levers, for not only is there motion of the limb as a whole but also between its component parts. The lengths of each of the several segments bear definite relations to the speeds developed and also to the loads they have to carry. those forms which, like the elephant, are mightly of frame, have a type of limb which is in marked contrast to that of a horse. To the former type has been applied the term graviportal (*i.e.*, weight-carrying), to the latter

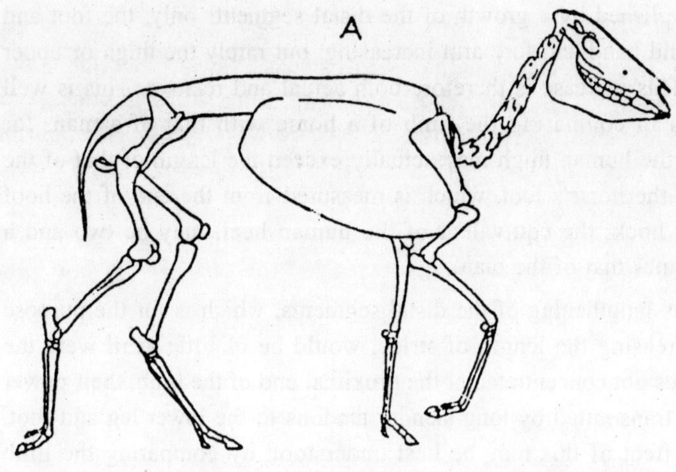

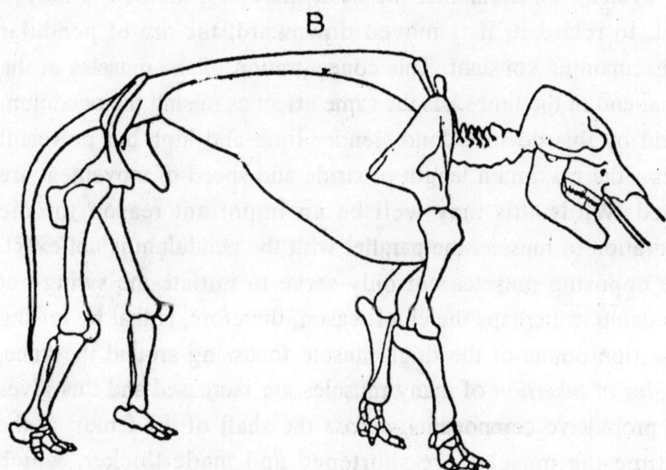

Fig.2.6. A, three-toed horse, **Hipparion (Neohipparion)***, showing cursorial adaptation for run or gallop; B, mastodon,* **M.americanus***, showingd graviportal (rectigrade) adaptation for walk or amble.*

cursorial, although both are adapted to increase their owner's travelling powers. In the graviportal type such as the mastodon, the foot is short and the thigh and shin relatively long, whereas in a cursorial form the foot elongates and the thigh is conservative.

Lengthening of Limbs

The lengthening of the limbs in cursorial types is usually thus accomplished by a growth of the distal segments only, the foot and shin and hand and fore arm increasing, but rarely the thigh or upper arm. This increase is therefore both actual and relative. This is well shown in comparing the limb of a horse with that of a man, for while the human thigh may actually exceed the length of that of the horse, the horse's foot, which is measured from the end of the hoof to the hock, the equivalent of the human heel, may be two and a half times that of the man.

This lengthening of the distal segments, which is for the purpose of increasing the length of stride, would be of little avail were the muscles not concentrated at the proximal end of the limb, their power being transmitted by long slender tendons to the lower leg and foot. One effect of this may be best understood by comparing the limb with a pendulum. The length of the pendulum determines the scope of swing, but the position of its center of gravity controls its speed or rate of beat. To accelerate the beat, therefore, the bob is moved upward, to retard it, it is moved downward, the arc of pendular motion remaining constant. This concentration of the muscles at the proximal end of the limbs has the same effect as raising the pendulum bob, and by this device—lond slender limb and high but powerful muscles—the maximum length of stride and speed of movement are obtained. While this may well be an important reason for the concentration of muscle, the parallel with the pendulum is not exact, for the opposing muscles not only serve to initiate the swing but also to damp it. Perhaps the chief reason, therefore, is that by raising the insertion points of the thigh-muscle focussing around the knee, the *angles of insertion* of many muscles are increased and this gives higher propulsive components, *across* the shaft of the femur; at the same time the muscles are shortened and made thicker, which increases their power and speed of contraction. It can readily be seen that a limit may be reached beyond which bone will not stand the strain to which it would be subjected, although bone is a wounderfully efficient material. Hence one would expect to find the greatest speed developed on the part of creatures of small to moderate size—the antelope of Africa or horses like the wild ass of Persia (*Equus onager*) the speed of which has been mentioned and which reaches a stature of but 11½ hands. The Mongolian wild ass or Kiang

(*Equus Kiang*), according to Roy Chapman Andrews, who chased one for 29 miles in a motor car; averaged 30 miles an hour for the first 16 miles and then when it began to slow down, still ran four more miles at a speed of 20 miles an hour. The modern race-horse is relatively small comparedb with some other breeds, and the limit of weight, size, and speed consistent with safety seems to have been approximately reached.

Ratios

Lengthening of limbs also implies, at any rate in a quadruped, the concurrent lengthening of neck and skull in order that the animal may readily reach the ground for food and drink. Hence the various parts of an animals's frame bear definite ratios to one another and this may also extend to individual bone proportions, definite "speed index" being recognizable. This makes it possible through the law of correlation to gain some insight into the habits of extinct and little-known forms through the study of comparatively fragmentary remains.

Bipedality

A two-footed mode of progression as an adaptation to speed has been repeatedly evolved among vertebrates, as follows :

Reptiles :

Lizards, several occasionally bipedal.

Dinosaurs, two evolutions.

Birds :

One evolution.

Mammals :

Marsupials, one evolution.

Rodents, three evolutions.

In all, eight or more times.

The erect posture of man was probably not originally a speed adaptation, nevertheless speed has always been a vital factor in human evolution, in all offensives and defensive operations. The human foot, which was originally a climbing sturcture, has been readapted for bipedal walking and running. The long thigh and shin of modernized man increases the stride materially in contrast to those of the gorilla and chimpanzee. The Neanderthal man had short stocky limbs as compared with the existing species, but doubtless could outrun any of the anthropoid apes.

Counterpoise

Some sort of counterpoise is always necessary in a semi-errect biped and the tail usually assumes this function. In the Kangaroo and in the dinosaurs it is a powerful organ and serves as a prop, like a third limb, when the creature rests without coming down on all fours. The tail may be comparatively short and heavy in larger forms or extremely long and slender in more lightly built creatures, on the principle that an ounce at the end of a sixteen-inch lever is as effective as a pound on one but an inch in length. Many dinosaurs and bipedal lizards have a long, attenuated tail. This is especially true of the dinosaur *Podokesaurus*, a Triassic form from the Connectucut valley, and of the Australian frilled lizard *Chlamydosaurus*. Among mammals the Kangaroos have a relatively short, heavy tail; the jerboa on the other hand has a very long one terminating in a tuft of hair, which through its resistance to the air adds effect to the counterpoise.

No existing birds have a long tail, that is, as regards the actual tail itself, although the feathers may be 'ong. These, as in the pheasants, may subserve a balancing function. The true cursorial birds, Ratitae, are practically tailless, but maintain their balance with ease, the heard and neck sufficient. The ostrich raises its wings as an aid in running, but with the practically wingless cassowary or the emu the head and neck alone must serve.

Shortening of Neck

In bipedal mammals there is a tendency toward reduction in the length of the neck, especially in the rodents such as the jerboa, in which cursorial adaptation is extreme and there is a remarkable cervical reduction associated with the shortened skull. There is of course no diminution in the number of neck vertebrae, for the number, seven, is with two, or three exceptions (sloths and manatee) constant among mammals; but the vertebrae themselves are shortened and tend to coalesce into a rigid mass of bone. Thus in the rodent *Pedetes* cervicals 2 and 3 are so closely articulated as to eliminate motion, in *Perodipus* the axis (2d cervical) and next two vertebrae are fused, while in *Dipus* (jerboa) all of the cervicals expect the atlas

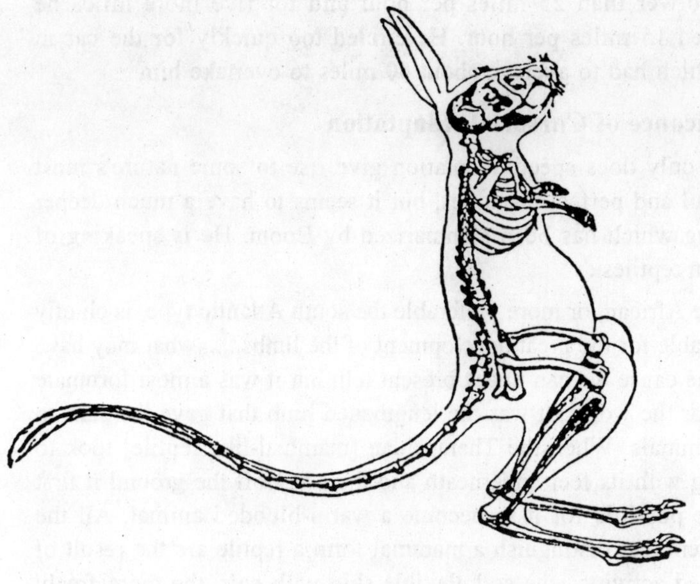

Fig. 2.7. Skeleton of jerboa.

(1st cervical) are coossified as in the whales. As we shall see, the shortening of the neck may be also an aquatic adaptation, since it occurs in the whales and sirenians.

Mental Precocity

Animals which depend upon speed for safety, as do the ungulates or the whales, cannot have helpless young. Such must either be brought forth in some secluded den or carried about by the mother. Carnivores and rodents have very feeble young, but they are kept hidden until able to shift for themselves. The Kangaroo, on the other hand, must carry her offspring with her and this undoubtedly proves a very heavy handicap to the race when competition with higher forms prevails, for the destruction of the mother means that of the young as well. With all other forms which depend upon speed for safety, the young animals must be able to keep up with the herd almost at once. Hence there is no period of helpless infancy, but the new-born deer on horse or camel, with its grotesquely long limbs, has the relative mental alertness of a very much older dog or rat, although the ultimate mental attainments of the ungulate may not be so great. As an illustration of precocity, Andrews describes an experience with a ten-day-old baby antelope. For four miles he seldom went solwer than 25 miles per hour and for five more miles he averaged 15 miles per hour. He circled too quickly for the car in fact, which had to average about 40 miles to overtake him.

Significance of Cursorial Adaptation

Not only does speed adaptation give rise to some nature's most beautiful and perfect machines, but it seems to have a much deeper meaning which has been summarized by Boom. He is speaking of Permian reptiles :

"The African, or more preferable the south Atlantic type, is chiefly remarkable for the great development of the limbs.......what may have been the cause we can not at present tell, but it was a most fortunate thing for the world. It was the lengthened limb that gave the start to the mammals. When the Therapsidan [mammal-like reptile] took to walking with its feet underneath and the body off the ground it first became possible for it to become a warm-blooded animal. All the characters that distinguish a mammal form a reptile are the result of increased activity—the soft flexible skin with hair, the more freely movable jaws, the perfect four-chambered heart, and the warm blood. It is further singularly interesting to note that the only other warm-blooded animals, the birds, arose in a similar fashion from a different reptilian group. A primitive sort of dinosaur took to walking on its hind legs, and the greatly increased activity possible resulted in the development of birds. Birds were reptiles that became active on their hind legs mammals are reptiles that acquired activity through the development of all four."

Back of all this lay the impelling natural cause. The earliest known mammals are late Triassie, the first recorded bird Middle Jurassic; the inference that both stocks arose in Permian time is justifibale from the degree of evolution which each class had attained by the time the actual record of their existence beings. Schuchert tells us that early in the Permian the climate of the lands seems everywhere to have been arid or semi-arid and that this condition lasted into Jurassic time. One characteristic of desert animals of to-day—the lizards, birds, gazelle, Persian ass—is *speed*, for the creature must fare widely for food and drink if he would fare well. Again we are told that during the Permian there was a period of extensive glaciation with a severity of climate, especially in the southern land masses, as great as, if not greater than, the polar one of Quaternary time, although, like the latter, the Permian glacial period had warmer interglacial intervals as well. The incentive for speed already given, rendering the development of warm blood possible, the devastating cold would soon place a premium upon such as did develop it and eliminate those which did not. From this fortunate relation of cause and effect might well have arisen on the one hand the primal mammal, making human evolution possible, and on the other hand the ancestral bird.

Walking

Human walking is unique. No animals walks as we do; althouh birds walk on two legs and apes sometimes do, thier styles of walking are quite different from ours. Most birds use flight rather than walking as their principal means of travel, and apes do most of their walking on four legs. Apart from ostriches and other fllightless birds, no other animal depends as much as we do on two-legged walking. Is there something particularly good about our peculiar walking style?

Walking: A Pendulum of Swinging Legs

The distinctive feature of human walking is that we keep each leg almost straight while its foot is on the ground. The sequence of photographs at the top of the page follows the course of movement

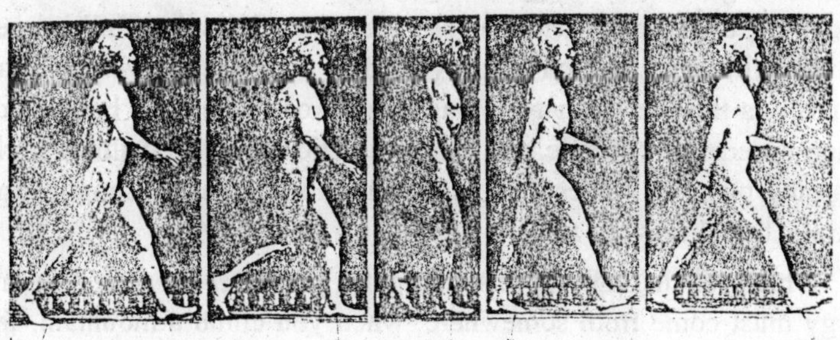

Fig. 2.8 : Stages of a walking stride.

through a single step. At stage (c) of a step, the supporting leg is straight and vertical, so the body is high. At stages (a) and (c), the legs are straight but sloping, so the body is lower. As a consequence of our leg's changing slopes, our heads bob up and down by about 40 millimeteres (1.6 inches) in the course of each step. Although this bobbing may seem like a trivial side effect, it actually may reduce the energy cost of walking. A comparison of speed and height over the course of a step shows why.

As we walk, our feet exert forces on the ground. The illustration shows the directions of the forces, which have been recorded by means of force plates, instrumented panels set into the floor that give electrical outputs indicating the downward, backward or forward, and sideways components of any force that acts on them. The records show that the force on each foot is always more or less in line with the leg. At stage (b) the foot is pushing forward as well as downward on the ground, so the body is not only being supported but is also being slowed down. At stage (d) the foot is pushing backward as well as down, so the body is being speeded up. thus the body is traveling relatiely fast at stages (a) and (c) of the stride and more slowly at stage (c). For example, the speed of someone walking moderately fast might fluctuate between 1.7 meters per second (3.8 miles per hour) at stages (a) and (c) and 1.4 meters per second (3.1 miles per hour) at stage (c).

To understand the implications of these movements and forces, we need to know about two kinds of energy : Potential energy and kinetic energy. Potential energy is the energy that matter has because of its height. That energy changes with heights is illustrated, for example, by hydroelectric schemes : as water flows downhill it loses potential energy, which is converted to electrical power. Kinetic energy is the energy that moving objects or fluids have because of their speed. For example, wind loses kinetic energy as it slows while passing over the blades of a windmill, and that energy supplies the power that drives the mill.

Whenever the body is raised it gains *potential energy.* That energy must come from somewhere; when you climb a mountain, for example, the potential energy you gain is supplied as work done by your muscles. The body also gains energy whenever it speeds up, and this *kinetic energy* must come from somewhere as well. When you accelerate at the start of a strint, your muscles do the work that increases your kinetic energy. You might conclude that the muscles must do quite a lot of work in the course of each stride and that they consequently uses a lot of metabolic energy. In walking, however, the body is high while it is traveling slowly (stage c) and low while it is traveling fast (stages a and c), so its potential energy is high while its kinetic energy is low, and viceversa. The same is true of a pendulum, which is highest when it stops at the end of a swing and lowest when it is moving fastest through the bottom of the swing. As the pendulum swings it converts potential energy to kinetic energy and back again. Energy is swapped back and forth between the two forms, and the pendulum will continue swinging for a very long time without any fresh input of energy. Similarly, very little work is needed from our leg muscles as we walk.

The pendulumlike quality of walking is a consequence of the straightness of our legs. It probably saves metabolic energy, but possibly not very much. If the changes in kinetic and potential energy were less balanced, our muscles would have to do more work at some stages of the stride to increase the total (kinetic plus potential) energy of the body; at other stages they would have to work like brakes, doing negative work to reduce the (kinetic plus potential) energy. While doing positive work they would use *metabolic energy* faster, but while doing negative work they would use it more slowly. The two effects might fairly nearly caneel each other our, but our knowledge of physiology is not precise enough for us to be certain. The pendulumlike quality of human walking probably reduces the "cost of work" element of the food energy requirement, but possibly not by very much.

There is another, possibly more important consequence of our straight-legged style of walking: it enables us to support our weight without the need for large forces in our leg muscles, thereby reducing the "cost of force" element of the energy requirement. When you stand with your knees straight, the line of action of your weight passes close to the knee joints and little tension is needed in your muscles to prevent the knees from collapsing under the load. If you stand with your knees bent, however, the line of your weight is farther from them and your muscles must exert more force. Similarly for walking : the straighter your legs, the less force your muscles need exert. for a practical demonstration, try taking a walk with your knees bent. You will feel the extra tension in the quadriceps muscles (at the front of the thigh), and you will find your steps unusually tiring.

Although our straight-legged style of walking is so economical of energy, only humans use it. Even our closest relatives, the apes, walk with their legs bent. Chimpanzees usually walk on all fours, on the soles of their hind feet and the knuckles of thier hands, but they sometimes walk on their two hind feet only, especially when carrying things in their hands. Even when walking bipedally they walk unlike us, with their knees bent and their back sloping. Gibbons usually travel through the treetops by swinging from their long arms, which are too long for quadrupedal walking. They sometimes walk along the upper surfaces of thick branches, moving on their hind legs alone, with their knees bent. In the wild they seldom or never descend to the ground, but in zoos they sometimes do—and again walk bipedally on bent legs.

Running : Bent Legs For Speed

Humans also move with bent legs—when they run. At the stage of a running step when the force on the foot is largest (stage c in the illustration on the next page), the knee is rather bent and the muscles must exert large forces, at a cost in metabolic energy. This extra cost suggests that running will be expensive of energy. If running is so expensive, and our straight-legged style of walking is so economical of energy, why do we ever run? Why don't we just walk faster?

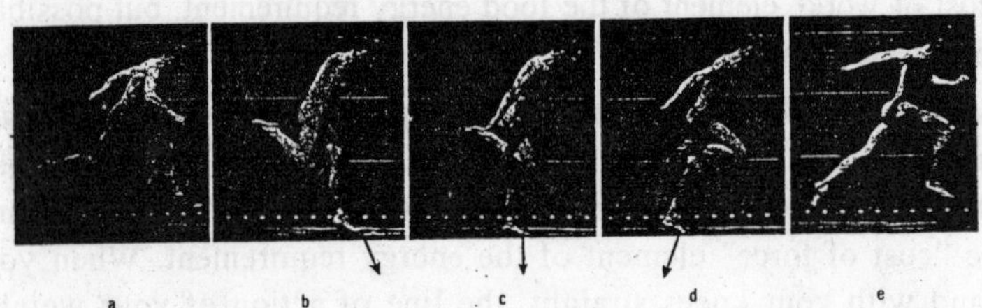

Fig. 2.9 : Arrows shows the directions of the forces on the ground during the stages of a running stride

Despite the apparent extra cost, adult people of normal size change from walking to running whenever their speed exceeds a highly predictable rate, about 2 to 2.5 meters per second (4.5 to 5.6 miles per hour). To try to explain why, we will analyze a simple model that seems to reduce walking to its essentials. The person shown in the diagram on the facing page is walking at a speed v, keeping the legs utterly straight while the foot is on the ground. The body moves along an arc of a circle centered on the foot, a circle whose radius is the length l of the leg. Here we need a simple formula from physics. An object moving in a circle is constantly changing its direction, turning toward the center of the circle. That change in direction affects the object's velocity, its speed in a specified direction . Because its direction is changing even though its speed may be constant, its velocity is also changing : in other words, the object has an acceleration toward the center of the circle. If the speed is v and the radius l, this acceleration is v^2/l. A force pulling toward the center of the circle is needed to give the object this acceleration and so prevent it from flying off at a tangent. In the case of the walker, the acceleration is downward toward the foot and the force causing it must be the weight of the body, so the acceleration cannot be greater than the gravitational acceleration g :

v^2/l cannot be greater than g.

so v cannot be greater than $\sqrt{gl}$.

The gravitational acceleration g is about 10 meters per second squared, and the length of an adult human leg from hip to sole is about 0.9 meter. Thus $\sqrt{gl}$ is about $\sqrt{10 \times 0.9} = 3$ meters per second. We can conclude that it is physically impossible to walk like the person in the diagram at speeds greater than 3 meters per second. That is the maximum possible walking speed, and people actually change from walking to running at the slightly lower speed of about 2 to 2.5 meters per second. The maximum speed is lower for children because their leg length l is shorter—and small children often have to run to keep up with their walking parents.

That seems to be a convincing explanation of the change from walking to running, until you look at walking races. The rules require the leg to be straight while the foot is on the ground, but athletes nevertheless manage speeds of 4 meters per second. The secret is a peculiar movement of the hips, which lowers the body's center of gravity a little at the stage of the stride when the leg is vertical. The body rises and falls less in each step than we supposed in the simple theory, so the accelerations that are needed at any particular speed are less than we calculated.

If you are not constrained by the rules of a walking race, it is best to change from walking to running at a speed a little less than the theoretical maximum. The reason is that the energy cost of

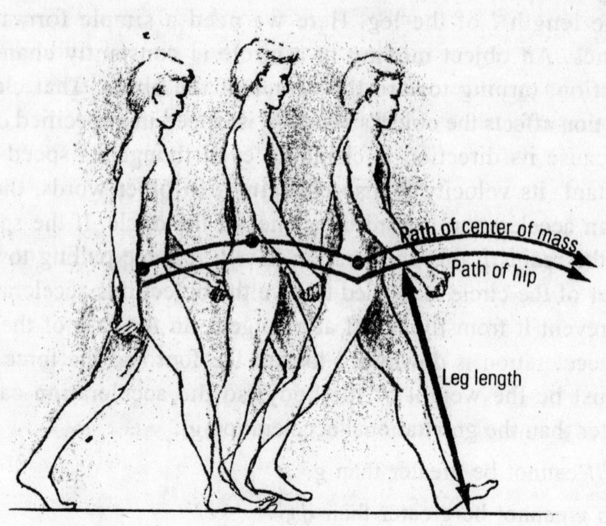

Fig. 2.10. : When a person walks on rigid legs, the body moves along an arc of a circle centered on the foot.

walking rises steeply at speeds near the limit. Physiologists have shown how the advantage shifts from one gait to the other by measuring the rates of oxygen consumption of people walking and running at different speeds. The subjects walk on a moving belt that keeps them stationary relative to the laboratory, while wearing face masks from which air is sucked through a hose to oxygen-analysis equipment. Plenty of fresh air for breathing is drawn in around the edges of the loose-fitting masks, but all the air the subjects breathe out is sucked away for analysis. the equipment measures the volume of air that passes through and the concentration of oxygen that has been removed by respiration, and from that the rate at which metabolic energy is being used. When energy consumption is plotted against speed for walking and then against speed for running, the resulting curves cross each other at about 2 meters per second. Below that speed, walking is the less energy consuming way of getting around; above it, running is more economical. At any particular speed we use the gait that needs less energy unless there is some special reason (such as the rules of a walking race) for doing otherwise.

Notice that both graphs give rates of energy consumption only for low to moderate speeds. The reason is that at higher speeds muscles work anaerobically and we build up an oxygen debt. Measuring oxygen consumption is a good way of finding out how much energy is needed for locomotion, but only at speeds at which the muscles are working aerobically.

The Spring in a Running Step

Runners bounce along in a motion that looks quite different from walking. Running also depends on a different energy-saving principle. When we walk, each foot is on the ground for more than half the stride, so there are stages when both feet are on the ground simultaneously. In contrast, when we run, each foot is on the ground for less than half the stride, so there are stages when both feet are off the ground. Thus running is a series of leaps, and the body is highest when the feet are off the ground. As in walking, the forces on the feet keep more or less in line with the legs, so the body is slowing down at stage (b) and speeding up at stage (d). Thus the body is higher and moving faster at stages (a) and (e) but is lower and moving more slowly at stage (c). Its kinetic and potential energies are highest at stages (a) and (e) and lowest at stage (c). There can be no question here of energy being swapped back and forth between the two forms in the pendulumlike manner of walking, but we will soon see how energy is saved by a different energy-swapping principle.

The basic principles of human running are the same as those of kangaroo hopping, even though the two forms of motion look very different. The kangaroo sets both hind feet down on the ground at the same time, but during each hop kinetic and potential energy rise and fall together as in running : they are highest in midair and lowest at the midpoint of the period of contact of the feet with the ground. Again, there is no question of a pendulumlike exchange of kinetic and potential energy. The same is true of the running gaits of quadrupedal mammals (dogs, horses, antelopes, and all the rest).

In all these gaits, the animal travels like a bouncing ball. When a ball hits the ground, it is brought rapidly to a halt, losing kinetic energy, which is largely converted into elastic strain energy as the ball is squashed out of shape. The ball then springs back to its original shape, and the elastic energy is converted back into kinetic energy as the elastic recoil throws the ball back into the air. Similarly, in the case of a running person or a hopping kangaroo, the kinetic energy and potential energy lost at each footfall are converted briefly into elastic strain energy and then returned in an elastic recoil.

Our bodies owe their bounce to springs that stretch to store elastic strain energy and recoil to return it. The most important of these springs are tendons, especially the tendons of muscles in the lower parts of the legs. As the connection between muscle and bone, tendons transmit the force from muscles to the moving joint. The force stretches the tendons whenever it increases and allows them to shorten whenever it falls. Tendons are not very obviously elastic : they stretch only a little before they break. In this respect they are more like ropes than like rubber bands. Yet the small amount of stretch is enough to save the muscles a considerable amount of work.

Tendons from different mammals and different parts of the body all have very similar properties, and all will stretch by about the same percentage of their length when under the same stress. Their elastic properties are best investigated in machines of the kind that engineers use to test the strength and elasticity of metals and plastics. At the top of the machine shown in the picture on this page there is a load cell, an electrical device that measures any forces that are exerted on it. At the bottom is a hydraulic actuator that can be made to move up and down and that is capable of exerting large forces. A tendon

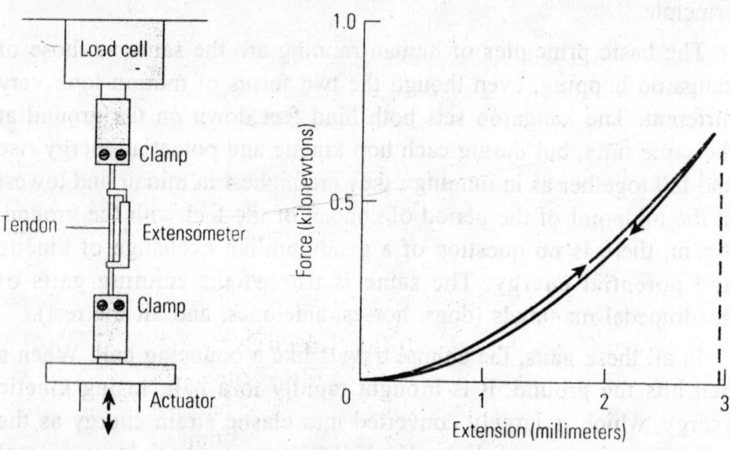

Fig. 2.11. : A dynamic testing machine used to measure the elastic properties of tendons (left), and the record of a test on a tendon from the hind leg of a wallaby (right)

dissected from an animal carcass is fixed in the machine, one end held in a clamp attached to the load cell and the other in a clamp attached to the actuator. Moving the actuator down stretches the tendon and moving it up allows the tendon to recoil. The machine can be adjusted so that the forces and the rates of stretching are about the same as they would be in running. The tendon would be moist inside the body and must be kept moist in the experiment because its elastic properties would change if it were allowed to dry out. The tendon amy be kept at the temperature of the living body, but the extra bother generally seems unnecessary because the properties of tendons at normal room temperature are almost exactly the same as at body temperature.

The graph shows the result of a test on a tendon from the hind leg of a wallaby, a small kangaroo. The record shows the force increasing as the tendon is stretched and falling as it is allowed to shorten, just as if a rubber band or a steel spring were being stretched and allowed to recoil. The line showing the force during stretching is slightly above the line for the recoil, so the record forms a narrow loop rather than a straight line. The loop in the record shows that the energy returned in the recoil is a little less than the work needed to stretch the tendon, which is inevitable because no material is perfectly elastic. (The energy that is not returned is lost as heat.) A tendon returns 93 percent of the energy, losing only 7 percent as heat; this percentage is good in comparison to rubbers and plastics. If it were a less good elastic material, more energy would be needed for running, because muscles would have to do work to replace the lost energy. Moreover, our tendons would heat up when we ran and might even get cooked.

The most important tendon for human running is the *Achilles tendon*, which you can feel through your skin, running behind your ankle to attach to the heel bone. To find out how much energy it can save, we need to know how much force acts on it and how much it stretches, because the strain energy stored in a spring is proportional to the force multiplied by the amount of stretch. It would be difficult to measure how much the tendon stretches when we run, but it is fairly easy to calculate that stretch.

From records of people running across a force plate at middle distance speeds, we find that each foot exerts a peak force on the ground of about 2.7 times body weight. The foot presses down with

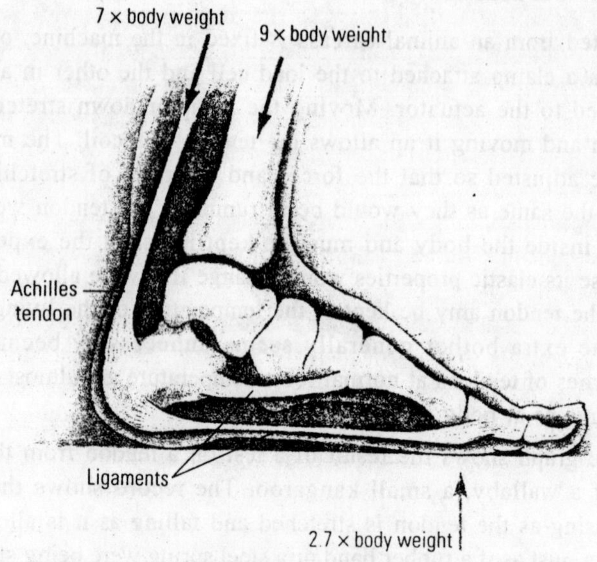

Fig. 2.12. : The forces on a human foot at the stage of a running stride at which they are greatest.

this force on the ground, and the ground pushes up on the foot with an equal, opposite force. This force is distributed over much of the sole of the foot, but for the purpose of calculation we can think of the entire force as acting at a single point, called the center of pressure. Sophisticated force plates tell us not only the size and direction of any force that acts on them, but also the position of the center of pressure. At the stage of a running stride the center of pressure is on the ball of the foot, close to the bases of the toes. The ankle joint is a freely movable pivot, so this force is 2.7 times body weight acting in front of it needs a balancing force behind it. (Similarly, the weight of a child on one side of a seesaw must be balanced by another child on the other side.) The balancing force on the foot is supplied by the muscles of the calf pulling upward on the heel bone through the Achilles tendon. The *Achilles tendon* is about 47 millimeters from the ankle joint, and the line of action of the ground force is about 116 millimeters from the ankle joint, so by the pricniple of levers the force in the tendon must be (116/47) × 2.7, or almost 7 times body weight. This force is about 3500 newtons, or a third of a ton, for a typical (50 kilogram) woman and 5000 newtons, or half a ton, for a 70-kilogram man.

The cross-sectional area of the Achilles tendon in adult men is about 90 square millimeters, so the 5000-newton force sets up a stress in the tendon of about 56 newtons per square millimeter (8000 pounds force per square inch). This stress is about half the stress that would be needed to break the tendon, and it is enough to stretch it by about 6 percent of its length. If we include the part of the tendon that runs up into the flesh of the calf muscles, the tendon is about 250 millimeters long. When stretched by 6 percent, it extends about 15 millimeters, enough to allow the ankle to bend through 18 degrees. If the tendon were not extensible, the muscle fascicles would have to lengthen and shorten this much more to allow the ankle to make the movements that are needed for running. About one third of the negative and positive work that would otherwise have to be done by the muscle lengthening and shortening is done by the passive stretch and recoil of the tendons. If the tendon were inextensible, all the kinetic and potential energy lost by the body in the first half of the step would have to be removed by the muscles doing negative work and lost as heat. It would then have to be replaced by the muscles doing positive work in the second half of the step. Because the tendon is extensible, however, one third of the energy that would otherwise be lost is stored and returned.

Because the muscles do less work, metabolic energy is probably saved, but even further savings are possible. If the muscle fascicles need not lengthen and shorten so much, the person or animal can make do with muscle fascicles that are shorter so fast, and the leg may be moved by slower, more economical muscle fibers. In either case, the "cost of force" elements of the metabolic energy consumption will be reduced.

In antelopes, horses, and related mammals, the tendons that serve as springs in running are very long and the muscle fascicles exceedingly short. Almost all the movement at the ankle joint, while the foot is on the ground in running, results from the stretching and recoil of the tendons, and the muscle fascicles lengthen and shorten very little. The most extreme example is the plantaris muscle of the camel. The plantaris is rudimentary or even absent in people, but in most other mammals it is one of the strongest muscles of the hind leg. It runs from behind the knee, down the shank, around the heel, and along the foot to the toes. In the camel, its muscles fascicles have almost disappeared. Those that remain are only about 2 millimiters long, buried in the tendon, and they can surely have no significant function. The tendon itself is about 1.3 meters (51 inches) long, continuous from the knee to the toes, and like all tendons must serve as a passive spring. Here is a "muscle" that can exert forces and allow joints to move without any metabolic energy cost.

An arrangement so economical must have a drawback; otherwise all mammals would take advantage of it. In this case, the penalty for saving energy is loss of agility. If a camel's leg is positioned so that the plantaris tendon is taut, the camel cannot bend its ankle joint unless it also bends the knee or the toe joints to slacken the plantaris. This loss of freedom of movement is one of the many reasons that camels are not good at climbing trees.

The Achilles tendon is the most important spring in the human leg, but films of barefoot runners suggest that there is another spring in the foot. The films show that while pressed against the ground the foot is considerably squashed : the ankle is forced about 10 millimeters nearer the ground than if the foot were resting lightly. The flattening of the foot is a consequence of its arched structure. We have already seen that large upward forces act on the ball of the foot and (through the Achilles tendon) on the heel. These upward forces are balanced by a downward force of nine times body weight at the ankle joint, where the tibia (the pricnipal bone of the lower leg) presses down on the joint. The two upward forces and the downward force is between partly flatten the arch of the foot, stretching some of the ligaments that connect the foot bones to one another.

My colleagues Robert Ker and Mike Bennett and I suspected that the arch of the foot might be a spring and wanted to test this idea experimentally. The machine that we had used for stretching tendons was also suitable for squeezing feet, but we could think of no way to use the machine safely on feet that were still in place on people's legs. Instead we used feet that had been amputated by surgeons becauses they were diseased or, in one case, because the knee had been damaged beyond repair in a traffic accident. The tibia was attached to the machine's load cell and the foot rested on two steel blocks, which in turn were supported by the actuator. When the actuator was made to rise, the foot was squeezed and the arch flatened. Rollers below the uppermost steel blocks allowed the slight lengthening of the foot that occurred when the arch flattened.

This simple experiment imitated quite well the forces that act on the foot during running. The upward pressure from the block under the ball of the foot imitated the force from the ground. The other block pressd directly on the heel bone (we had removed the fatty pad of the heel, to expose the bone), and its upward push imitated the upward pull of the Achilles tendon. Finally, the downward push from the load cell imitated the force at the ankle joint of the living foot.

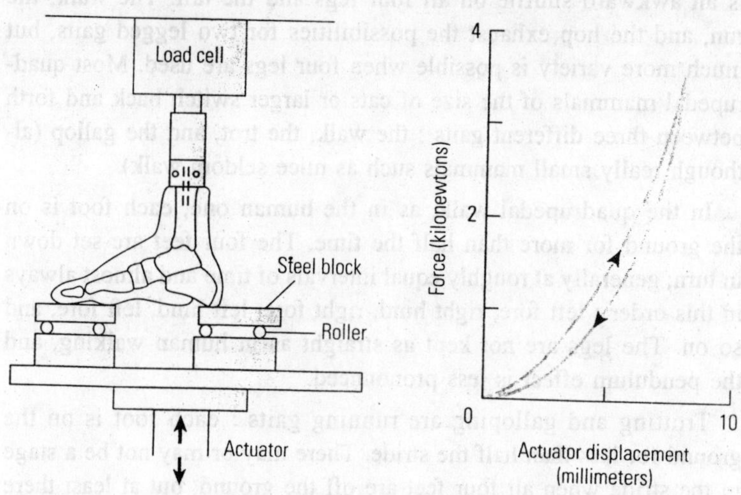

Fig. 2.13. : The experiment that demonstrated the spring in the arch of the human foot (left) and a typical result (right).

The experiments confirmed that the foot is indeed a reasonably good spring. It compressed under load and recoileds immediately when the load was reduced. The records showed loops that were wider than those in the experiments on tendon, indicating that more of the energy was being lost as heat, but most of the energy (78 percent) was returned.

It has been estimated that one third of the kinetic and potential energy that the body loses and regains in a running step is stored in the Achilles tendon and returned in its elastic recoil. The experiments on feet showed that they would store and return a further one sixth of the kinetic plus potential energy. Together, these two springs halve the work that the muscles have to do $\left(\frac{1}{3}+\frac{1}{6}=\frac{1}{2}\right)$. The tendon springs of animals such as horses, camels, and antelopes are probably even more effective.

Four-Legged Gaits

Most animals that travel on two legs have a slow gait and a fast one. People and some birds (chickens, for example) walk to go slowly and run to go fast. Other birds, such as American robins, walk at low speeds, but at higher ones they hop. Kangaroo also have two gaits (the hop is the faster one), but in their cases the solw gait is an awkward shuffle on all four legs and the tail. The walk, the run, and the hop exhaust the possibilities for two legged gaits, but much more variety is possible when four legs are used. Most quadrupedal mammals of the size of cats or larger switch back and forth between three different gaits : the walk, the trot, and the gallop (although really small mammals such as mice seldom walk).

In the quadrupedal walk, as in the human one, each foot is on the ground for more than half the time. The four feet are set down in turn, generally at roughly equal intervals of time and almost always in this order : left fore, right hind, right fore, left hind, left fore, and so on. The legs are not kept as straight as in human walking, and the pendulum effect is less pronounced.

Trotting and galloping are running gaits : each foot is on the ground for less than half the stride. There may or may not be a stage in the stride when all four feet are off the ground, but at least there are stages when both fore feet, or both hind feet, are off. In trotting, the feet move in diagonally opposite pairs, the left fore with the right hind and the right fore with the left hind. Camels and some longlegged breeds of dog use the pace, a gait that is superficially trotlike but in which the two left feet move together and the two right feet move together, instead of the diagonal pattern. Perhaps these animals prefer the pace because if they trotted, the long legs might get in each other's way at the stage when the fore leg swings back and the hind leg of the same side swings forward.

In walking and trotting the left and right feet of a pair are set down at equal intervals; for example, if the right fore foot is set down a quarter of a second after the left, the left is set down a quarter of a second after the right. In a gallop, the intervals are unequal. The two feet of a pair are set down in rapid succession, and then a longer interval follows before the first is set down again. In a full gallop, the two hind feet are set down and then the two fore. The canter, which is sometimes recognized as a distinct gait, is a slow gallop in which the first fore foot is set down at the same time as the second hind.

The footfall pattern of galloping makes it possible for the animals to lengthen the stride by bending and extending the back. While only the fore feet are on the ground, the back bends, pulling the hindquarters forward. While only the hind feet are on the ground, the back straightens again, pushing the forequarters forward. Thus the body moves farther forward while each pair of feet is on the ground than it would if the back remained rigid. That enables the animals to travel faster or to travel more economically at the same speed.

Fig. 2.14. : Four stages of a stride of a camel (Camelus ferus).

The principal muscle that straightens the back is connected to the skeleton of the hindquarters by a sheet of tendon (the technical term is *aponeurosis*). This aponeurosis has elastic properties like other tendons and serves as an energy-saving spring, but only in galloping. To see why a spring might be useful, we have to think more about the animal's kinetic energy. It has seen how the body as a whole decelerates and reaccelerates while the feet are on the ground, but we have ignored the movements of the legs, which swing back while their feet are on the ground and forward again while they are off. Kinetic energy is associated with this movement whenever the legs are moving *relative to the body's center of gravity*. At the end of its forward swing and again at the end of its backward one, each leg has to stop and start swinging the opposite way : it has to lose and regain kinetic energy twice in each stride. The faster an animal runs, the faster the legs have to swing and the larger the swings in the kinetic energy. At the stage of the stride when the back is most bent, the fore legs have been swinging back and are about to swing forward and the hind legs have been swinging forward and are about to swing back. Both pairs of legs have to be stopped and started moving again, so kinetic energy has to be lost and regained. That could be accomplished entirely by muscles doing negative work to stop the legs and then positive work to reaccelerate them, but less metabolic energy is needed if some of the kinetic energy is stored as elastic strain energy in the aponeurosis and returned by its elastic recoil. That is what seems to happen.

It has already been observed how, for people, walking is more economical than running at speeds below 2 meters per second, whereas running is more ecoknomical at higher speeds. Dan Hoyt and Richard Taylor of Harvard University showed similarly for ponies that each of the three gaits—walk, trot, and gallop—was the most economical in the range of speeds at which it is used. Just as in the experiments with people, they had the ponies run on a moving belt while the air they breathed out was collected through face masks and analyzed. It is fairly easy to to train ponies (and many other animals) to run on a moving belt, but Hoyt and Taylor achieved the more difficult feat of training the ponies to walk, trot, or gallop on command. These ponies could be made to gallop at speeds at which they normally would have trotted and to trot at speeds at which they normally would have trotted and to trot at speeds at which they would have preferred to gallop. By analyzing the use of oxygen Hoyt and Taylor obtained a graph of energy consumption per unit distance plotted against speed. The walking and trotting curves cross at 1.7 meters per second, telling us that walking is more economical below that speed and trotting above. Similarly, the trotting and galloping curves corss at 4.6 meters per second; above that speed the advantage shifts to galloping. To travel as economically as possible, the ponies should have changed from walking to trotting at 1.7 meters per second and from trotting to galloping at 4.6 meters per second.

To find out whether the ponies did that, Hoyt and Taylor filmed them moving around their paddock. The two men did not chase the ponies or disturb them in any way, but simply allowed them to move as they chose. They found that the ponies did indeed select the most economical gait, walking below 1.5 meters per second, trotting at 2.8 to 3.8 meters per ssecond, and galloping above 5 meters per second. Furthermore, the ponies generally avoided speeds near the intersections in the graph : they accelerated quickly from walking well below 1.7 meters per second to trotting well above it, and from trotting well below 4.6 meters per second to galloping well above it.

The reason is that ponies need to use less energy per unit distance near the middle of the speed range for each gait and must use more near the transition speeds. For example, a pony that travels at 4.6 meters per second would use 340 joules per meter if it moved steadily at that speed (whether trotting or galloping), but it could cover the same distance in the same time for less energy if it alternated between trotting at 3.5 meters per second and galloping at 6 meters per second using (in each case) only about 300 joules per meter.

People similarly avoid speeds near the walk-run transition. Ultrarunning is the sport of racing over a distance of 100 miles. The best competitors cover the distance in about 13 hours, at a speed of about 3.4 meters per second, running all the way, but many others average speeds of around 2.2 meters per second. These latter competitors walk part of the way at lower speeds and run the rest at higher speeds, avoiding the uneconomical transition speed. Not surprisingly, different animals change gaits at different speeds: for example, short-legged cats make the changes at lower speeds than do long-legged giraffes.

Walking, Running, and the Design of Ships

Imagine that small animals were exact scale models of large ones. To say the same thing in more technical terms, imagine that animals of different sizes were geometrically similar to one another. If one animal were twice as long as another, it would also be twice as wide and twice as high, and all its bones would be twice as long and have twice the diameter. That is obviously not the case: a 2-kilogram cat is not an exact scale model of a 250-kilogram tiger, nor is a 20-kilogram gazelle an exact scale model of a 800-kilogram buffalo. However, it is more nearly the case than you might suppose, as a simple argument will show. The masses of geometrically similar animals would be proportional to the cubes of their lengths (our imaginary animal that was twice as long, twice as wide, and twice as high as another would have 2 × 2 × 2 × = 8 times the volume and so be 8 times as heavy). In other words, the lengths of geometrically similar animals would be proportional to the cube roots of their masses, or to (body mass)$^{0.33}$.

From similarity of shape we move to similarity of movement, for by extending the idea of geometric similarity we can arrive at the idea of dynamic similarity. Two shapes are geometrically similar if

Fig. 2.15. : The galloping movements of cats and rhinoceroses are remarkably similar.

they could be made identical by uniform changes in the scale of length. Two motions are dynamically similar if they could be made identical by uniform changes of the scales of length, time, and force.

To understand what that means, imagine you had a film of a small cat running and another of a tiger running. You could change the sizes of the images on a screen by moving the projectors closer or farther away. You could also change the time taken for each stride by running the pojectors faster or slower. If the animals were running in dynamically similar fashion, you would be able to make the two films seem identical.

The movements of cats are remarkably like those of tigers using the same gait. Indeed, there are close similarities of movement even between less similar animals; for example, between cats and rhinoceroses. Would it be reasonable to suggest as a rough approximation that different mammals using the same gait move in dynamically similar fashion?

There is a physical principle that says that movements that are affected by gravity cannot be dynamically similar unless their Froude numbers are equal:

$$\text{Froude number} = \frac{(\text{speed})^2}{\text{gravitational acceleration} \times \text{length}}$$

The principle was first used by William Froude, a Victorian navel engineer who was making tests on small-scale models before building ships to new designs. He wanted to know how much power is needed because ships push waves along in front of their bows. To find out what the waves would be like around the real ships, Froude needed to know how fast he should propel the models to produce the same pattern of waves. He showed that the model speed should be adjusted to make the Froude number the same as for the ship. The wave patterns around model and ship would then be dynamically similar. As always when the Froude number is applicable, gravity is important here—in this case because it tends to flatten the wayes. The same principle has since been applied to other situations in which gravity is important.

The speed and the length that are used to calculate the Froude number can be defined in various ways to suit different kinds of movement. When considering walking and running, it seems sensible to use the forward speed of the body and the length of the legs:

$$\text{Froude number} = \frac{(\text{speed of locomotion})^2}{\text{gravitational acceleration} \times \text{leg length}}$$

For example, camel's legs are about nine times as long as those of cats, so the Froude numbers of these animals are equal when the camel is travelling three times as fast as the cat (the 3 squared in the dividend will cancel the 9 in the divisor).

Many observations have confirmed that the movements of different-sized mammals are fairly nearly dynamically similar when the mammals are traveling with equal Froude numbers. For example, in dynamically similar movement, animals of different sizes would have equal relative stride lengths (stride length divided by leg length). Measurements from films confirm that a graph of relative stride length against Froude number is more or less the same for mammals as different as dogs, camels, and rhinoceroses—and even for bipeds such as people and kangaroos. Recall that the camel and the cat differed in their speed of movement at equal Froude number: the camel traveled three times faster. We can expect a camel to change gaits at about three times the speed at which a cat makes the corresponding change. Speed measurements from films show that this is aproximately true if you count the peculiar pace of camels as equivalent to the trot of cats. More generally, mammals change from

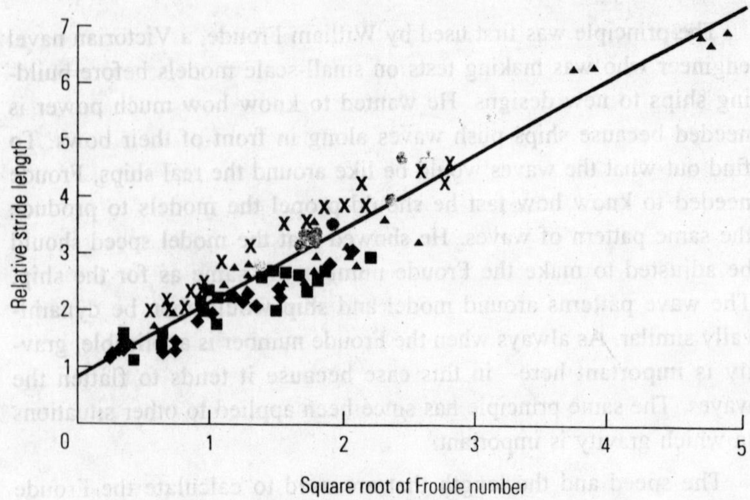

Fig. 2.16. : Animals of different sizes have about the same relative stride lengths when traveling with equal Froude numbers, × humans, ▲ kangaros, ■ dogs, ■ rhinoceroses.

walking to trotting of pacing at a Froude number of about 0.5 (1.0 meter per second for a cat, 2.9 for a camel) and from trotting to galloping at a Froude number of about 2.5 (2.2 meters per second for cats, 6.5 for camels).

This generalization fits well with a conclusion reached earlier in the chapter, that straight-legged walking is impossible when the speed v is greater than the square root of the gravitational acceleration g multiplied by leg length l —when v is greater than $\sqrt{gl}$. This is equivalent to saying that stiff-legged walking is impossible at Froude numbers v^2/gl greater than 1. The change from walking to running is actually made at a rather lower Froude number, but the example may help to show why Froude numbers are important.

Energy Costs and Size

Although animals as different in size as cats and camels move in a similar manner, very tiny and very large mammals do not. If you look at the whole rangde of land mammals from shrews to elephants, you will see that their movements are not quite dynamically similar; smaller mammals run on bent legs and larger ones run with their legs much straighter. This has major consequences for the forces the muscles have to exert and for the energy they use.

The definition of dynamic similarity says that dynamically similar movements can be made identical by adjusting the scales of length, time and *force*: in dynamically similar movements all forces are scaled up or down in the same proportion. An animal's weight is one of the forces that acts on it when it runs. Thus if different-sized animals moved in dynamically similar fashion, their muscles would exert forces proportional to their body masses. Each of the muscles of a 200-kilogram lion, for example, would have to exert 100 times as much force as the corresponding muscle of a 2-kilogram cat. The masses of geometrically similar animals are proportional to $(\text{length})^3$, but the cross-sectional areas of their muscles are proportional only to $(\text{length})^2$, or $(\text{body mass})^{2/3}$. The cross-sectional areas of their muscles would be only $100^{2/3} = 22$ times as much in the lion as in a geometrically similar cat. We divide force by cross-sectional area to obtain the stresses in the muscles, which would be $100/100^{2/3} = 100^{1/3} = 4.6$ times as much in the lion as in the cat. That seems bas enough for lions, whose muscles would have to work much nearer their limits of strength than the muscles of cats have to do, but cats and lions are far from the extremes of mammal size. If a 3-gram shrew were scaled up to the size of a 3-tonne elephant and still ran on bent legs like a shrew, the stresses in its muscles would be increased 100 times.

Elephants and other large mammals avoid the need for impossible muscle stresses largely be keeping their legs much straighter than small mammals. (We say when we discussed human walking how straight legs reduce muscle forces.) Andy Biewener of the University of Chicago has studied the posture of different-sized mammals and the dimensions of the muscles and their positions of attachment; he has concluded that when mammals ranging at least from 90-gram chipmunks to 300-kilogram horses use similar gaits, the peak stresses in their leg muscles are about the same.

It seems obvious that elephants cannot run on bent legs, but why don't shrews run on straighter ones? The forces in their muscles would be less, so they would use less metabolic energy. The most plausible reason suggested so far is that an animal with its legs bent is immediately ready to accelerate or jump, but one standing on straight legs cannot pounce or jump out of the way until it has first bent its legs. Small mammals run on bent legs, and large ones keep their legs straighter; the difference is a matter of size.

Richard Taylor's laboratory is an old missile site in the words outside Cambridge, Massachusetts. There is plenty of room in the surrounding paddocks for the ponies that trotted or galloped on command in the gait experiments, and there is also room for more exotic animals. At various times Taylor has kept kangaroos, gazelles, cheetahs, and even young lions. Indoors he has moving belts to suit animals of all sizes, from 100-kilogram ponies down to tiny chipmunks less than a thousandth of that mass. There he and his colleagues have measured rates of oxygen consumption of a very wide range of mammals (and also of birds and lizards) walking and running at different speeds. They also shipped moving belts and oxygen analyzers to kenya, where they measured mammals ranging from a 600-gram mongoose to a 240-kilogram eland (a large antelope). One of their most ambitious projects was to measure the oxygen consumption of a moving elephant, but even they were daunted by the prospect of building a moving belt strong enough to support the animal. Instead they made the measurements in a zoo while the elephant walked along paths and the oxygen analyzer traveled alongside on a golf cart.

Not surprisingly, this ambitious program of research showed that animals in general (like people and ponies) use oxygen faster when running fast than when walking or running slowly. Also (again not surprisingly), large animals use oxygen faster than small ones traveling at the same speed. More interestingly, the research produced a general equation relating the energy used by a running animal to its speed and body mass. It can be described by an equation,

$$P_v = P_o + Cv$$

P_v is the rate of consumption of metabolic energy when running at speed v and P_0 is the rate of consumption when standing still. C_v is the extra rate of energy use (over and above the standing rate) for running at speed v; therefore C is the extra power (rate of energy use) divided by speed. In other words, C is the energy cost per unit distance:

$$C = \frac{\text{energy cost}}{\text{distance}}$$

C is a constant for each individual animal, but it is different for different species, especially for species of different sizes. You might expect locomotion to be twice as costly for a large animal as for a smaller one hald its mass; in other words, you might expect C/m to be constant (where m is body mass). However, the real situation is more complicated. C is larger for larger animals, but not in proportion to their masses: c/m is *smaller* for larger animals.

The graph on this page shows C/m plotted against body mass. The scales have been made logarithmic so that the distance along the x-axis from 10 grams, to 100 grams, for example, is the same as from 100 grams to 1 kilograms or even from 10 kilograms to 100 kilograms. Plotted in this way, the data points all lie close to a line that has a slope of –0.32. This tells us that C/m is proportional to (body

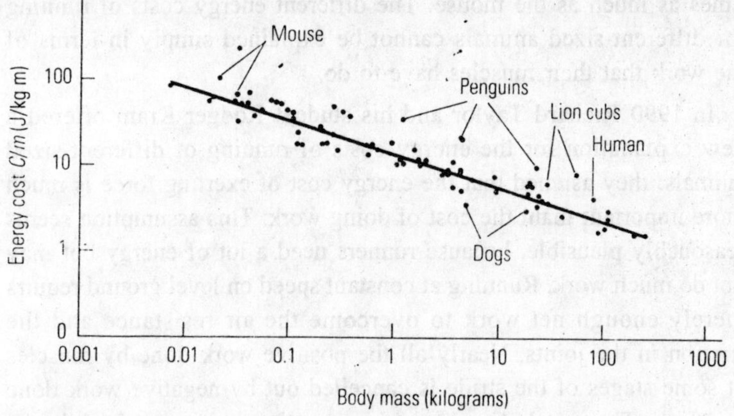

Fig. 2.17. : The energy cost of walking or running for various mammals, plotted against body mass.

mass)$^{-0.32}$, so the cost of travelling a unit distance, C, is proportional to (body mass)$^{0.68}$. Every time you increase body mass by a factor of 10, for example, you increase the energy cost per unit distance by a factor of $10^{0.68} = 4.8$. Some obviously ungainly animals such as penguins have a higher value of C than the line predicts, but the data are probably not accurate enough to say much about which animals move economically and which do not. Their main value is that they show the general relationship between C and body mass.

Imagine similar animals of different sizes, moving with dynamically similar gaits. The forces involved would be proportional to their body weights, and the distances covered would be proportional to their leg lengths. Thus the work (force times distance) done in corresponding movements would be proportional to body weight multiplied by leg length. The distance traveled in a stride would also be proportional to leg length, so work per unit distance would simply be proportional to body mass—that is, to (body mass)$^{1.00}$. If muscles of different-sized animals worked with the same efficiency (and there is no very obvious reason why they should not), metabolic energy used per unit distance would be proportional to (body mass)$^{1.00}$.

It is actually approximately proportional to (body mass)$^{0.68}$. The discrepancy is serious because we are dealing with a very wide range of body masses. If energy cost were proportional to (body mass)$^{1.00}$, a 1.5-tonne elephant would use 75,000 times as much energy per unit distance as a 20-gram mouse. The energy cost is actually proportional to (body mass)$^{0.68}$, and the elephant uses only about 2000 times as much as the mouse. The different energy costs of running for different-sized animals cannot be explained simply in terms of the work that their muscles have to do.

In 1990 Richard Taylor and his student Rodger Kram offered a new explanation for the energy costs of running of different-sized animals: they asumed that the energy cost of exerting force is much more important thant the cost of doing work. This assumption seems reasonably plausible, because runners need a lot of energy but may not do much work. Running at constant speed on level ground requirs merely enough net work to overcome the air resistance and the friction in the joints. Nearly all the positive work done by muscles at some stages of the stride is cancelled out by negative work done at others. The metabolic energy consumption is increased while the muscles are shortening (doing positive work) and reduced while they are lengthening (doing negative work) and reduced while they are lengthening (doing negative work), but the total energy used in a complete stride may not be much more than if the muscles had exerted the same forces, without either lengthening or shortening. It may be possible to explain the energy cost of running almost entirely in terms of the cost of force production.

The energy cost of force production is expected to be

$$\text{cost of force} = \frac{\text{force} \times \text{fascicle length} \times \text{time}}{\text{economy}}$$

This esquation tells us the energy used in a given time, but we would like to know the energy per unit distance, the qunanity C that we have been discussing. Divide both sides of the equation by distance and remember the speed is distance/time:

$$\frac{\text{cost of force}}{\text{distance}} = \frac{\text{force} \times \text{fascicle length} \times \text{time}}{\text{economy} \times \text{distance}}$$

$$= \frac{\text{force} \times \text{fascicle length}}{\text{economy} \times \text{speed}}$$

Kram and Taylor argued that the less time the feet stay on the ground at each footfall the faster (and so less economical) the muscles must be. For example, a mouse whose paws each stay on the ground for one twentieth of a second in each step needs faster, less economical muscles than a horse whose hooves each stay on the ground for half a second. They suggested that economy might be proportional to ground contact time:

$$\frac{\text{cost of force}}{\text{distance}} \text{ is proportional to } \frac{\text{force} \times \text{fascicle length}}{\text{ground contact time} \times \text{speed}}$$

Ground contact time multiplied by speed is the distance the animal's body travels while one particlar foot is on the ground; this is called the step length:

$$\frac{\text{cost of force}}{\text{distance}} \text{ is proportional to } \frac{\text{force} \times \text{fascicle length}}{\text{step length}}$$

If animals of different sizes were geometrically similar to each other and used dynamically similar gaits, fascicle length and step length would be proportional to each other. As a consequence, the cost per unit distance of generating the required muscle forces would simply be proportional to the forces themselves. The forces would be proportional to body mass, and so the cost per unit distance would be proportional to body mass. That conclusion is identical to the one we reached when assuming that the cost of running was the cost of work. We do not seem to have made much progress toward understanding the energy costs of different-sized animals.

Animals of different sizes, however, do not use gaits that are precisely dynamically similar: small mammals like mice run on strongly bent legs, and large ones like elephants run on much straighter legs. In geometrically similar animals, cross-sectional areas would be proportional to $(\text{length})^2$ or to $(\text{body mass})^{0.67}$. This argument suggests that the forces in leg muscles, and so the energy cost per unit distance (C), should be proportional to $(\text{body mass})^{0.67}$. This conclusion is almost exactly right: measurements of oxygen consumption already described show that C is approximately proportional to $(\text{body mass})^{0.68}$.

This explanation of the energy cost of running is impressive and very attractive. It assumes that the energy cost of running is predominantly the cost of force rather than of work, and it assumes that economy is precisely proportional to ground contact time. Our knowledge of muscle physiology is not yet good enough for us to be sure whether these assumptions are sound.

The Kram and Taylor theory seems to explain why many animals that run well have rather long legs for their body mass. For example, a 60-kilogram gazelle has hind legs 0.80 meter long, whereas a 60-kilogram warthog has hind legs only 0.55 meter long. The longer legs of the better runners enable them to increase step length, and the theory tells us that the cost per unit distane should be proportional to 1/step length.

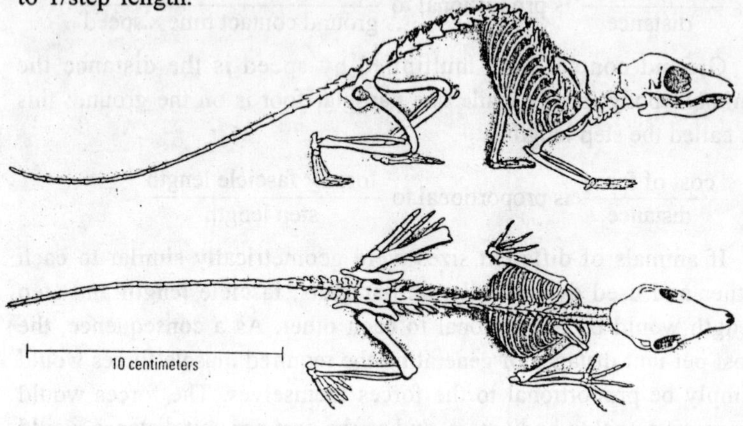

Fig. 2.18. : Small mammals such as this tree shrew run in a crouched position, with their legs strongly bent.

A similar argument provides a further explanation of the advantage of galloping. It has already seen how the work that the muscles have to do, swinging the legs backward and forward, is reduced in galloping by the spring action of an aponeurosis in the back. In addition, bending and extending the back enables the body's center of gravity to travel farther while a foot remains on the ground: it increases the step length and thereby (by our new argument) should reduce the energy cost of running. Greyhounds and cheetahs are fast runners with shorter legs than antelopes of the same mass, but with exceptionally flexible backs. This flexibility suggests that they should be economical runners, but measurements of oxygen consumption while running do not show much difference between cheetahs and other mammals of similar size.

Antelopes, horses, greyhounds, and cheetahs seem to be among the fastest running animals, but again it is difficult to be sure. Many of the animals running speeds that can be found in books are merely subjuctive impressions based on the observer's experience of road traffic. Others are speedometer readings made by driving alongside a running animal in a vehicle, but they too are often unreliable. For example, if the animal tries to swerve away from the vehicle it will be traveling on the inside of a bend and the vehicle on the outside: the vehicle therefore must travel faster than the animal to keep alongside. The only large animals for which maximum speeds have been reliably measured are greyhounds and racehorses. Times given in the sporting pages of newspapers show that most greyhound races are won at 15 to 16 meters per second (34 to 36 miles per hour) and that most horse races are won at 16 to 17 meters per second (36 to 38 miles per hour). These animals have been bred for speed and seem likely to be faster than most wild animals, which have been selected for in the course of evolution not only for speed, but also for other qualities. A good athlete who runs 100 meters in 10 seconds is averaging only 10 meters per second, and the peak speeds of the world's best sprinters are only about 12 meters per second.

Many books will tell you that cheetahs can run at 70 miles per hour (31 meters per second). Such claims seems to be based on a popular article that described a tame cheetah running 80 yards in 2¼ seconds. Unfortunately, according to a later article, the enclosure where the test was made is only 65 yards long, so the speed that is claimed is probably much too high. A later record of 56 miles per hour (25 meters per second) is still astonishingly high, although more believable.

Sprawling Gaits

Birds and mammals run with their feet close under the body, so the lines of footprints made by their left and right feet are close together. Fossil footprints of dinosaurs show that they walked in the same way, but modern reptiles run with their feet well out on either side of the body, so that the lines of left and right footprints are well apart. This manner of running is usually described as a "sprawling" gait. Reptiles move diagonally opposite feet together, the left fore with the right hind and the right fore with the left hind. This sequence is the same as in mammalian trotting, but lizards also use it for slow gaits. They bend their bodies from side to side as they run,

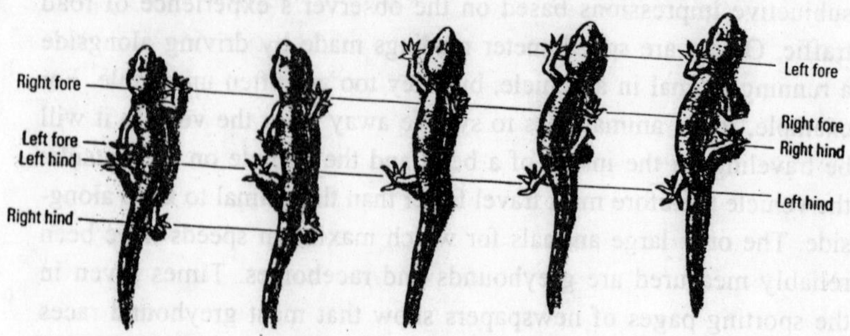

Fig. 2.19. : Lizards use a sprawling gait and bend their backs from side to side as they run.

timing the bending so that it increases the length of their steps.

Until recently, zoologists assumed that the sprawling style of running was uneconomical because it required large forces in muscles. Only when the rates of oxygen consumption of running lizards were measured was it realized that they run as economically as mammals. The energy cost per unit distance (*C* in the equation on page 43) is about the same as for mammals of equal mass, and the metabolic rate while standing still (P_o) is considerably less than for mammals. Why then did posture change in the course of evolution, from the sprawling stance of early reptiles to the mammalian stance with the feet tucked in under the body?.

Dave Carrier of the University of Michigan showed that they stop to breathe; running and breathing use the same muscles in different ways, and so the two activities cannot be performed simultaneously. In running, the side-to-side movements of the body require the muscles to the left and right sides of the trunk to contract alternately. In breathing, however, the muscles of the two sides must act together to enlarge the ribcage and draw air into the lungs, or to compress it and drive air out. Mammals have no such difficulty: indeed, the movements of galloping seem to help breathing. The mammalian back, bending up and down instead of from side to side, works like a bellows. When it bends it squeezes the body cavity, driving air out of the lungs, and when it straightens it draws air in. Records of the flow of air through galloping horses nostrils show that they take one breath per stride, breathing out as the back bends and in as it extends again.

The sprawling gait can be fast as well as economical. As a general rule, lizards can sprint about as fast as mammals of equal mass. Quite small (50-gram) lizards have been timed at the remarkably high speed of 8 meters per second (18 miles per hour).

In contrast, tortoises are notoriously slow. Their slow muscles are very economical. If you try to ride a bicycle very slowly, you will probable wobble and fall off. Similarly, walking very slowly requires more precise control of the forces on the ground than walking faster or running. The reasons are not the same as for slow bicycling (there is nothing in walking comparable to the stabilizing gyroscope action of bicycle wheels), but the effect is the same: slow movement requires more precise control.

Whatever the speed of walking, the feet exert fluctuating forces on the ground, and the body is generally not in equilibrium. At one stage the forces on the feet may total more than body weight and accelerate the body upward, but at another they may be smaller and let the body fall. The body will rise and fall in the course of a stride but this may not matter, unless the vertical movements are so big that the belly bangs on the ground. Similarly, temporarily unbalanced horizontal forces may make the animal speed up and slow down during each stride, or veer from side to side. Imbalance between the forces exerted by different legs may tilt the animal, causing it to pitch and roll. One can think of the rising and falling, speeding and slowing, pitching and rolling as unwanted movements, but walking remains effective if they are not too large.

When an animal goes fast, footfalls follow each other in rapid succession, and an unwanted movement started by an unbalanced force at one stage of a stride can soon be corrected. If the animal is moving slowly, however, the unwanted movement movement will continue for longer before there is an opportunity to correct it, and it may go too far. If the animal is moving very slowly, force fluctuations must be kept small.

The problem is especially severe for a low-slung animal, such as a tortoise that supports its shell close above the ground. A small loss of height or a tilt through a small angle may make it hit the ground, which presumably would be unsatisfactory. A quick calculation will make the problem more obvious. Imagine that an animal's legs suddenly stopped exerting any force, so that the trunk was completely unsupported. If the animal were a tortoise with its belly initially 5 centimeters from the ground, it would hit the ground after only 0.10 second. For a tortoise of this size, each stride would last about 2 seconds, so the shell would hit the ground if the animal was unsupported for a mere twentieth of a stride. However, if the animal were a dog with its belly 40 centimeters from the ground , it would take 0.28 second to fall. The dog's strides would each last about 0.5 second, so in its case the falling time would be half a stride period. The tortoise could tolerate only a tiny unsupported fraction of a stride (or smaller force fluctuations for larger fraction), but the dog's feet could be off the ground for quite a lagre fraction of a stride without ill effects.

In theory it is possible for a tortoise or other quadruped to keep perfect equilibrium through each stride without rising, falling, pitching, or rolling at all. Three feet are enough to support a body in stable equilibrium: a three-legged stool is stable, but a two-legged one is unstable and will fall over. The tortoise could remain stable by moving one foot at a time, always leaving three on the ground for support.

If you sit on a three-legged stool and lean over too far to one side, it will fall over. All is well so long as the center of gravity (of you and the stool combined) is vertically over the "triangle of support" formed by the stool's three feet, but if the center of gravity moves outside the triangle, you will toppl:. Similarly, an animal standing on three feet cannot remain in equilibrium if its center of gravity is not over the triangle of support. The problem disappears if the feet are big enough and can be placed directly under the center of gravity (birds and people can stand on one foot), but animals with small feet need three, suitably placed.

It turns out that it is enough for a quadruped to move one foot at a time: to keep the center of gravity always over the triangle of support, it must move its feet in a particular order. This order, shown in the diagram on this page, is actually used by dogs, horses, and most other quadrupeds whenever they walk. Tortoises, which have far more need to keep close to equilibrium, also move their feet in this order, but unlike large mammals they do not move their feet at equal intervals. Each fore foot is set down only slightly before the diagonally opposite hind foot, so the footfall pattern is very like that of lizards (which move diagonally opposite feet together), and there are times when only two feet are on the ground.

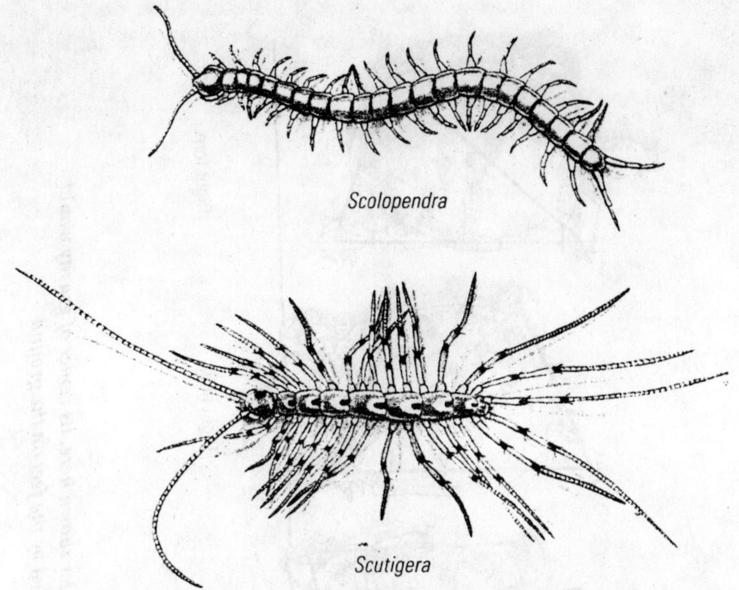

Fig. 2.20. : Scolopendra, a short-legged centipede, moves its legs in sequence from front to back. Scutigera, with long legs, moves them in teh reverse sequence. Legs whose feet are on the ground are tinted.

At first sight this gait seems faulty, but Alan Jayes and I were able to show that, for an animal with slow-acting leg muscles, it would allow steadier walking than if the feet moved one at a time. The reason is that the "ideal" gait in the diagram can keep the animal in constant equilibrium only if the forces exerted by the other feet can be changed instantaneously, whenever a foot is lifted or set down. If the muscles are incapable of rapid adjustments, the animal can keep closer to equilibrium by moving its feet in diagonally opposite pairs.

Keeping close to equilibrium is easier with six or more legs. Insects generally move their six legs in two groups of three, the two groups taking turns to provide the triangle of support. Each group consists of the front and rear legs of one side of the body and the middle leg of the other.

Centipedes may have to use gaits that keep their many legs out of one another's way. *Scolopendra* is a short-legged centipede that moves the legs of each side of the body in sequence from front to back: leg 1 before 2 before leg3, and so on. In a group of adjacent feet that are on the ground, the ones in front are at a later stage of

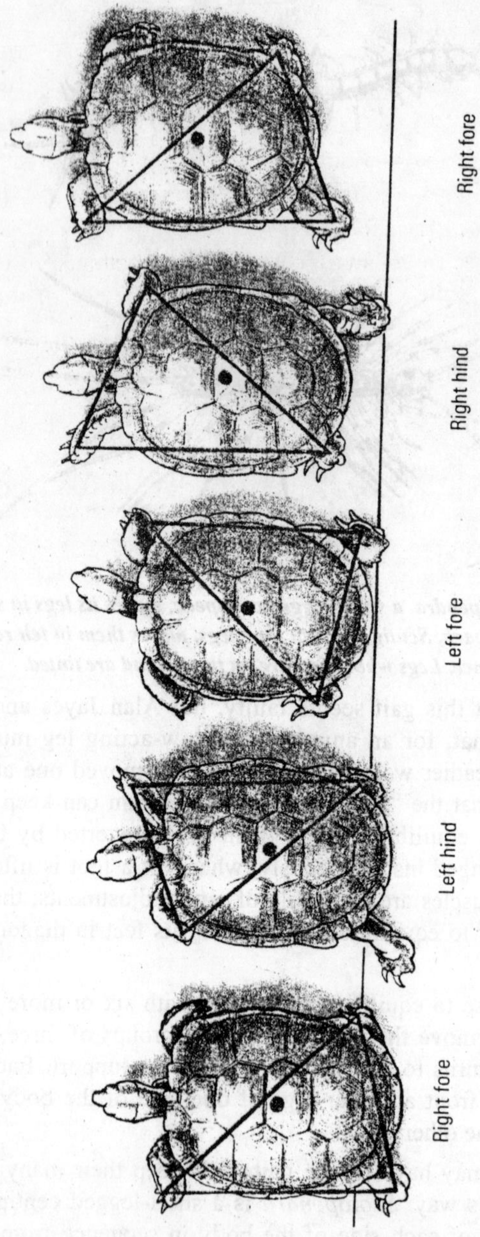

Fig. 2.21.: If a tortoise moves one foot at a time in the order shown here, its center of gravity would always be over the triangle of support provided by the feet on the ground.

the step than the ones behind, so the feet on the ground form clumps, as the diagram on this page shows. This gait presents no problem for *Scolopendra*, but if the legs were much longer they would cross over each other so that the foot of leg 2 was set down in front of foot 1, and foot 3 in front of foot 2. It is hard to see how the centipede could do that without getting into a tangle. Another diagram shows how *Scutigera*, a long-legged centipede, avoids the difficulty. It moves its feet in reverse sequence, starting at the rear. The result is that the feet are spread out instead of being clumped or crossed.

Insects, centipedes, and other arthropods use sprawling gaits, keeping their feet well out on either side of the body. For reasonably large land animals, sprawling gaits are optional: lizards and crabs use sprawling gaits, but birds or mammals of similar mass do not. For small animals such as insects, sprawling is probably essential: if they did not stand with their feet well out to either side of the body, they would be apt to be blown over by gentle breezes. The reason is that the force of the wind on an animal is proportional to its surface area, but the weight that helps to stabilize the animal is proportional to its volume. For geometrically similar animals of different sizes, smaller ones have larger ratios of area to volume and so are more likely to be blown over.

CHAPTER 3
MOVEMENT IN WATER

The aquatic realm covers a large area of the world. It is one of the most important habitat in which a large number of animals dwell.

Fishes are the primarily aquatic forms which have evalved from more primitive aquatic progenitors and never had a terrestrial ancestry. Their adaptation to aquatic medium is perfect and they do not have any inconvenience due to the fact that they are primitive gill-breathing vertebrate. But those which are secondarily adapted to the aquatic mode of life, feel much inconvenience especially to breathe in water. Fishes perform their respiratory function in the watery-medium but in secondarily adapted animals, this process is carried on by the air which passes through the nostrils into the lungs.

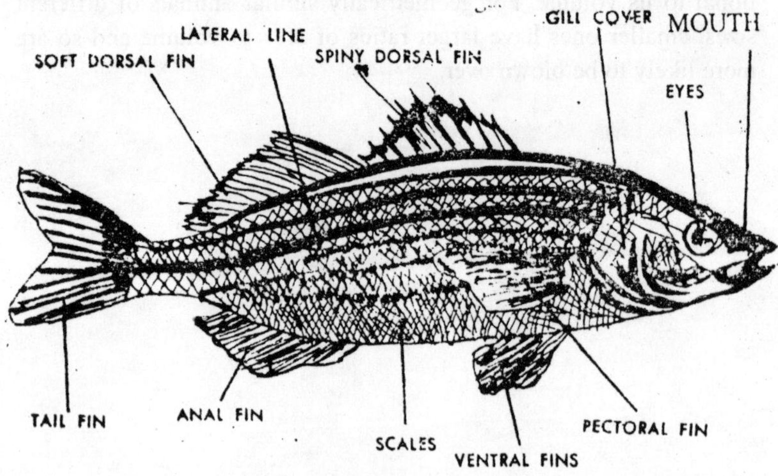

Fig. 3.1 : A Typical Bony Fish

Fishes are the primitive gill-breathing vertebrates, which are often provided with accessory respiratory organs but still retain gills as the chief organ for respiration throughout their life. The dense medium in which they live, except for a few, has exerted profound influence upon them and hence they have acquired similarity in form, in such a way that everyone can recognize a fish easily.

1. Body contour :

The contour of an aquatic vertebrates is very important and has been moulded according to the needs. The head, body and tail of aquatic vertebrates are compressed into a beautifully curved stream-lined form, the entire surface is accurately rounded without any protuberances which would be an obstacle in the swift passage of the animal through water. They have a sub-conical head. The edges of jaws and gill-covers fit precisely and even the eyes conform accurately to the curvature of the head. When the typical swiftly-swimming fish e.g. Spanish mackeral (*Scomber-omorous maculatus*) is viewed from the front, it is found that is tis a perfect ellipse and the fins which are so prominent when seen from the side are hardly visible because they are reduced to thin keel-like lines.

2. Locomotion :

The process of locomotion is performed by lateral undulations of the flexible body, which is provided with fins. The pectoral fins act as balancers and the caudal fins serves as a rudder in changing the direction of movement. Therefore, fish are accessory locomotive organs. The body is thrown into a series of curves which begin at the tail region. The static water is enclosed in the incurved places which is pressed upon and as a result, causes the (orwar) movement of the fish. In an ordinary fish, there is some lost motion or "slip", for the water is not sufficiently resistant to oppose the thrust of the body but in the larval eels, which are thin, ribbon-shaped animals and whose great relative depth presents a large lateral surface, the swimming is so precise that if a pencil is held in one of the hollows, the body of the fish passes without touching it. It moves forward without any loss of motion due to the fact that its lateral surface is sufficiently great. The fish develops unpaired fin-folds of the skin, which are stiffened by fin-rays made up of cartilage or elastic bone and thus increase of resistant surface is obtained in the fish through it. These unpaired fins may by more on less continuous from the head along the middle line of the back, around the tail and forward along the underside the body, as far as the vent, or they may be broken into a number of distinct fins. These unpaired fins may be dorsal, caudal and anal and out of these three, the caudal is the most important propelling organ.

The fish has also paired or lateral fins, i.e., pectoral and pelvic fins which correspond to the fore and hind-limbs of thee higher vertebrates. The pectoral-fins lie just benin the gill-apparatus at the shoulder, while the position of pelvic-fin is very much variable, though they normally led on either side of the vent. Pectoral-fins have additional functions also, as they as stabilizers and help to check the animal's way.

3. Swim-bladder :

All fishes above the sharks may have a swim-bladder, which is a further adaptation to the aquatic life. It is a hollow organ filled with air or gas. It is largely hydrostatic in functions, as it maintains the fish at a certain depth of floatation. If the animal tends to sink, the body is compressed slightly by the contraction of muscles and as a result, the swim-bladder or air-bladder is also compressed. Due to this, the bulk is lessened where-as the weight remains constant, the specific gravity is increased with the resultant loss of buoyancy and ultimately, the fish sinks down. When the fish wishes to come up, it repeats the process just the reverse described above.

Respiration and not hydrostatic, was the primal function of swim-bladder in certain ancestral fishes. This organ is homologous with the lung of terrestrial vertebrates and there can be no doubt that the swim bladder or air bladder of certain ancient fishes actually evolved into a lung. In this way, the progenitors of all higher vertebrates arose.

Secondary aquatic adaptation of vertebrates:

Secondarily adapted aquatic forms were the lung-breather vertebrates, which through the series of circumstances, i.e., in hostile lands, where there was a scarcity of food and severe competition, were forced to return to the water, the primal habitat in which their remote ancestors lived. There is always the handicap of lung breathing but otherwise the adaptation of certain to the more extreme is little short of being marvellous.

When the lung breathers returned back to the aquatic habitat, they readapted the externals of fish life. They kept in a varying degree the higher brain and the more efficient methods of aeration of the blood and locomotion.

1. Amphibious vertebrates :

Amphibians are those vertebrates which spend a part of their time on land and a part in the water. They are actually terrestrial animals which exhibit partial aquatic adaptation only, and rarely extend beyond the possession of webbed-feet and a laterally compressed swimming-tail. The tail may bear a fin-like expansion long its upper margin. Sometimes, some of the wrist and ankle-bones lack ossification. The class Amphibia represents the transitional forms in the original landward migration.

Among vertebrates and in the above class Amphibia, we find many instances e.g., the Galapagos lizards, *Amblyrhnous cristalus* which are terrestrial from choice but aquatic from necessity. These swim to the surface-line and dive for the sea-weeds, upon which they fed. In this case, a flattened swimming tail and a slightly compressed body are their aquatic adaptations. Another example are the Cretaceous Hadrosaurs, which died in the open and therefore, the fossils of the whole animals are found, showing detailed morphology. We find a wonderful battery of teeth, which is an adaptation to a very harsh heybage which have been preserved within these animals. These Hadro-saurian dinosaurs had a splendid swimming tail, webbed hands and feet which were employed in the aquatic or at least amphibious life. These were defenceless and devoid of either weapons or armours. One would, therefore, expect in the Hadro-saurian dinosaurs, a great degree of aquatic adaptations than the Galapagos lizards show.

The marine turtles have undergone a long step farther in the fact that both food and safety is present in sea-water for them but the females come to the shore for the purpose of egg-laying. In some species of turtles, males never come to the shore.

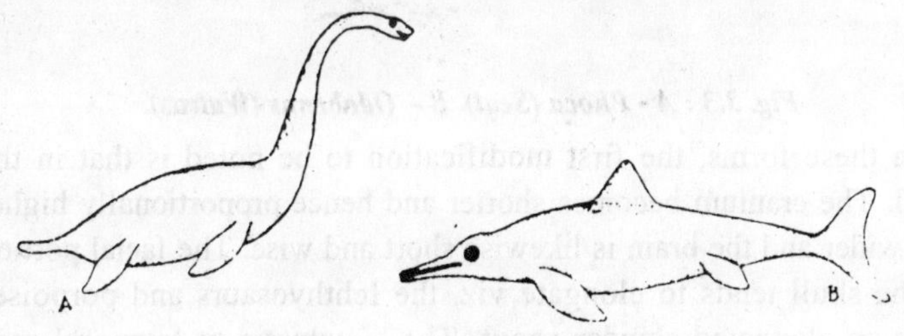

Fig. 3.2: A - Muraenosaurus (Amphibious). B - Ichihyosaurus (Marine aquatic)

Icthyosaurs of Mesozoic era represent the final stage in the aquatic adaptation of reptiles. The perfection for their life conditions equalled that of the modern whales. These reptiles gave birth to the young alive and hence, they did not need to go ashore for egg-laying purpose.

There are 27 orders of secondarily adapted water-inhabiting

vertebrates, some of which are exclusively aquatic.

Among aquatic lung breathers, the order Proganosauria of class Reptilia, are the first in point of time, because the fossil remains are found embedded in the Permian age rocks not very long ago but relatively after the reptiles were established.

2. Body contour :

As in the primarily adapted aquatic forms, here also the following aquatic adaptations are found:

(a) Body contour becomes stream-lined.

(b) The neck construction disappears.

(c) The tail enlarges.

This assumption of fish-like form is best seen in the fully aquatic orders e.g. Ichthyosauria, Cetacea, Sirenia, Pinnipedia and to a lesser extent in several other group (Fig. 3.3).

Fig. 3.3 : A - Phoca (Seal). B - Odobenus (Walrus).

In these forms, the first modification to be noted is that in the skull. The cranium becomes shorter and hence proportionally higher and wider and the brain is likewise short and wise. The facial portion of the skull tends to elongate viz. the Ichthyosaurs and porpoises have an elongated slender snout. The zygomatic or temporal arch of the skull in the Cetacca is reduced almost to a vestige.

The neck shortens to a great extent and the mobility of the neck is lost in the swifter, tail-driven forms. But the number of cervical vertebrae in whale is seven, which is the standard mammalian number; but in manatee among Sirenia, only six cervical vertebrac are present, which is one of the three exceptions to the standard mammalian number. These vertebrac may be fused into a solid, compressed mass of bone. The vertebrae have become secondarily simplified. The centre of the vertebrae and zygapophysis includes the simplification. Several processes may become greatly reduced in the trunk region ad it may be elongated in the tail region to provide greater surface for the attachment of muscle.

The sacrum is more or less reduced in direct ratio to the loss of supporting or propelling function on their part. Thus in Ichthyosaurs, Cetaceans and Sirenians, the actual identity of the sacral vertebrac is lost.

The chest of truly aquatic forms has become cylindrical. The chest is modified in such a way, as to bring the internal cavity higher towards the back. This ensure a greater stability of floatation and increased lung capacity and is accomplished by the ribs, which tend to become highly arched dorsally.

In whales, the articulation of ribs and vertebrae is very much loose which allows a great mobility of chest for the rapid respiration, which is essential after prolonged submergence. The diaphragm has become horizontal in position in Sirenia and Cetacea. The nostrils are often capable of being closed. The eyes have also become adapted to the aquatic vision which because of the denser medium, require a different curvature of the lenses.

3. Locomotive Mechanism :

In whales and lchthyosaurs, fleshy, fin-like expansion of the body-wall has occurred like fishes, there fins may be dorsal and caudal but there secondarily aquatic forms often go back to first principles ad readopt the old undulatory or wriggling movement of their ancient ancestors to which fins, etc. are only subsidiary. Together with it, goes the elongation of the body, multiplication of segments and the loss of limbs.

Thus, two methods of propulsion are seen among these aquatic forms. According to Williston, these are known as "Oar propulsion" and "tail propulsion".

Oar propulsion type of locomotion: It is met with in turtles and plesiosaurs. In these forms, the limbs are nearly equal in size and act as oars in propelling the organisms.

Tail propelled type: This is included in undulatory type of locomotion. This type of locomotion found in the Ichthyosaurs, whales and sirenians. In these forms, the hind limbs tend to disappears until no external vestige is detectable ad the tail is responsible for the locomotion of the animal. Unlike fishes, unpaired fins of secondarily aquatic vertebrates are never supported by fin rays and in this case it is supported by masses of dense connective tissue.

The caudal fin of the marine mammals in horizontal instead of being vertical as in reptiles. By the development of tail, flukes with extremely powerful musculature ad vertical movements, are propelled forward. The medium dorsal fin of the whales serves as a stabilizer.

4. Limbs :

Webbed feet are the first aquatic adaptation. One of the aquatic adaptations is the development of flexible paddles which is of great aquatic utility. In whales, the forelimbs have modified to form paddles having no trace of digit. Due to the development of paddle, the mobility of various joints has been lost and the entire skeleton of the limb has been enclosed by the skin in a single mass. As a further modification, we notice hyperphalangy (increase in the number of phalanges) and hyperdactyly (increase in the number of digits). The length of the fore-limb is much reduced, due to which, the paddle is held very close to the body - a special modification in round headed dolphin.

Clobicephalus The length of two or three digits may be very great while the other may be much reduced.

5. Integument :

Reduction in armouring of hair, skin-glands, muscles and nerves is very characteristic of aquatic animals. In Ichthyosaurs, loss of armour has taken place. In marine mammals, loss of hair has taken place because in water, hair are of little value.

The salivary glands are also reduced in those aquatic forms which devour the food under the water. This is possible because their secretions would be diluted very much in watery medium and after dilution, it will have no great digestive value and also partly due to the mechanical function of lubrication to aid in swallowing is eased by the water, which is taken with the food.

6. Mouth armament :

The power of mastication has been lost, except in Walrus and seacows because the jaws are no longer used for this function and so the coronoid process and other areas for muscle attachment are greatly reduced.

The teeth are entirely absent in the case of Baleen-whale and the function of teeth is taken by baleen or whale bone. This whale feeds on plankton and baleen or whale bone serves as a straining apparatus. In sperm-whale, total loss of teeth from one jaw has occurred, the marine reptiles have simple prehensile teeth which help them to retain the slippery prey.

7. Percocity :

Mental percocity is necessary in the gregarious aquatic animals as among the cursorials and they show an ability to keep up with the mother soon.

8. Speed :

The aquatic animals can move very fast and this speed is maintained by the highly efficient propelling tail. Porpoises are known to keep pace with a 39 - knot destroyer, going ahead of the craft with the utmost ease.

9. Size :

The water-borne animals are larger than animals living in the other habitats, possibly because of the energy exhausted by the terrestrial animals in overcoming gravity might have been turned into a growth force. For example, sulphur - bottom whale may be 87 feet long as compared with the largest recorded terrestrial animals, the elephants.

From the foregoing account, it becomes clear that the fishes are primarily adapted aquatic animals in having a streamlined body and gills for respiration. They have paired (pectoral ad pelvic) and unpaired fins which aid in swimming.

The secondarily adapted (amphibious and aquatic) animals which are derived from terrestrial ancestors also show modifications in body contour, limbs, integument and likc fishes in the development of fins, which are structurally different from those of fishes.

The movement of fishes

All fish can move in water, but they do not all move in the same way. Their shape is important for swimming purposes but their internal structure is even more important. It is this that limits them to particular movements. If you watch a fish swimming slowly, you can tell from its quick, sudden movements, that may seem the same to the human eye, whether it is a goldfish, a herring or a cod. It is useful to have a cinecamera to record the movements of fishes so that you can watch their swimming in slow motion. Two important observations can be made: the main part of the movement is not carried out by the fins but by the tail and latter part of the body: the movement is the result of a series of rhythmical flexures of the body. Analysing these flexures further, we see that they are made up of waves running along the body. The extent of these waves increases little by little until they reach as far as the head and the tail. The 'motor' that makes the fish swim is the series of muscular segments on either side of the vertebral column. These muscular segments, or myotomes, are an important part of a fish's body. An inactive fish, such as the well-known goldfish, has muscle equal to two-fifths of its total weight. A tunny, on the other hand, can have three-quarters of its weight made up of muscle. The myotomes are in close contact with the vertebral column and placed in such a way as to exercise their force by pulling crosswise on the joints that separate the vertebrae. The myotomes are also placed as if they were embedded in each other and usually slope from head to tail. This results in the flexure of the fish's body from one part to the other, according to whether the right or left side of the muscles is contracted. Swimming is thus an alternate contraction or distension of the fish's muscles. It is obvious that when the myotomes on the right side contract, those on the left side distend.

Fish also need to maneuver in their surroundings, to be able to perform turns and stops, and to make certain movements in reverse. These are carried out by other fins. If you watch goldfish in an aquarium, it is easy to surprise them carrying out little movements forwards, backwards and sideways, near the walls of the tank or some other obstacle. These fishes reveal considerable ability for maneuvering: their pectoral fins are the most efficient instruments. These fins are most important to the stickleback in the mating season. This little freshwater fish builds a nest where his mate will lay her eggs. The nest is made of plants joined together by a viscous secretion from his kidney. The stickleback uses its mouth in the building, but it could not do this without its pectoral fins. After the eggs have been fertilised the fins are used as a fan to create a current of water that keeps the eggs well-oxygenated. The unpaired fins, such as the dorsal and the anal, act as rudders and are also used to prevent vertical oscillations. Many species equipped with particularly well-developed dorsal fish can fold them back until they disappear, so that they can reach higher speeds in rapid swimming.

Swimming styles

We have seen how the flexing movement impels the fish : now let us look at the way the waves of flexure spread out in the water, producing the forward impetus necessary for swimming. By analysing the movements of various species in slow motion, we can see their different ways of swimming. It varies from the style of the tunny, in which only the caudal fin moves, to that of the mackerel (2) and the shark (3), which displays the waves of flexion far more, and to the style of the eel, in which the waves convulse the whole animal (4). Both the eel and the tunny move in their surroundings without any difficulty, but the eel has virtually no caudal fin while the tunny has a very well developed one. To understand how different types of fish move, we must analyse the forces that create the movement. The undulations of the eel-shaped body produce a series of movements in the water which are obliquely directed to both sides of the fish's rear.

This oblique movement can be divided into two components, one

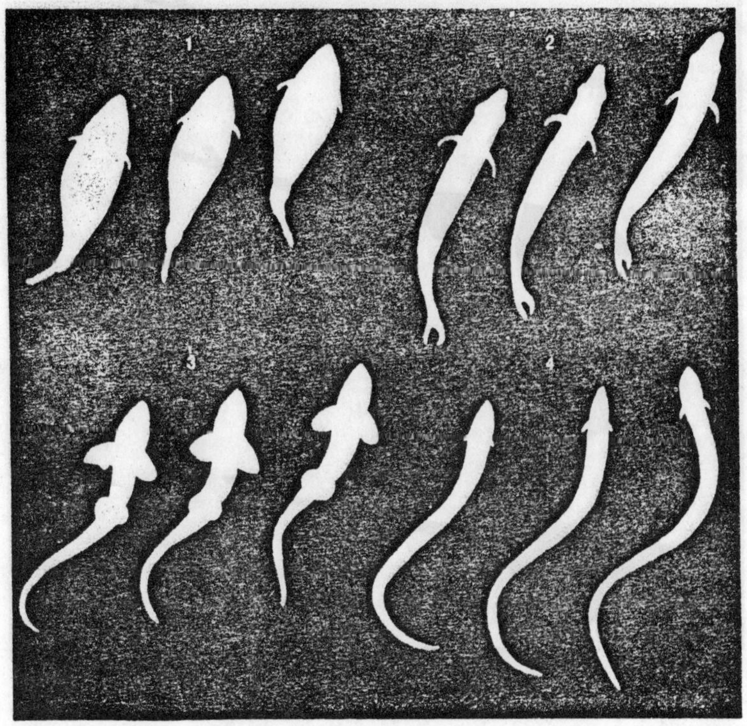

Fig. 3.4. Swimming movements of fishes

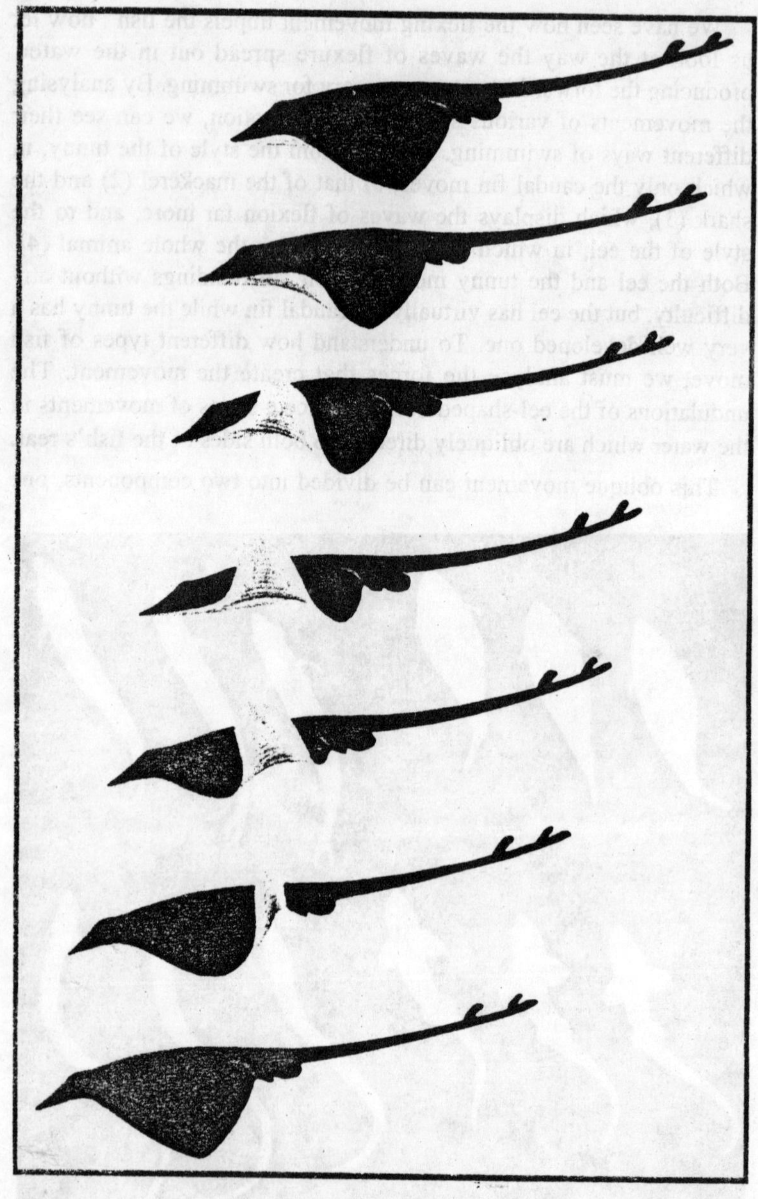

Fig. 3.5. Swimming of a ray

perpendicular to the fish and the other directed to the rear : the perpendicular forces cancel each other out, because those on one side compensate those on the other, but the rearward force makes the fish go forward. The swimming of spindle-shaped fishes is not very different : they reduce the oscillations of the body, but the impetus is supplied equally by the rear part and the caudal fin. If we watch the oscillating movement of the caudal fin, we shall see that the force it exerts on the water is directly oblique to the rear and, as in the case of the eel, can be divided into two parts. The tail and the

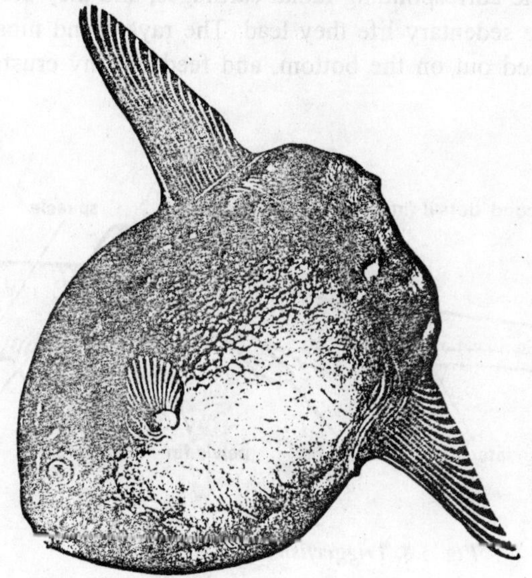

Fig. 3.6. Sun-fish Mola mola

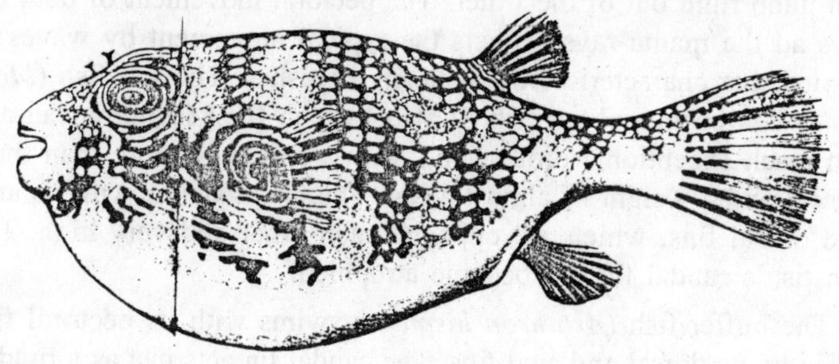

Fig. 3.7. Puffer-fish Arothron lispidus

body contribute to the movement to a different extent : in the eel it is only the body, in the tunny only the tail. We know that the tunny is one of the fastest fish, while the eel is rather slow in its movements.

It is obvious that the caudal fin makes the most important contribution to fast swimming. A special swimming style is adopted by the rays and the devil-rays or manta-rays, which are propelled by the impetus they receive from large pectoral fins which are like great wings. The rays are equipped with a series of pectoral muscles that move the corresponding radial cartilages, and they are perfectly suited to the sedentary life they lead. The rays spend most of their time stretched out on the bottom, and feed on tiny crustaceans or

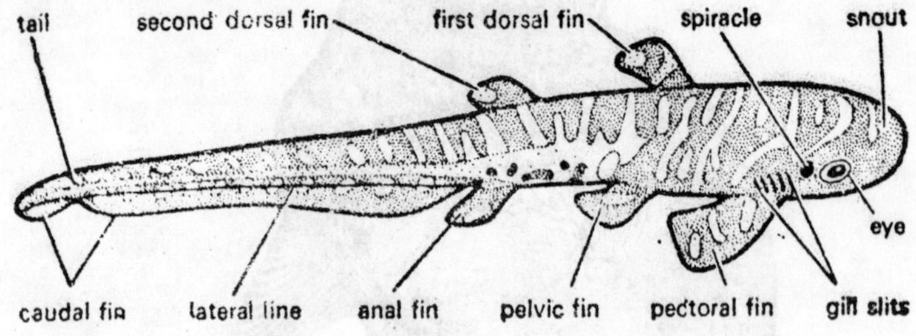

Fig. 3.8. Triggerfish Balistes carolinensis

molluscs. Their flat body is adapted to this life-style ad allows them to disguise themselves by covering themselves with sand flung up by their pectoral fins. The manta-rays, which lead a more active life, can jump right out of the water. The pectoral movement of both the rays ad the manta-rays reflects the typical movement by waves of flexion that characterises the majority of fishes. The sun-fish (*Mola mola*) is a cosmopolitan fish. It also lives in the Mediterranean and can reach exceptional dimensions, with a height of more than three meters and a weight of almost a ton. This giant swims with its anal and dorsal fins, which are opposite each other and very high. The sun-fish's caudal fin has become atrophied.

The buffer-fish (*Arothron hispidus*) swims with its pectoral fins aided by its dorsal and anal fins. The caudal fin acts just as a rudder. The puffer-fish swims slowly, but it can protect itself by taking in water so as to increase its size greatly. The trigger-fish (*Balistes carolinensis*) moves like the puffer-fish, using its anal and dorsal fins. It also has a characteristic, spiky dorsal fin that it can erect when it likes but which plays no part in swimming. Surgeon-fishes (Acanthuridae) move by using their pectoral fins like oars. The trunk or box-fishes (Ostraciontidae) are literally enclosed in a rigid, bony box that prevents any movement of the body except for the tail, although pectoral, anal and dorsal fins move. The strangest of all fishes is perhaps the seahorse (*Hippocampus*), which moves by quickly vibrating the dorsal and pectoral fins. Its movement is slow, but adapted to the life it leads, usually attached to algae by its prehensile tail.

Fast fishes

Best known in America, deep-sea fishing is carried out with a rod and fishing-line on board boats built for this purpose that cruise slowly in fishing areas. The fishermen hope to catch pelagic fish such as sharks, tunnies and bonitos. Once hooked, these fish fight actively and swim fast, and by measuring the amount of line they draw out in a given period of time, it is possible to work out the

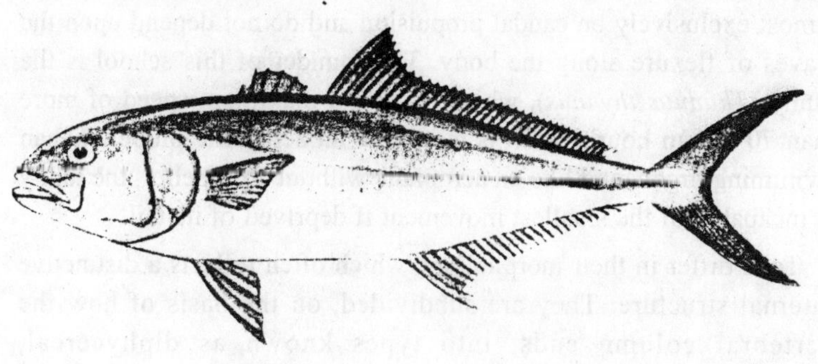

Fig. 3.9. Amberjack

speed at which they make off. The wahoo (*Acanthocybium*), a relation of the tunny, is just under two meters long, but when in flight it pulls out the first hundred meters of line at a speed of more than 65 km an hour. Sharks too are subject to intensive deep-sea fishing. Their speed is plainly slower, but even they can maintain 40 km an hour. All the speed records are held by fishes with a crescent-shaped tail.

Considering tunnies, sharks, bonitos, and amberjacks, we can see

Fig.3.10. Blue shark

an obvious similarity in the shape of the caudal fin. Plainly evolution has caused very different species to adopt the most efficient form of prospulsion. All these fast fish use a swwimming style based almost exclusively on caudal propulsion and do not depend upon the waves of flexure along the body. The founder of this school is the tunny (*Thunnus thynnus*), which reaches a maximum speed of more than 70 km an hour. Its tail is more like an aeroplane propellor than swimming organ, and like an aeroplane without a propellor, the tunny is incapable of the smallest movement if deprived of its tail.

Tails differ in their morphology, which often reflects a distinctive internal structure. They are subdivided, on the basis of how the vertebral column ends, into types known as diphycercal, heterocercal, hemocercal, and gephyrocercal. The diphycercal tail fin is typical of eels and is characterised by complete symmetry between the dorasl and vertral parts. This kind of tail can also be seen in many fish embroys. Sharks have heterocercal tails, characterised by an asymmetry that, in the case of the thresher shark, is considerable, with the longer lobe of the tail reaching a length of around two metres. Heterocercal tails have a characteristic feature, linked to the particular structure of the two caudal lobes, of which the lower is far more flexible. The vertebral column continues into the tail as far as the apex of the upper lobe. The sturgeon and all the other Acipenseriformes have heterocercal tails. The majority of bony fishes have homocercal tail that seems to be symmetrical. But by analysing the structure of the bony parts of the tail, we can see the asymmetry of the vertebral column. Angler fishes and many other

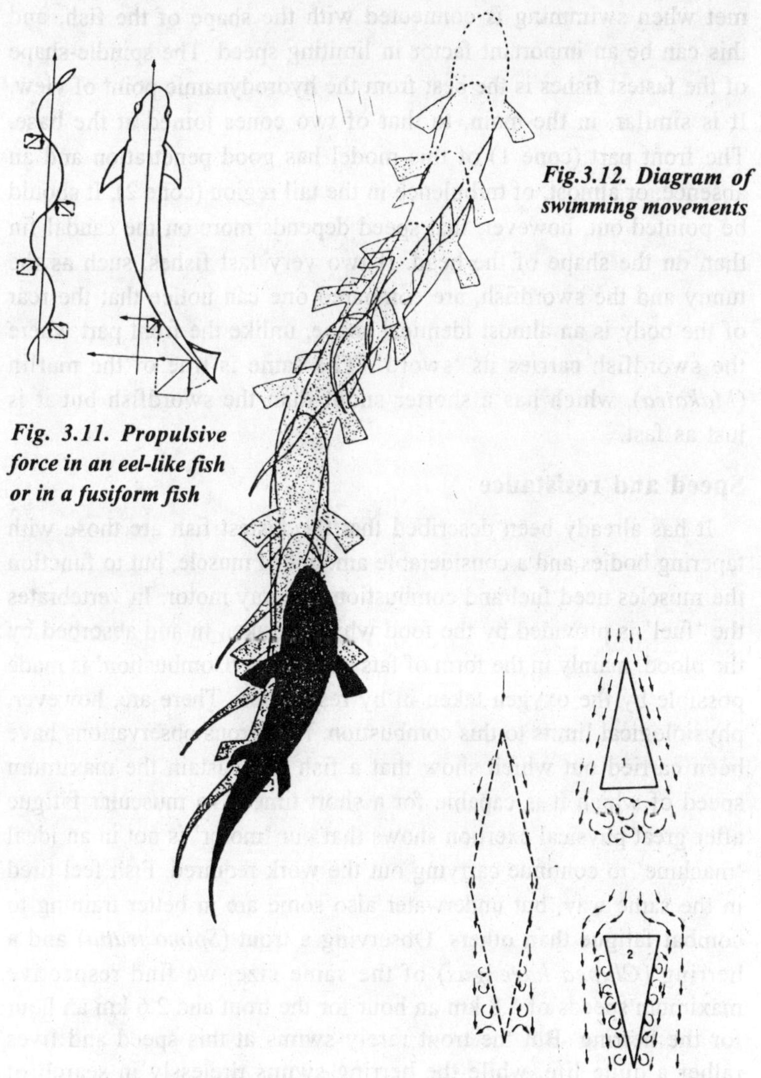

Fig.3.12. Diagram of swimming movements

Fig. 3.11. Propulsive force in an eel-like fish or in a fusiform fish

Fig. 3.13. Hydrodynamic shapes of fishes

benthonic fishes have a gephyrocercal tail, which is very reduced. The tails that do the most to increase speed are the homocercal and the heterocercal, which have the greatest elongation (that is, the relation between the height of the tail and its surfaces). There is , however, another element that has not been dealt with : the resistance met when swimming is connected with the shape of the fish, and this can be an important factor in limiting speed. The spindle-shape of the fastest fishes is the best from the hydrodynamic point of view. It is similar, in the main, to that of two cones joined at the base. The front part (cone 1) of this model has good penetration and an absence, or almost, of turbulence in the tail region (cone 2). It should be pointed out, however, that speed depends more on the caudal fin than on the shape of the head. If two very fast fishes, such as the tunny and the swordfish, are compared one can notice that the rear of the body is an almost identical shape, unlike the front part where the swordfish carries its 'sword'. The same is true of the marlin (*Makaira*), which has a shorter snout than the swordfish but it is just as fast.

Speed and resistance

It has already been described that the fastest fish are those with tapering bodies and a considerable amount of muscle, but to function the muscles need fuel and combustion like any motor. In vertebrates the 'fuel' is provided by the food which is taken in and absorbed by the blood, mainly in the form of fats and sugar. 'Combustion' is made possible by the oxygen taken in by respiration. There are, however, physiological limits to this combustion. Numerous observations have been carried out which show that a fish can sustain the maximum speed of which it is capable for a short time. The muscular fatigue after great physical exertion shows that our 'motor' is not in an ideal 'machine' to continue carrying out the work required. Fish feel tired in the same way, but underwater also some are in better training to combat fatigue than others. Observing a trout (*Splmo trutta*) and a herring (*Clupea harengus*) of the same size, we find respective maximum speeds of 5.8 km an hour for the trout and 2.6 km an hour for the herring. But the trout rarely swims at this speed and lives rather a quite life, while the herring swims tirelessly in search of food. There thus exists a fairly loose relationship between the maximum speed, used by the fish only in times of necessity, and its cursing speed which can be maintained for a period of time. To flee from the

Fig.3.14. Herring Clupea harengus

jaws of a predator, quick acceleration is more necessary than speed. Some fishes have amazing performances : carp, minnows, pike, cover the first five centimetres, starting from scratch, with an acceleration of 40 m/sec^2. Over such a short time, their acceleration equals that of a racing car.

The Heavy Fishes

When a dogfish attacks its prey, the onlooker cannot help noticing the particular tactics it uses. Having singled out its prey, the dogfish begins to moves round it in circles until with a sudden dart it is on top of it. This technique is imposed by the structure of the dogfish. The diagram opposite shows that the dogfish's tail is heterocercal (asymmetrical) and that when it moves it generates two forces : one propulsive (1) and the other elevating (2). The propulsive force pushes the animal forwards, while the elevating force tends to lift the dogfish's tail upwards. As a result of this elevating force the snout would tend to turn downwards, but dogfish are able to maintain a completely horizontal position. They counterbalance the impetus of the tail with an upward thrust of the large pectoral fins (3). Their function is very like that of aeroplane wings : the pectoral fins like wings, produce this thrust only when the dogfish is moving, But dogfishes are slightly heavier than water and if their movement stops, they start to sink to the bottom, like a plane whose engine has stopped. Forced to act as wings, the pectoral fins are hardly mobile. Dogfish find it impossible to stop suddenly and must always be on the move; they can only rest on the bottom where the sea is shallow. Thus their best hunting ground is in the open sea where, on the surface of the water, they find their food and must spend their lives. This force is one reason for their great voracity : they are forced into a life of action and so their need for fuel is continuous.

Weightless Fishes

As everyone knows, the goldfish (*Carassius auratus*) spends much of its time motionless in the water if it is not disturbed, and can stay like this without fatigue. How is it able to maintain this stable position in liquid without using its fins? Archimede's principle must compensate the weight of the fish exactly. But the density of the water varies and the weight of the fish is not constant either. Obviously something must exist that is able to adapt Archimede's principle to the fish's need at any given moment. Given that Archimede's principle depends on the volume of water displaced by the fish, it will achieve perfect hydrostatic equilibrium by modifying its volume, even just a little.

The Swim-Bladder

The specific gravity of a fish is generally in the region of 1.076.

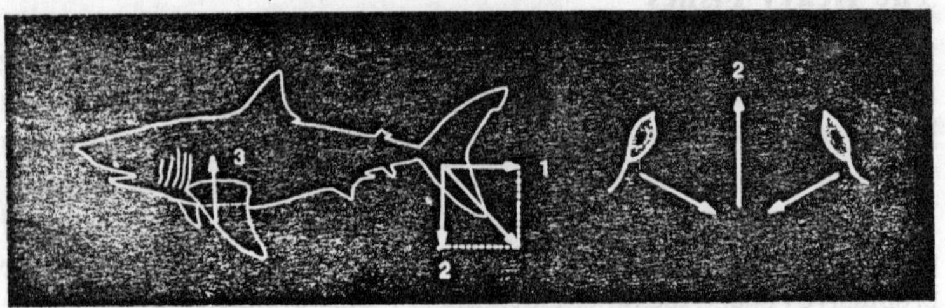

Fig. 3.15. Propulsive forces in a dogfish

A freshwater fish (the density of freshwater is 1) that displaces 100 ml of water will receive an upward thrust of 100 gr. The volume of the fish multiplied by its specific gravity gives its own weight, which equals 107.6 gr. The result is a downward thrust of 7.6 gr. In order not to sink, the fish will therefore have to increase its volume by 7.6, without changing its weight. This is possible because bony fishes are equipped with a swim-bladder.

This is usually an elliptical-shaped sac, situated beneath the vertebral

Fig. 3.16. Starry smooth-hound shark **Mustelus asterias**

column and filled with gas. It is derived from the oesophagus, to which in some types of fish it is linked by a canal called the pneumatic

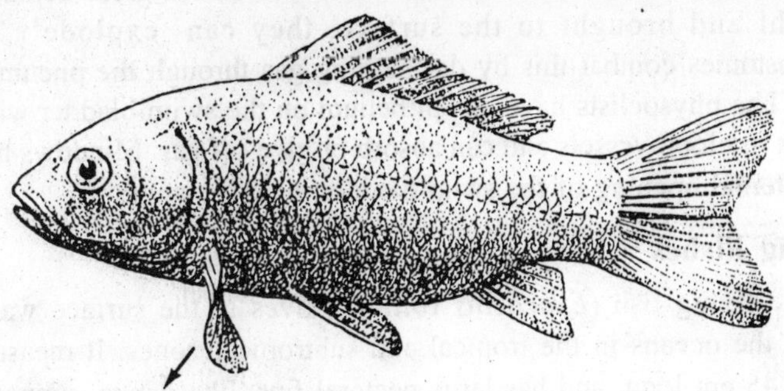

Fig. 3.17. Movement of the pectoral fin of a cyprinid which is staying still in the water.

duct. Its size differs in marine and freshwater fishes : in fact sea-water, with its density of 1.026, offers more support. In freshwater fishes the volume of the swim-bladder is around 7% of the total volume, in marine fishes it is 5%. The swim-bladder has many forms and can even disappear, as in flat fishes and blennies. If the pneumatic duct is present, the fish is a physostome; if the swim-bladder is separated from the oesophagus, the fish is a physoclist. A major problem, related to whether the swim-bladder is linked to the oesophagus or not, concerns rapid changes of depth in the water. Changes of depth vary the external pressure, which is felt by the whole fish and by its swim-bladder. There is more danger in coming up than in doing down : coming up, the fish runs the same dangers as

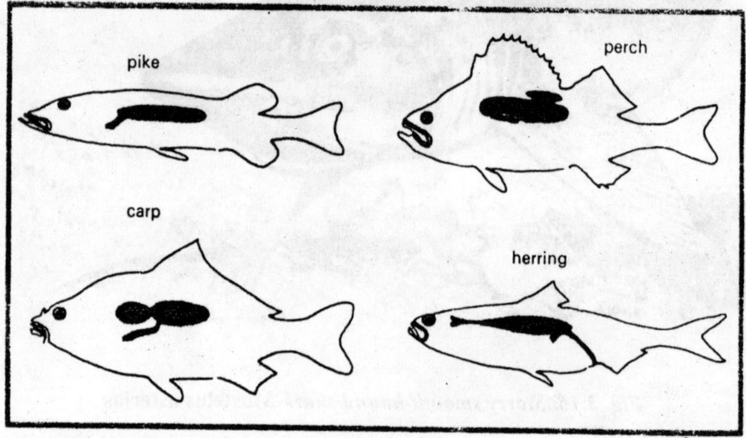

Fig.3.18. Swimbladders

a diver, the risk of embolism. If fish who live at a great depth are caught and brought to the surface, they can 'explode'. The physostomes combat this by discharging gas through the pneumatic duct. The physoclists have an epithelium on the swim-bladder which is rich in blood vessels and can absorb or give off gas. Minnows have an external opening of the swim-bladder, near the anus.

Flying Fishes

The flying fish (*Exocoetus volitans*) lives in the surface waters of all the oceans in the tropical and subtropical zones. It measures up to 45 cm long, and has large pectoral fins, like wings, supported by strong, flexible rays. When it is frightened, it leaps out of the water and glides along with the help of its pectoral fins. Three phases can be distinguished in the flight of the flying fish : the sprint, takeoff, and flight. Before leaving the surface of the water, the flying fish increases its velocity to 25 or 30 km an hour, pushing itself forward with its tail, anal and dorsal fins. When it reaches the right speed for flight, it opens its pectoral fins and bends its body at an angle of 15 degrees to the horizon, so that its head and pectoral fins are out of the water but its caudal fin is still immersed and continuing to supply power. In this takeoff phase, the caudal fin is beating rhythmically about 50 beats a second. In order to provide a greater impetus, the caudal fin is modified so that the lower lobe is much longer than the upper one; this helps the fish to keep its tail in constant contact with the surface of the water. Final takeoff from the water is assisted by the pelvic fins.

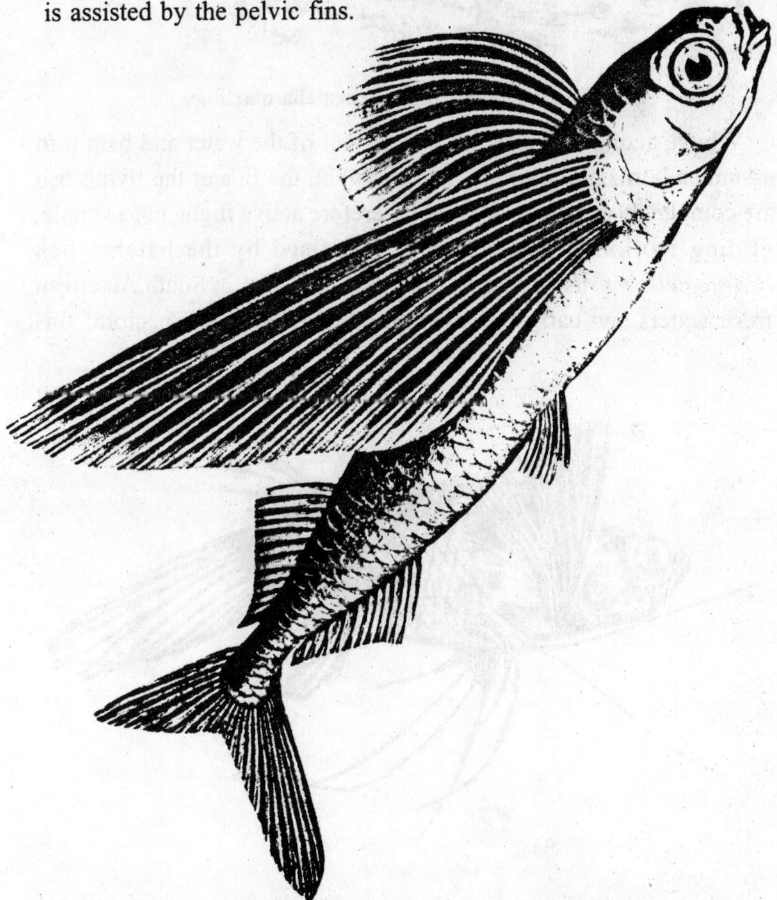

Fig.3.19. Flying fish **Exocoetus volitans**

***Fig.3.20. Hatchet fish* Carnegiella marthae**

Which, acting as wings, lift the tail out of the water and help it to assume a horizontal position. During flight, the fins of the flying fish are completely motionless. It is not therefore active flight, but a simple, gliding motion. Active flight is practised by the hatchet-fish (*Carnegiella marthae*), and its relatives. They live in South American fresh waters and carry out short flights, beating their pectoral fins

***Fig.3.21. Freshwater butterfly-fish* Pantodon buccholzi**

like wings. The butterfly-fish (*Pantodon*), which lives in African streams, behaves in a similar way to these fishes.

Fishes That Walk

Many species adapted to a benthonic life have developed pectoral or pelvic fins suited to their habit of remaining motionless on the bottom for much of the time. Many benthonic fishes, however, do not use their fins just to stay still but, like the angler fish (*Lophius*), are able to make little movements by using their pelvic fins. Even more specialised are the appendages of certain gurnards. These fishes have pectoral fins with the lower rays free; these act as tactile and locomotory organs as the fish moves about on the bottom in search of food. The blennies, which have scaleless rather slimy skins, have pectoral fins with the lower rays that are free of membrane at their extremities. With these 'digitate' fins, they can literally walk about on the sea bed. These small fishes live on the rocky bottoms covered with vegetation in all temperate and tropical seas; they are numerous in the Mediterranean where, besides rocky bottoms, they also live on sandy bottoms with artificial rocks such as the breakwaters of ports. The tompot blenny (*Biennius gattorugine*) is one of the largest blennies and measures up to 30 cm.

Other fishes are capable of movement and, unlike the blenny, do not only move about on the sea bed. The mudskipper (*Periophthalmus Koelreuteri*), for example, lives in the littoral zones of the Indian Ocean, the Pacific and West Africa. Its habitat is the mangrove swamps, trees whose roots go down from the branches and which grows along the coast and can even be half submerged in brackish lagoons. In these surroundings, which are sometimes flooded by the sea, the mudskippers often remain on land and move about by literally walking on their strong pectoral fins, jumping about and overcoming obstacles by thrusting upwards with their tails. Their body is adapted to bear the dehydration caused by being permanently out of water. They protect themselves from excessive dehydration by keeping part of their body immersed in the pools of water left by the high tide. Among the many adaptations to various life styles, there is the extreme case of the garden-eels, eel-shaped fishes, which spend their life in a hole in the sand, from which part of their body emerges just to catch the plankton on which they feed. At the least sign of danger, they immediately withdraw into their holes.

CHAPTER 4

LOCOMOTION IN THE DEEP SEA

The resistance to the movement of a macroscopic animal through water is determined by its size and velocity, and the density and viscosity of the water. The density of seawater increases steadily with increase in pressure whereas viscosity changes in a biphasic manner, reaching its lowest value at about 500 atm. Neither the viscosity nor density of seawater changes more than 4 per cent as a result of naturally occurring pressures so deep water offers virtually the same resistance to macroscopic animals as does shallow water.

In the case of the movement of microscopic animals and moving parts such as cilia, it is conceivable that pressure may affect the behaviour of water at their surfaces. The small size and the complex shape of microscopic animals, especially small planktonic Crustacea, further complicates the issue. For example, in the swimming of the shallow water copepod *Labidocera trispinosa* Vlymen (1970) has argued that in the 'hop and sink' progression typical of copepods, the bursts of rapid acceleration are not an energetically expensive form of rapid acceleration are not an energetically expensive form of progression. In fact, it is argued that the energy cost of movement through the water is negligible in the animal's total energy budget. Vlymen points out how the animal derives benfit from rapid movement in escaping predators without paying a high cost. It is difficult to imagine such a state of affairs in macroscopic animals where rapid acceleration involves a mass of muscle which undoubtedly requires significant energy from the animal's metabolism. Some bathypelagic fish appear to have relinquished powerful muscle and live a quiescent and economic life.

To a first approximation then, the deep sea does not present mid–water animals with special mechanical problems of locomotion. This is also true of those animals which crawl over, or burrow in, deep sea sediments where the nature of the sediment determines the problems of locomotion. This some what unexciting observation is worth making if only because the floor of the deep sea, covering twice the area of the terrestial environment, is populated by such creatures. Nevertheless, these same benthic organisms may well be interesting in respect of their buoyancy.

Animals tend to sink because protein–based tissues and skeletal meterials are denser than seawater. Lipids, certain body fluids, and gases in bulk are lighter than seawater and are deployed in ways which vary in degree of refinement to bring animals close to neutral buoyancy. At depth, buoyancy devices have to generate expansion against significant hydrostatic pressure, and it will be argued here that uplift is obtained in one of two ways. The increase in partial molar volume of a solute (or its displacement of water) either involves a phase change or a low density arrangement of water around solute particles. Because the problem of deep sea buoyancy forces us to think of molecular events, it is worth noting the molecular forces which act to make tissues denser than seawater in the first place. The case of skeletal salts is simple; these are dense because of their atomic constituents and close packed arrangement. What of proteins?. In Chapter 2, mention was made of how a protein, β–lactoglobulin, changed in volume and density on dissolving in water. In the dry crytalline state the specific volume (reciprocal of density) is 0.802 which diminishes to 0.751 in aqueous solution. Thus the solid state protein is less dense than the dissolved substance. Part of the change in density is attributed to the internal pressure of water and part to the electro–striction of the water by charged groups on the protein. β–lactoglobulin is probably typical of many proteins. Ovalbumin in the dry state has a density 1.2655, casein 1.318, horse serum globulins 1.279–1.312, and horse serum allumin 1.27–1.28. When the last three proteins enter aqueous solution at room temperature their density was found to increase by 5–8 per cent, doubtless due to the forces already mentioned.

So far as is known, buoyancy devices in marine animals appear to compensate for heavy tissues as if protein–based tissues have a density of 1.3 compared to the density of normal seawater which is 1.026. Buoyant proteins are an intriguing possibility however.

Neutral buoyancy is likely to be of advantage to an animal for two reasons. It can save energy and it can allow the animal to hover inconspicuously. The energetice of neutral buoyancy have been quantified in a most illuminating way but the value of being able to hang silently in the water is less easily quantified. Table 4.1 summarises

Table 4.1: **Buoyancy mechanisms in some deep sea animals**

	Species	**Depth (m)**
	A. Bulk phase mechanisms	
Gas		
(a) Gas-filled swimbladder,	Bony fish	
examples from deep water with	*Bassogigas profundissimus*	5610–7160
well-developed swimbladder	*Bassozetus toemia*	4570–5610
presumed functional at depth.	*Nematonurus armatus*	2600–3600
	Lionurus carapinus	>2000
(b) Gas-filled shell	Cephalopods	
	spirula	<1000
(c) Lipids, regression of	Bony fish	
swimbladder in adult	*Lampanyctus mexicanus*	Mesopelagic
(d) Hydrocarbon squalene	Elasmobranchs	
in liver	*Centroscymnus coelolepis*	~ 1500
	B. *Aqueous Solutions*	
(e) Ammonium-rich body fluid	Squid, e.g. Heliocranchia pfefferi and many others	(Shallow water)
(f) Absence of SO_4^{2-}	Beröe sp.	(Shallow water)
None		
(g) Reduced heavy tissues	Bony fish	
	Gonostoma elongatum	~1000

a number of buoyancy mechanisms found in marine animals.

Deep–sea life

The sea embraces about one half of the entire globe and it forms a special biotype by its unique ecological features in the biosphere.

Bionomic Features

Briefly speaking, there are four factors such as cold, quiescence, darkness and the total absence of living green plants beyond the light zone and this is the summary of conditions in the deep–sea.

Deep sea fauna consists of water breathing organisms only and so the higher Arthropods and all the vertebrates above fishes are debarred from such a habitat. Among invertebrates, sponges, corals, echinoderms (Brittle stars, stalked crenoid, sea–cucumbers and starfish, etc.). Bryzoans, Brachiopods, tube dwelling annelids, barnacels, ostracods and pelecypods, decapods (among Mollusea) are chief deepsea fauna.

Vertabrate fauna include sharks, rays, chimeroids (among Elasmobranchs) and many teleosts. The deep–sea fishes live under great pressure below the water at great depths. When released from this terrific pressure, *i.e.,* taken out from the sea in which they are adapted, they are spoken of as weak and flabby of flesh. The jaws are relatively great. There great jaws and the armament of cruel mouth are characteristic features of deep–sea fishes.

Another characteristic feature of deep–sea fishes is their very small size but they can swallow a fish bigger than themselves.

Adaptive Character

The deep–sea animals are frail or weak because of the absence of water–currents at these depths. They have small amount of carthy matters in their bones. The flesh is generally thin and flabby. They have simplified colours red colour dominates over the others. Some deep sea fishes are blind, some have telescopic eyes *e.g. Gigantura,* while other have eyes like concave mirrors. Either of the two latter types of eyes, helps in absorbing the greatest possible volume of light–rays. To compensate for the loss of vision, tactile organs are well developed. Almost all deep–sea animals are luminiscent. Many deep–sea fishes have lost their mastication powers as they live on decaying matters. Others are predatory with powerful jaws.

Perhaps the most striking features of both deep–sea and of the cave–dwellers also is their changelessness, *i.e.,* after they have adapted in the peculiar environment, their evolution gets practically ceased as they diminish their individual activity and consequent metabolism.

Deep–sea modifications, met in the animals, are the following:

1. *Frail and weak body :* The animals are frail and weak having thin and flabby flesh and with very little earthy matter in their bones, but they possess relatively great jaws and cruel mouth armanent *e.g.,* silver sharks, *Gastrostomus.* In invertebrates, long legged crabs, delicate long stalked crinoids, fragile siliceous glass sponges and hexactinellids attached to soft oozo through glassy rope like root spicules are the deep sea forms.

2. *Simple body colour:* Diversified colour of animals of the light zone are lacking from the deep–sea forms. They are red or brown, pearly grey or black, although some fishes have scarlet fins. But *Macrurus filicauda* has black belly and is silvery on the top.

3. Eyes, Some deep–sea fishes are *blind (Ipnops)* belonging to the

family Amblyopsidae; *Oncirodes* (anglers), others have *telescopic eyes,* yet others have *eyes like concave mirrors* to absorb the greatest possible volume of light rays.

In carbs eyes are not found and in pecten these are reduced.

4. *Presence of tactile organs:* As a compensation for the loss of vision, deep–sea forms possess long recrers and slender attenuations of the fins. In *Bathypterois,* one fin ray of the pectoral fins is long and acts as a sensory filament. In *Stylophorus paradoxus,*caudal fin a produced into a long filament, which is sensory.

Crustaceans also possess long antennae. These may be 8 to 10 times the length of the body. A crustacean isopod (*Munnipsis longicornis*) has antennae 8 times longer then the body.

Lateral line system is very well developed in deep sea fishes.

5. *Presence or luminescent organs:* Luminescent organs are found either over the entire body or on the belly or localized on highly modified organs. These are large oval bodies or irregularly elliptical shape, siuated on the head near the mouth or smaller round globular bodies placed symmetrically in series along the side of the body and tail. Teeth and mouth may also be luminous. It is due to the dark habitat, *e.g., Ipnops, Linophryne, Oneirodes*–deepsea fishes. Luminescence is also found in crustaceans, cophalopods, starfishes, many coelenterates and annelids. Luminescence is useful for th: recognition of sea and for attracting the prey.

6. *Loss of power of mastication:* Many deep–sea fishes loose the masticatory power since these feed on decaying ooze, while others have powerful rapacious jaws, *e.g., Cetomimus.* Most of the deep–dea fishes have very large mouths, and long sharp teeth and enormous stomach, capable of devouring large animals,*e.g., Saccopharynx and Euryph-arynx.*

7. *Sexual dimorphism:* In the abyssal dark none, it is very difficult to search the mate. In angler fish (*Photocorynus spiniceps*) the males are very small and remain attached to a process on the head of the female.

8. *Perental care:* Many deep–sea fishes take care of their youngs, while other produces youngs in large numbers.

Vertical distribution of animals. Life exists at all depths, *i.e.,*from pelagic to abyssal zone, but the numbers of individuals decrease as they descend to greater and greater depths. The abyss is comparatively rich

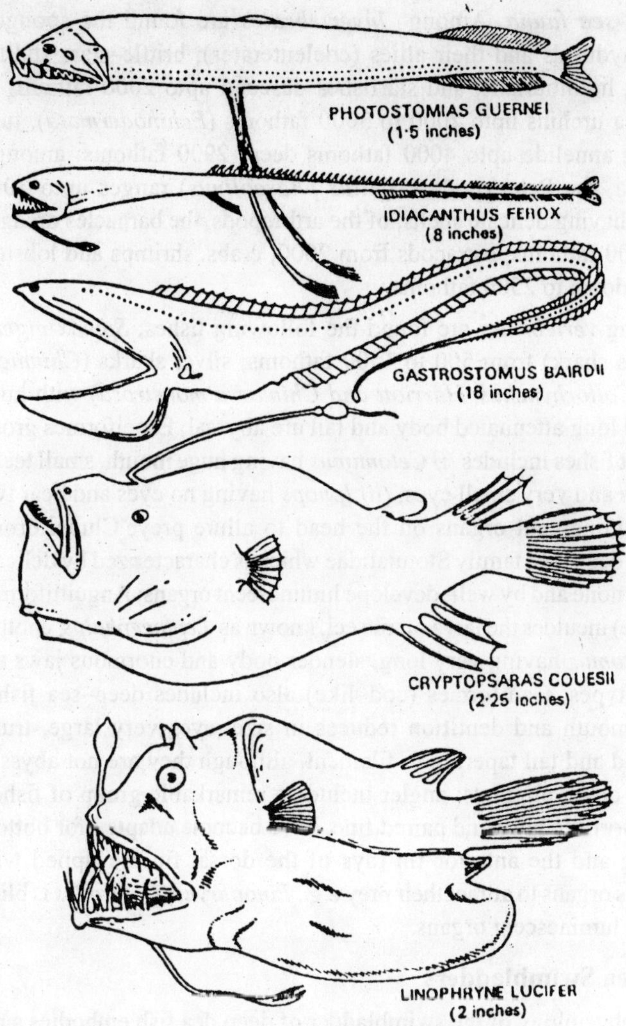

Fig. 4.1 : Deep sea fishes

in species, but poor in number of individuals. According to Sir John Murray, the number of bottom–dwelling forms or various depths are as follows:

Pelagic region Down to	200 meters,	about	4,200 species.		
"	2000	"	"	600	"
"	4000	"	"	400	"
over	5000	"	"	150	"

Deep–sea fauna. Among *Invertebrates* are found the sponges, corals, hydroids and their allies (coelenterates); brittle–stars stalked crinoids, holothurians, and starfishes descend upto 2000 fathoms or more; sea urchins upto 2000 to 3000 fathoms (*Echinodermata*); tube dwelling annelide upto 4000 fathoms deep; 2900 fathoms; amongst Mollusca, the Pelecypoda (*Mytilus phaseolinus*) ranges upto 3000 fathoms having delicate shells; of the arthropods, the barnacles dredged from 3000 fathoms; ostracods from 2000; crabs, shrimps and lobsters ranging down to 2500 fathoms.

Among *vertebrates* are found the following fishes; *Spinax niger* (luminous shark) from 500 to 1500 fathoms; silver sharks (*Chimaera affinis, Collorhynchus, Harriott and Chimaera monstrosa*) with huge eyes and long attenuated body and tail are abyssal; Esociformes group of teleost fishes includes *(i) Cetomimus* having huge mouth, small teeth, no scales and very small eyes, *(ii) Ipnops* having no eyes and bear two large, luminescent organs on the head to allure prey; Clupeiformes includes deep–sea family Stomiatidae which is characterized by delicate scales or none and by well–develope luminescent organs; Anguilliformes (ece–like) inculdes the larva of true eel, known as *Leptocephalus,* another *Gastrostomus* having very long, slender body and enormous jaws are abyssal types; Gadiformes (cod–like) also includes deep–sea fishes having mouth and dentition reduces in size, eyes very large, trunk shortened and tail tapering to filament, although they are not abyssal, they are down migrants; angler include a remarkable group of fishes, having horrible jaws and paired fine, have become adapted for bottom crawling and the anterior fin rays of the dorsal fin are tipped with fuminous organs to attract their prey *e.g., Linophryne, Oneirodes* is blind but with luminescent organs.

Deep Sea Swimbladders

The physiology of the swimbladder of deep dea fish embodies such a range of physiological and physical phenomena that it could well be the centre–piece of a book on deep sea physiology. For two good reasons the topic is only dealt with in summary here; first, because the author has no practical experience in this field and second, because the subject is well dealt with in the literature, having attracted the attention of some brilliant experimenters and lucid writers. Accordingly, for the sake of completeness there follows only a brief description of the structrure and function of the swim–bladder of shallow water fish and a few comments on the special problems of its function at great depth.

In some fish swim–bladder connects by a duct to the alimentary canal, but in all deep sea fish it is a closed sac. The gas gland is perfused by capillaries which form a long coonter current rete which is normally buried in the swimbladder wall. The gas gland is the point from which gas leaves solution as a gas phase. The oval is another perfused patch which permits dissolved gas to be transported away from the bladder. Note the absence of a rete there. Gas exchange between oval and bladder is determined by muscular control over the extent to which the oval is exposed to the gas contents. The wall of the swim–bladder is highly impermeable to gases and compliant. Its volume in relation to the body it buoys is fairly constant, for example in shallow water fish it lies in the range of 3—6 per cent of the body volume. The relative constancy is, of course, due to the approximately uniform denisities of fish and seawater. In fish which lay down a lot of lipid during their life span, the swimbladder may be relatively small.

Marshall's (1972, 1960) thorough studies and reviews of deep sea swimbladders have yielded a number of important conclusions. About one third of all mesopelagic species of fish possess a swimbladder of a size suited for buoyancy purposes. Others, insluding such common animals as *Cyclothne* spp. and *Stoimias* spp. do not possess functional swimbladder in the adullt state.

Bathypelagic fish living deeper than 2000 m do not possess gas filled swimbladders.

Fish living near the deep sea floor (benthopelagic) possess a gas filled swimbladder irrespective of depth. Marshall states that of the 16 species known to live in this way at depths greater than 3500 m, 11 possess an apparently functional swimbladder.

Benthic fish generally do not have a functional swimbladder, irrespective of depth. In this context the celebrated case of *Basogigas profundissimus* is often quoted. A single specimen was retrived from a trawl which collected sedentary animals from the floor of the Sunda trench (7160 m,1.5°C). On retrieval it was found to have an undamaged swim–bladder, which is difficult to reconcile with the suggestion that it contained sufficient gas at a pressure of 700 atm to give it useful buoyancy. Further, there is no evidence that the trawl was deployed and finally, no evidence to show that the animal was even alive when it was caught.

One might expect at least two pieces of evidence to establish the maximum depth at which swimbladders function as true buoyancy

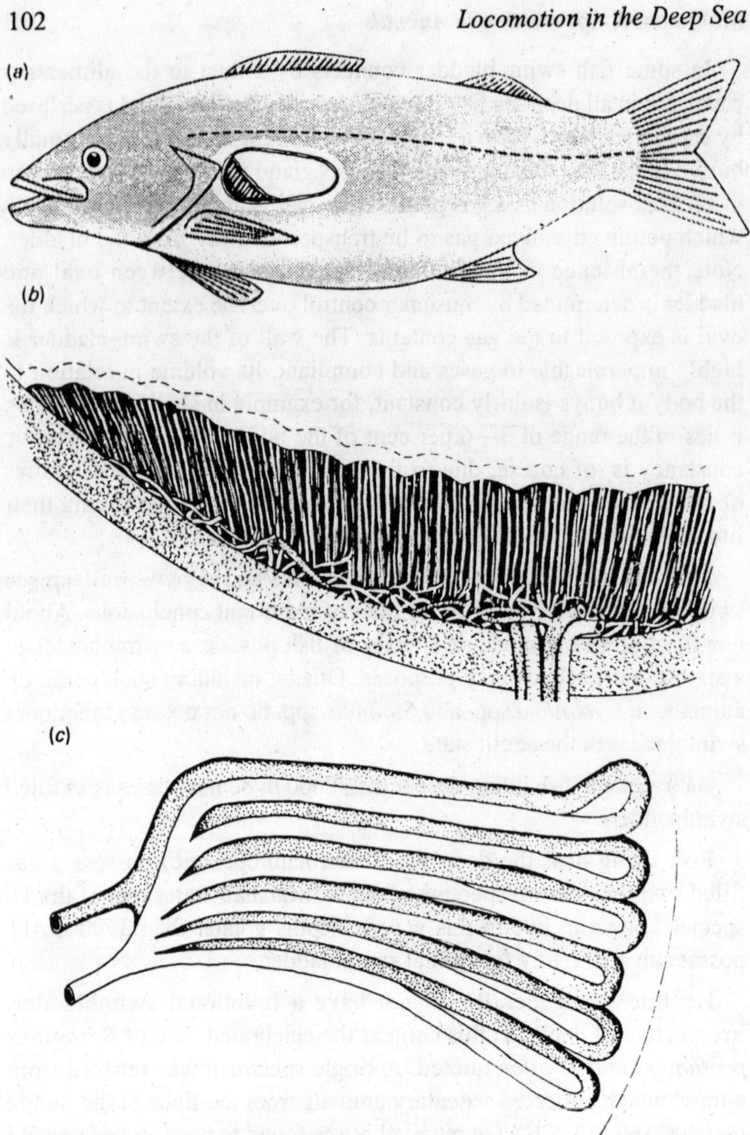

Fig.4.2: Diagram of a swimbladder.* (a) *shows position within fish body,* (b) *the gas gland and* (c) *the rete. See text.

organs. It is essential to know that the fish in question actually possesses a fully developed swimbladder and not much lipid, and secondly it is important to observe the animal floating at depth. The present evidence suggests that useful upthrust might be obtained from a swimbladder in fish at depths to about 5000 m. The rat tail *Coryphaenoides* has been filmed 'floating or swimming slowly', presumably at a depth of 2000m.

Pressure is not the most important limiting factor in the distribution of gas filled swimbladders; an adequate food supply appears an equally important prerequisite to their inclusion in the repertoire of fish physiology.

The density of atmospheric gases at deep sea pressures and temperatures is less than that of the liquid hydrocarbons or lipides. Oxygen, the major constituent of the gas mixtures in the swimbladders found at septh, has a density of approximately 0.45cm^{-3} at 400 atm. The upthrust obtained from a given volume of swimbladder gas measured at environmental pressure declines markedly in the bathypelagic zone. Obviously the region where change in depth affects a given gas volume most is the top few hundred metres.

Functioning of the gas filled swimbladder at depth

Before considering some of the special featureas of swimbladder function at abyssal depths a brief description of the situation in shallow water is given. Figure shows the rete through which blood flows in counter cureent fashion. The solubility of atmospheric gases physically dissolved in the blood and the association of oxygen with haemoglobin decreases as the blood enters the effetent capillaries where a higher lactate level prevails. The partial pressure rises in accordance with Henry's Law, causing gas to diffuse across to the afferent rete and, by a continuous process, builds up a high partial pressure of gas in the blood at the swimbladder end of the afferent rete. This high partial pressure sustains that of the gas in the swimbladder.

Taking the rate of blood flow and the trans–rete diffusion of gas into account, an equation defining the partial pressure of gas in the blood at the gas gland end of the afferent capillaries has been derived by Enns:

$$P_L = P_0 \exp\left(\frac{K}{F} L \frac{\Delta\alpha}{\alpha}\right) \qquad (2)$$

where P_L = partial pressure of stated gas at the gas gland end of afferent capillaries, P_o = partial pressure of stated gas at the entrance to afferent capillaries, K = trans–rete diffusion of gas (Q (ml min^{-1} length^{-1} pressure difference^{-1}), divided by α the solubility coefficient of the stated gas in distilled water), F = blood flow rate (ml min^{-1}), L = length of rete, and

$$\frac{\Delta\alpha}{\alpha} = \frac{\text{gas solubility in distilled water} - \text{solubility in solution}}{\text{solubility in distilled water}}$$

The important point to notre is that K is a very large figure and it dominates the exponential term; $\Delta\alpha / \alpha$, or the salting out effect, is a small figure.

Table 4.2 A: Some properties of substance which contribute upthrust in marine animals (all data at atmospheric pressure unless otherwise stated)

Substance	Specific gravity (see Table 4.2 B)	Source	Up-thrust (g ml^{-1})	Volume giving 1g upthrust (ml)
Cod liver oil[1]	0.925	Cod liver	—	—
Triglycerides[2,3]	0.91–0.92	Fish body oil, *Euphausia*	0.106	10
Diacyl glycerol ethers[2,3]	0.89–0.91	Elasmobranch liver, e.g. *Dalatias, Squalus, Acanthias*	—	—
Squalene[1]	0.856	Elasmobranch liver, e.g. *Centrocymnus*	0.166	6
Wax esters[2,4]	0.857	(Albatross stomach oil)	—	—
Pristane[5]	0.78	Bony fish, whales, *Latimeria*, small amounts only, no buoyancy function demonstrated	0.246	4
n-pentane[4]	0.626	Not found in organisms, for comparison only	—	—
Oxygen at 400 atm[7]	Density approx. 0.457 g cm^{-2}	Swimbladders	0.706	1.4
Dry oxygen at 2 atm total pressure 0°C[1]	Density approx. 0.0028 g cm^{-3}	—	—	—
Ammonium-rich body fluids (isosmotic)[8]	Denisty 1.010 g cm^{-3}	Body fluid, *Heliocranchia*	—	—
Hypotonic body fluid of bony fishes[9]	1.013	Watery body fluid, *Gonostoma*	—	—
Seawater lacking SO_4^{2-} (isosmotic)[10]	1.022	Beroe, diatoms	—	—

Specific gravities of lipides normally refer to the substance at room temperature compared to water at 4°C.

The critical temperature of oxygen is—118.8°C, of nitrogen —147°C and of argon —122°C. Carbon dioxide has a critical temperature of 31°C and a critical pressure of 72.8 atm. At its critical temperature and pressure its density is 9.46 g cm^{-2}.

Table 4.2 B.

Pressure above atmospheric (bars/atm)	Specific gravity of seawater (35 ‰ °C)
0	1.028
100	1.032
500	1.050
1000	1.070

Plausible figures for $K = Q/\alpha$ used by Enns *et al.* for nitrogen are 65/0.02 = 3250 ml min^{-1}cm^{-1} at 5°C.

Taking $\Delta\alpha/\alpha$ as 0.00153 (a value based on the afferent–efferent difference in lactate in shallow water fish), F as 1 ml min^{-1}, and L as 1 cm, then P_L equals 112 atm when P_o is 0.78 atm as it is in the case of nitrogen. Thus the generation of 100 atm of nitrogen is broadly accounted for, provided leakage through the swimbladder wall is small, which it is in shallow water conditions. It therefore appears that a shallow water swimbladder only requires a long rete if it is called upon to buoy a fish at considerable depth. To reinforce that over–simplifying assertion we can consider the permeability of the swimbladder wall of shallow water fish to very steep partial pressure gradients. Table shows the permeability of the wall of two swimbladders from shallow water fish. The low permeability of the conger eel bladder is attributes to the presence of guanine crystals and is unlikely to be upset by a high ambient pressure. Kutchai and Steen argue that oxygen will leak through the wall of the conger swimbladder at a rate equal to a plausible rate of secretion when the partial pressure of the gas equals 250–fold that of the ambient water, i.e., at 250 atm or 2500 m.

Table 4.3: Oxygen and carbon dioxide permeabilities of swimbladder walls in ml (cm² min atm µ m⁻¹) compared with values reported for connective tissue and for water. Readings at STP.

	O_2 permeability	CO_3 permeability
Swimbladder of common eel[a]	0.0106	0.308
Swimbladder of conger eel[b]	About 0.001	About 0.03
Connective tissue from frog rectus abdominus muscle	0.115	—
Central tendon of dog diaphragm	—	2.65
Water	0.34	9.75

The equilibrium pressure of dissolved gas rises with increase in hydrostatic pressure in an exponential manner. This some what surprising state of affairs has already been discussed in Chapter 4. The experimental verification of the theromodynamic arguments which yield this conclusion is limited, but in the absence of data to the contrary we will assume, with Enns *et al.* (1967) that dissolved gas activity rises 14% per 100 atm all the way to the greastest depths. These workers make the point that if we regard the sea as equilibrated with nitrogen at normal atmospheric pressure and consider a swimbladder being inflates with nitrogen, then at both 3000 m depth and a hypothetical depth of 13300 m, the ratio of P_{N_2} in the bladder to P_{N_2} in solution in the sea will be 246. The situation with oxygen is more complecated but it is clear that the exponential ries in gas equilibrium pressure with hydrostatic pressure helps rather than hinders the maintenance of gas in a swimbladder at great depth.

Special features of deep sea swimbladders — the rete

L, in equation (2) is increased in deep sea fish. Table shows data selected from Marshall (1972) to illustrate this point and the original papers should be consulted to verify the generalisation. The rete in Table are uncorrected fot body size or swimbladder size but the correlation with living depth is real and clearly connected with the working of the countercurrent multiplier.

Table 4.4: *Length of swimbladder rete in deep sea fish.*

Species	Length of rete (mm)	No. of rete	Range of depths where most commonly found (m)
Bassogigas profundissimus	15	1	
Bassozetus taenia	25	2	4570—5610
Nemotonurus armatus	25	5	2600—3600
Lionus carapinus	25	6	> 2000
Coryphaenoides guntheri	20	4	1000—2000
Synaphobranchus Kaupi	10	2	800—2000
Hymenocephalus italicus	6	2	300—800
Malacocephalus laevis	4	2	150—600
Vinciguerria attenuata	0.8	—	150—600
Myctophum punctatum	2	—	150—600

The energy required to maintain the high partial pressure of gas within the swimbladder has been estimated by Alexander (1972). If the partial pressure of gas (oxygen or nitrogen) in the ambient seawater is

P_W and the partial pressure in the swimbladder is $P_{S,}$ then the work of compressing the gas us RT In (P_S/P_W); R and T are the gas constant and absolute temperature repectively. Diffusional losses must be matched by the secretion of gas to maintain a steady state. Let D represent the number of moles of gas diffusing out of the swimbladder, a quantity which can be compuuted from permeability data and measurements of a particular swimbladder. Thus the steady state power requires to maintain a swimbladder volume constant at depth is DRT In (P_S/P_W). Using permeability data from the conger eel swimbladder and arbitrarily assuming that gas secretion proceeds with an efficiency of 5 per cent, Alexander calculates a 1 g fish would expend energy equivalent to respiring 35 cm^3 O_2 kg^{-1} $hour^{-1}$ at 100 atm pressure to maintain its swimbladder full of oxygen. This is 17 times the amount of energy required at 10 atm pressure, and would appear to be rather a high of expenditure to apply to a real fish.

The efficiency of gas secretion in even shallow water fish may be much higher than 5 per cent. Alexander's argument shows that swimbladders may become metabolically very demanding at great depth and we might reasonably anticipate adaptations in deep sea swimbladders to improve their efficiency.

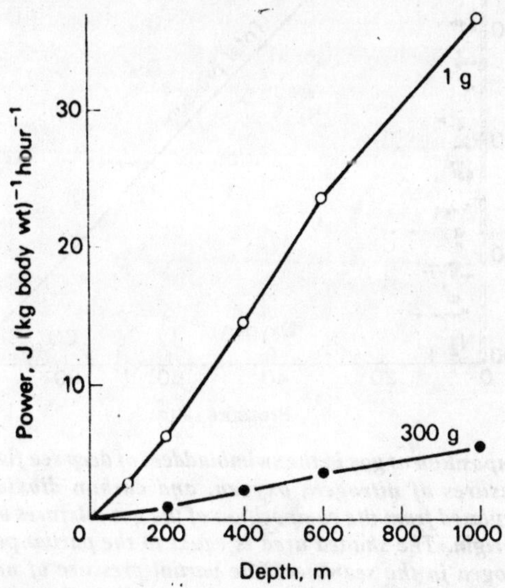

Fig. 4.3: Graphs of estimated power required to replace oxygen diffusing out of the swimbladder, against depth. Separate graphs are presented for a 1 g and a 300 g fish. An ambient P_{O_2} at the following depths is assumed :0.2 ATA down to 200m, 0.02 ATA at 400–1000 m (approximately).

Composition of the gas in deep sea swimbladders

Oxygen is the major gas in deep sea swimbladders, but nitrogen and argon are usually present at partial pressures in excess of that prevailing in the ambient seawater. A small amount of carbon dioxide is also present. Fig. 4.4 shows the results of Scholander and Van Dam (1953) who analysed the swimbladder contents from many different species. These results are consistent with a gas secreting machanism which works in a physical rather than a selective chemical way. High concentrations of oxygen were also found in the swimbladders of hatchet fish retrieved from the oxygen minimum layer at a depth of about 800 m. Under these conditions the P_{O_2} of the gas was 10000 times the ambient P_{O_2} which may still be accounted for by the 'salting out' mechanism of gas secretion. Note also the volume of gas present in these animals.

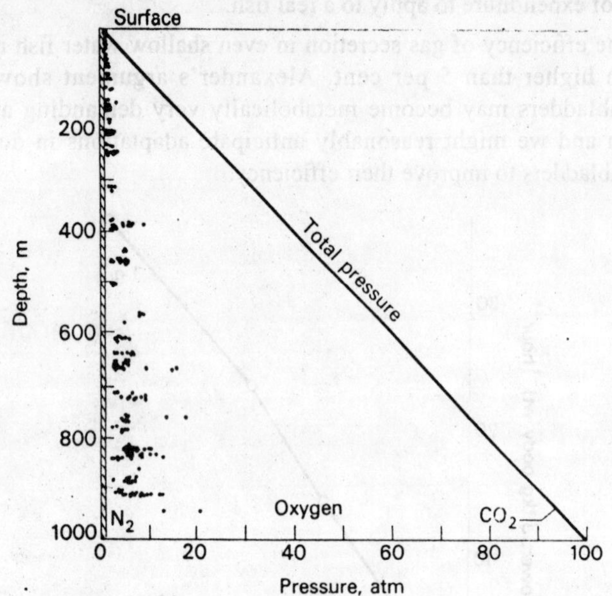

Fig. 4.4 : ***Composition of gas in the swimbladders of deep sea fish. The partial pressures of nitrogen, oxygen, and carbon dioxide have been calculated from the composition of the gas mixtures and the depths of origin. The shaded area is equal to the partial pressure of the nitrogen in the seawater. The partial pressure of nitrogen in the swimbladder is to the left of each point, that of oxygen to the right. The partial pressure of the carbon dioxide is usually less than that represented by the thickness of the diagonal line. The over–all picture suggests a linear increase in the partial pressure of the nitrogen with depth.***

Table 4.5: The volume and oxygen content of the swimbladder gasses from oceanic fish.

	% O_2	Swimbladder gas as % volume of inflated fish at the surface
Surface pelagic species		
Exocoetidae		
Exocoetus monocirrhus	33	3.5
Cypselurus xenopterus	23–25	
Hirundichthys sp.	33	3.1
Hemiramphidae		
Hemiramphus sp.	29	
Deep water, fishes taken at the surface		
Myctophidae		
Myctophum evermanni	74–91	3.2–5.0
M. spinosum	77–91	4.0–5.5
M. affine	90	
Fishes taken at depth (all below 100 m)		
Sternoptychidae		
Argyopelecus lychnus	72–92	32–65
A. affinis	67–93	42–55
Sternoptyx obscura	71–88	
Vinciguerria lucetia	78–90	32–52
Ichthyococcus sp.	77–83	
Melamphaidae		
Melamphases macrocephalus	73–87	37–65
M. nigrofulvus	67–77	56
M. opisthopterus?	71	
M. nycterinus	54–93	48–54
M. cristiceps?	61–81	
Myctophidea		
Lampanyctus omostigma	89	53
Lampadena bathyphila	85	

Deleterious effects of high partial pressures of gases

At partial pressures in excess of about 1 ATA oxygen is toxic to intact animals and impairs the function of many *in vitro* preparations.

Nitrogen at partial pressures of about 4 ATA exerts a depressive action on the mental performance of men, and at slightly higher pressures narcotises a number of physiological systems. The gas gland of the swimbladder works at elevated partial pressures of both these gases, and in particular, the gas gland of deep sea fish functions at a P_{O_2} of 100 or more atmospheres. D'Aoust (1969) has brought this into sharp focus by subjecting the epipelagic fish *Sebastodes miniatus* to a partial pressure of oxygen which only its gas gland would normally encounter. Fig. 4.5 demonstrates the fact that the gas gland functions normally in a P_{O_2} lethal to the whole animal. it is not yet clear if a gas gland which tolerates a few atmospheres partial pressure of oxygen would remain unaffected by a PO_2 of 100 atm. Intuitively, one might suppose further adaptive changes would be necessary. In this connexion pressure itself may play a role in modifying the potency of the gas .

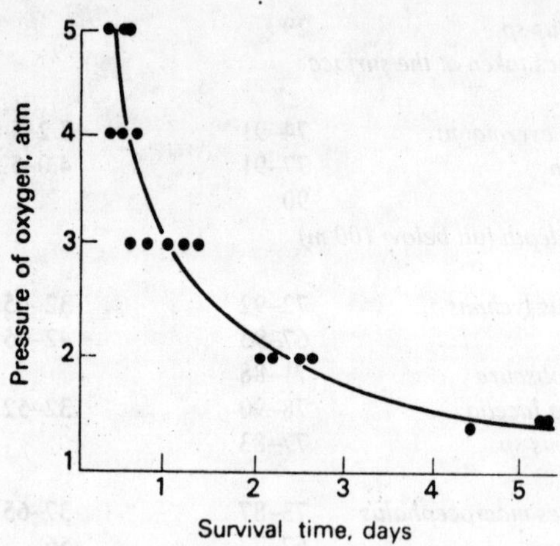

Fig. 4.5: Effects of an elevated P_{O_2} on* Sebastodes miniatus. *Survival time (horizontal scale) is plotted against oxygen pressure. In order to rule out the effects of hydrostatic pressure, the lowest partial pressure of oxygen (1.5 ATA) was also administered to fish in the chamber as compressed air so that the total pressure was 7.5 ATA. There was no significant difference in the survival time of fish treated with these two gas mixtures.

The sensitivity of gas glands to high pressures of nitrogen has not been studied. In general, there is at present little evidence to show that they are exposed to partial pressures much greater than 20 atm, but such pressures are, of course, mildly narcotic in some systems. Hydrostatic pressure ca n enhance or antagonise the narcotic action of nitrogen in a number of preparations.

In a shallow water species, gas bubbles have been seen in a live gas gland. Lipids rich in cholesterol and phospholipids have been found within the swimbladder of *Coryphaenoides acrolepis* and *Antimora rostrata* and it is conceivable that these substandes are connected with the secretory process. The role and metabolism of lipids both inside and outside swimbladders of deep sea fish is obscrure.

$\Delta\alpha/\alpha$ at great depth

A generalised salting out of dissolved gases could give rise to a multiplying effect in the rete sufficient to generate gas pressures in excess of those needed in the deep sea. In addition, a shift in the equilibrium $HbO_2 \rightleftharpoons Hb + O_2$ could also cause oxygen to leave the blood and enter the swimbladder. From solubility measurements carried out by Enns *et al.*(1967) the ratio $\Delta\alpha/\alpha$ for argon and nitrogen can be computed for different molarities. It would be interesting to know how hydrostatic pressure affects these data.

If lactate released into the efferent capillary blood decreases the solubility of nitrogen in a manner comparable to NaCl then quite high partial pressures of nitrogen may be slowly generated. Lactate also shifts blood pH which might affect the equilibrium shown above. Indeed Scholander and Van Dam (1954) have shown that acidified blood will dissociate off oxygen at quite high pressures (Root effect), but probably not at a P_{O_2} of more than 40 atm. Perhaps other substances can releade oxygen from oxyhaemoglobin against higher partial pressures. Fig. summarises a recent view of the gas secreting process. Oxygen may be driven from the aqueous and haemoglobin components of blood by two different machanisms. A pH effect on the oxygen binding capacity of haemoglobin dissociates the oxygen off into solution. A salting out effect drives it from solution to diffuse into the swimbladder. The first machanism is limited to a P_{O_2} of 40 atm, the second is not limited by a naturally occurring pressure.

The problem of the rate of gas secretion is important. For example, Enns *et al.* (1967) suggest that a hypothetical 100 g fish at 200 atm might take 174 days to fill its swimbladder with oxygen. In short, in swimbladders working at great depth the equilibrium situation is broadly explicable. The rate at which buoyancy equilibrium is achieved in deep sea fish is not known but extrapolation from those systems which have been studied suggests that it may be achieved only very slowly unless supplementary mechanisms exist.

Table 4.6: Solubilities of nitrogen and argon in salt solutions at 20.48°C.

	Nitrogen			Argon		
NaCl conc. (mole cm^{-3})	Solubility (cm^3 cm^{-3} atm^{-1})	$\Delta\alpha / \alpha$[a]	(cm^3 $mole^{-1}$)	Solubility (cm^3 cm^{-3} atm^{-1})	$\Delta\alpha / \alpha$[a]	(cm^3 $mole^{-1}$)
0	0.01545	—	—	0.03375	—	—
0.0953×10^{-3}	0.01495	0.0324	3.40×10^2	0.03271	0.0308	3.23×10^2
0.1594×10^{-3}	0.01454	0.0589	3.69×10^2	0.03195	0.0533	3.34×10^2
1.784×10^{-3}	0.00827	0.464	2.60×10^2	0.01901	0.437	2.45×10^2
5.259×10^{-3}	0.00293	0.810	1.54×10^2	0.00722	0.786	1.495×10^2

[a] $\frac{\Delta\alpha}{\alpha} = \frac{\text{(solubility in distilled water)} - \text{(solubility in solution)}}{\text{solubility in distilled water}}$

Buoyancy by a Gas Filled, Rigid Shell

Certain cephalopods achieve neutral buoyancy means of a gas filled shell. The pressure within the shell is atmospheric or less and the shell with–stands the ambient pressure. *Nautilus* and *Spirula* are mesopelagic animals living at pressure of less than 100 atm. Their coiled and chambered shells fail catastrophically at pressures of 60 and 170 atm respectively. The mechanism by which the shell, which is initially filled with water, is evacuated during the growth of these animals, is not fully understood; nor is it clear how the siphuncle, the perfusing stem running the length of the shell, withstands the ambient hydrostatic pressure and remains waterproof. An osmotic pressure gradient maintained by the active transport of salts appears one of the forces which holds back the water pressure, but such a mechanism in its simple version cannot account for the way in which *Spirula* withstands hydrostatic pressures greater than the osmotic pressure of its blood (about 20 atm).

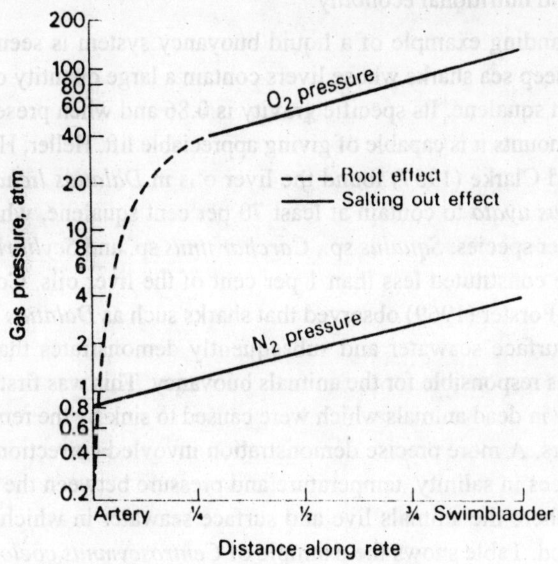

Fig.4.6: Summary of the development of gas pressure along the swimbladdeer rete.

Lipids

Some inactive myctophids (lantern fish) show an interesting reduction in their swimbaldders and an accumulation of lipid material during their life. Capen (1967) has studied example of this progress in some detail and found both a 'cottony' tissue growing inside and an oily tissue outside the swimbladders of, for example, *Lampancytus mexicanus*. Fig. shows the swimbladder regressing and the lipid tissue increasing, and Table shows the relationship between the volume of the swimbladder volume and size of the whole animal during growth. Table summarises data showing how swimbladder volume varies in a number of deep sea fish. The adults of a number of mid–water fish of the deep scattering layer may be slightly negatively buoyant in sea–water and it may be that gas filled swimbladders are too slow to adjust during vertical migrations.

Other myctophids including the genus *Lumpanyctus* sp. contain interesting lipids. Certain species contain large amounts of wax esters which probably provide significant upthrust (e.g. *Stenobrachius*); others, e.g. *Diaphus,* contain negligible amounts. Presumably,. the proportion of these substances present in an animal is related to its buoyancy and nutritional economy.

An outstanding example of a liquid buoyancy system is seen in a number of deep sea sharks whose livers contain a large quantity of the hydrocarbon squalene. Its specific gravity is 0.86 and when present in sufficient amounts it is capable of giving appreciable lift. Heller, Heller, Springer and Clarke (1957) found the liver o is in *Dalatias licha* and *Centrophorus uyato* to contain at least 70 per cent squalene, whereas in many other species, *Squalus* sp., *Carchar inus* sp. and *Scyliorhinus* sp. squalene constituted less than 1 per cent of the liver oils. Corner, Denton and Forster (1969) observed that sharks such as *Dalatias licha* floated in surface seawater and subsequently demonstrates that the squalene was responsible for the animals buoyancy. This was first seen qualitatively in dead animals which were caused to sink by the removal of their livers. A more precise demonstration invovled corrections for the differences in salinity, temperature and pressure between the deep sea water where the animals live and surface seawater in which they were weighed. Table shows the example of *Centroscymnus coelolopic* which, in Plymouth seawater, reqyured the addition of 18 g to achieve neutral buoyancy. At the animals normal depth (about 1500 m) the salinity was greater than that of Plymouth water and a corresponding correction was made by Corner *et al.* in the following way. If the volume of the animals is *V*, and the surface seawater density in which the animal floated with an upthrust of 18 g is d_a, then the change in

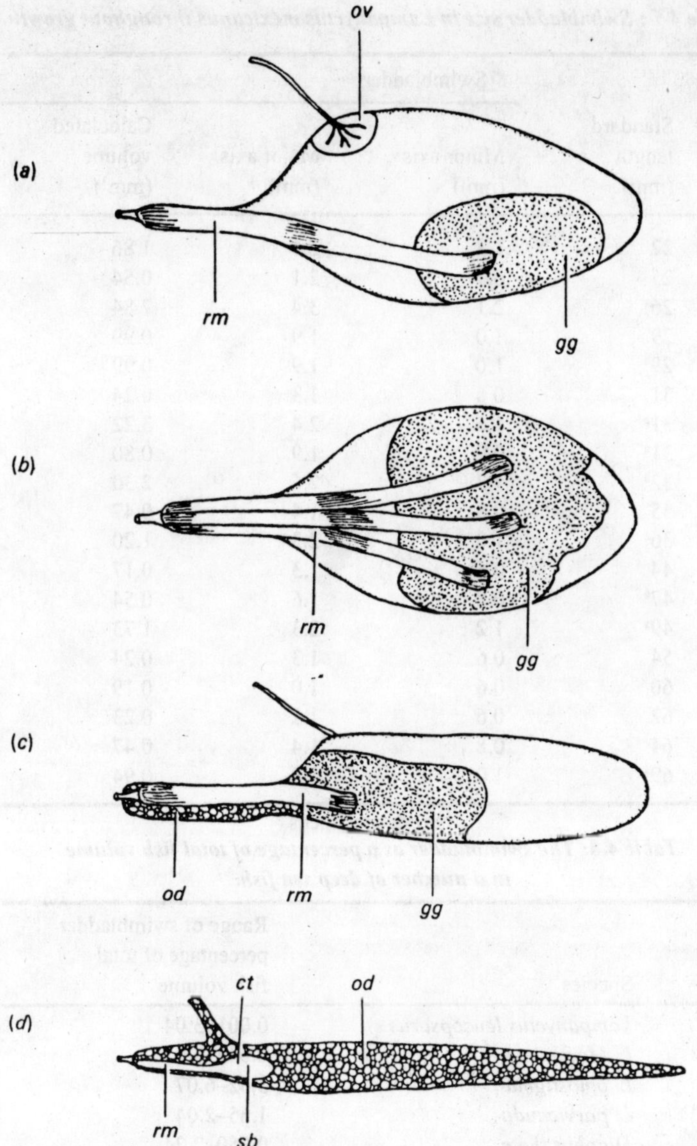

Fig. 4.7: Growth stages in oil–filled tissue surrounding the* Lampanyctus mexicanus *swimbladder. Partially diagrammatic. Rate mirabile and gas gland shown as if swimbladder wall were transparent.* (a) *and* (b)*, 22 mm fish, left lateral view and ventral view, respectively;* (c) *25 mm fish, left lateral view;* (d) *60 mm fish, leftlateral view; in* (d) *the swimbladder,* sb*, is filled with cottony tissue, and the oval is not apparent.* (a) *and* (b) *enlarged 21½ times;* (c) *23, and* (d) *7½ times.* ov*, oval;* rm*, rete mirabile;* gg*, gas gland;* od*, oil droplets;* sm*, swimbladder;* ct*, cottony tissue.

Table 4.7 : Swimbladder size in Lampanyctus mexicanus throughout growth.

Standard length (mm)	Swimbladder Minor axis (mm)	Major axis (mm)	Calculated volume (mm^3)
22	1.3	2.1	1.86
25	0.7	2.1	0.54
26[a]	2.1	3.4	7.84
29	1.0	1.9	0.99
29[a]	1.0	1.9	0.99
31	0.6	1.3	0.24
31[a]	1.6	2.4	3.22
31[a]	0.9	1.9	0.80
32[a]	1.3	2.6	2.30
35	0.8	1.4	0.47
36[a]	1.0	2.3	1.20
44	0.5	1.3	0.17
47[a]	0.8	1.6	0.54
49[a]	1.2	2.3	1.73
54	0.6	1.3	0.24
60	0.6	1.0	0.19
62	0.6	1.2	0.23
64	0.8	1.4	0.47
69[a]	1.0	1.8	0.94

[a] Fresh specimens

Table 4.8: The swimbladder as a percentage of total fish volume in a number of deep sea fish.

Species	Range of swimbladder percentage of total fish volume
Lampanyctus leucopsarus	0.001–3.04
L. Mexicanus	0.032–6.03
L. omostigma	3.62–6.07
L. parvicauda	1.65–2.04
Diaphus theta	0.050–2.24
Myctophum aurolaternatum	0.57–0.91
Argyropelecus pacificus	0.68–0.81
A. lychnys	0.53[a]
A. hawaiensis	1.02[a]
Merluccius productus	0.61–1.44

[a] Only one specimen examined

Table 4.9: Lipids of Myctophidae

Species	Mean standard length (cm)	Mean wt (g)	Tissue	Lipid (%) Fresh wt.	Dry wt.	wax entera in total lipid (%)
Hygophum reinhardtii	3.9 (3)[a]	0.58 (3)[a]	Whole fish[b]	3.34	14.0	10 (est.)[f]
Symbolophorus evermanni	6.5 (2)	3.7 (1)	Whole fish[c]	3.1	13.4	10 (est.)[f]
			Muscle[d]	1.16 ±0.3	5.1 ±0.7	—
S. californiensis	No data	13 (1)	Whole fish	4.3	19.8	Trace[f]
Tarletonbeania crenularis	No data	0.43 (2)	Whole fish	2.1 ±0.3	11.4 ±1.8	5 (max.)[f]
Diaphus theta	No data	21 (1)	Whole fish	15.8	57.0	Trace[f]
Stenobrachius leucopsarus	7.0 (17)	4.0 (17)	Whole fish	15.6 ±1.4	56.4	90.9
			Muscle	18.5	54.7	90.5 ± 1.7
			Viscera[e]	14.4	63.9	90 (est.)[ff]
Triphoturus mexicanus	6.0 (4)	1.85 (4)	Whole fish	14.5 ±0.8	54.1	82.2
			Muscle	16.9	57.8	74.5
Lampanyctus ritteri	7.8 (5)	5.2 (5)	Whole fish	14.2 ±0.3	50.6	58 (min)
			Muscle	11.8	51.1	86.7
			Liver	—	—	30 (est.)[ff]
			Viscera[e]	9.5	35.4	—

[a] No. of fish is parentheses [b] Three adult females [c] One male adult [d] Nine pooled individuals
[e] Includes gut contents [f] From thin-layer chromatograms.

that upthrust will be $V(d_a - d_b)$ g when the animal is transferred to a different salinity seawater of density d_b. The effects of temperature and pressure were allowed for separately. The coolor deep sea water caused a change in the upthrust prosuced by the squalene which was related to the remperature coefficient of expansion of oil and seawater. The upthrust in grams of oil floating in water changes with temperature according to αW, in which W is the weight of oil and α is $d_B/d_2 - d_A/d_1$ where d_A and d_B are seawater densities at high and low temperatures and d_1 and d_2 are densities of oil at the respective temperatures. Oils are caused to contract a lot more than seawater by a temperture decrease and in the buoyancy balance sheet it is seen that this is the largest single adjustment.

Table 4.10: The buoyancy balance sheet of Centroscymnus coelolepis (fish ♂).

	Weights in air (g)
Whole animal	5260
Liver	1550
Weight of liver, % total	29%
Weight of oil in liver, % liver	82%
	Weights in Plymouth seawater (g)
Whole animal (18 g upthrust)	– 18.0
Correction for salinity difference between Plymouth and deep sea water (whole animal)	–9.3
Correction for temperature difference between Plymouth water and deep sea water :	
(i) liver oil	+ 11.5
(ii) animal minus liver oil	+ 4.1
Correction for pressure difference between Plymouth water and deep sea water:	
(i) liver oil	+ 2.6
(ii) animal minus liver oil	—
Total net correction	+ 8.9
Calculated weight of fish in its natural environment	– 9.1 g

The negative buoyancy of the animal lacking its liver is affected by the cooler deep sea water to the extent indicated in Table 4.10 Corrections for the effect of pressure on the buoyancy of squalene and the negative buoyancy of the animal lacking a liver are both small. The compressibility of squalene has not been directly determined but it probably lies in the range typical of oils generally which is slightly in excess of that of seawater. It would be interesting to know the melting points of squalene and lipids at high pressure and low temperatures. The buoyancy correction is given by $10^{-5} \times (P_B - P_A) \times$ (density of seawater) $\times$ (volume of oil) where P_A and P_B are 1 atm and the normal ambient pressure of the animal respectively. The difference between the bulk compressibilities of protein and seawater is even smaller and Corner *et al.* quite reasonably discounted it in the buoyancy balance sheet. The balance sheet for the individual given in Table 4.10 is quite representative and shows that squalene, the dominant liver oil with no apparent metabolic function serves a buoyancy role in life.

How much swimming effort does a shark have to make in order to move its huge squalene–filled liver through the water? Indirectly this question in answered by Alexander's theoretical studies (1972) which show that, to a first approximation and in small fish, the energy cost of moving a mass of lipid through the water is much less than the cost of providing an equivalent amount of hydrodynamic lift in the absence of the lipid. In many sharks the pectoral fins generate uplift and Corner *et al.* (1969) showed that in neutrally buoyant sharks the pectoral fins are significantly smaller than the fins of similar animals which are negatively buoyant.

The metabolic cost of synthesising squalene has been estimated at 0.7 cal g^{-1} compare to 0.5 cal g^{-1} for oleic acid, but more upthrust is obtained from squalene.

The control of the synthesis of squalene is presumably connected with any control the animals might have over their buoyancy. Figures from Corner *et al.* (1969) suggest that the mass of squalene would have to be controlled to within 1 per cent in order to keep the animal within 0.1 per cent of neutral buoyancy.

Malins and Barone (1970) have demonstrated a possible basis for the regulation of the upthrust provided by the liver of deep water sharks. They used *Squalus acanthias* whose liver does not contain large amounts of squalene. Although the whole animal is not neutrally buoyant, the liver does contain significant amounts of diacyl glyceryl ethers and triglycerides. 3 kg animals which were loaded with 100 g weights showed a change in the mixture of liver lipids over 2 days, with diacyl glyceryl ethers present in higher proportions. The specific gravity of the diacyl glycerol ethers is lower than that of the triglycerdes so the changed ratio would to doine extent, offset the effect of the weights.

The naturally occurring lipid which might provide most buoyancy in animals, pristane, has a specific gravity of 0.78 but is not found in sufficient quantities to exert much effect. Blumer, Mullin and Thomas (1963) have shown that starving *Calanus* do not utilise pristane, which only constitutes 1–3 per cent of the animal's lipid. The energy cost of negative buoyancy in copepods may well be negligible in the overall economy of the animal anyway.

No Buyancy Device

A conditin of near neutral buoyabcy is reached in some bathypelagic fishes by a reduction of the skeleton and muscles. The buoyancy balance sheet of *Gonostoma elongatum* for example reads as follows, The weight of protein in water is 1.1 g per 100 g of fish, and the weight if skeletal material is also 1.1 g per cent. Upthrust is provided by the dilute body fluids, 1.2 g per 100 m fish, and from lipid to the extent of 0.5 g per cent. Thus without bulk lipid or gas buoyancy devices this fish weight only 2.2–1.7 g or 0.5 g per 100 g in seawater, which approximates a fifth of the weight in water of a normal shallow water fish with its swimbladder removed. This ingenious economy is probably closely related to the animal's nutritional economy and feeding habits.

Aqueous Solutions

Certain marine animals, representatives of which may live in the deep sea, reduce their weight in water through the regulation of their ionic contents. The coleomic fluid in a variety of squid and a few Crustacea is isosmotic but lighter than seawater due to the presence of ammonium ions. *Gigantocypris mulleri* is one of the deep sea crustaceans which buoys itself in this way.

Other invertebrates appear to exclude SO_4^{2-} ions, thereby reducing weight. The partial molar volume of SO_4^{2-} in aqueous solution is large and negative; it is an intensely hydrated ion and notably heavy. The prevalence of these buoyancy devices in the deep sea is not known. Only a limited uplift is obtained from a large bulk of 'light' fluid.

Some Conclusions About Buoyancy

The act of generating upthrust either involves motive power or molecular expansion. The deep sea has little effect on the resistance of water to muscular propulsion but exerts a variety of effects on buoyancy mechanisms. Some of the buoyancy devices employed in shallow water are adaptable to great depths. The buoyancy chamber of the cephalopods *Spirula* and *Nautilus* appears severely limited, and a theoretical study of its depth limitations would be an interesting sequel to the full elucidation of the buoyancy mechanism in these animals.

Other buoyancy systems work less well at increasing depths but are not subject to such an abrupot limitation by pressure as is the case in *Spirula.* Bulk phase bouyancy utilising gas or lipid molecules, is achieved by a phase change and the accompanying expansion of the system is due, in the case of lipids, to the formation of hydrophobic bonds. The energy required to generate the gas pressure required in deep sea swimbladders is not particularly demanding, but the work required to sustain adequate buoyabcy against diffusional losses, or to vary buoyancy daily, seems high.

Aqueous systems are different as no phase change is seen. Uplift is obtained by the selective maintenance of ions in which the process of molecular expansion has already taken place. Heavy and heavily hydrated ions are excluded. Although little upthrust is obtained by light aqueous solutions, deep sea animals may well be found to make extensive use of this type of buoyancy device.

The energy cost of generating upthrust at depth has yet to be investigated. Alexander's theoretical approach points the way. Making plausible assumptions, Alexander (1972) calculates that in fairly shallow water upthrust by muscular propulsion at constant depth uses about 12 times the energy which upthrust from a swimbladder would cost. Upthrust obtained from buoyant lipids is intermediate in its energy cost. At the 1000 m (100 atm) depth level the energy saving advantage of a buoyant swimbladder seems greatly reduced, but the widespread occurrence of swimbladders at depths greater than this suggests otherwise.

As we descend into the depths Nature has increasing difficulty in providing neutral buoyancy to save energy which, in turn, is becoming increasingly scarce. A more satisfactory description these interacting forces and solutions to them can only come from physiological study of specific cases. Barham (1971) has suggested how two types of mesopelagic myctophids may be distinguished. There is an active type, exemplified by *Myctophum,* which have muscles, and in those cases which possess a swimbladder it is well developed and perhaps capable of rapid gas exchange. It is these active myctophids which migratre all the way to the surface at night and are menbers of the deep

Barham has also made the interesting observation that these suspended animals may obtain upthrust from their opercular pump mechanism.

The biochemistry, physiology and behaviour of mesopelagic fish have a lot to tell us about how different design–solutioins to a common physical environment are arrived at. The deep sea elasmobranchs, far less numerous than muctophids but rather more robust as experimental animals, may prove good bathypelagic animals to study.

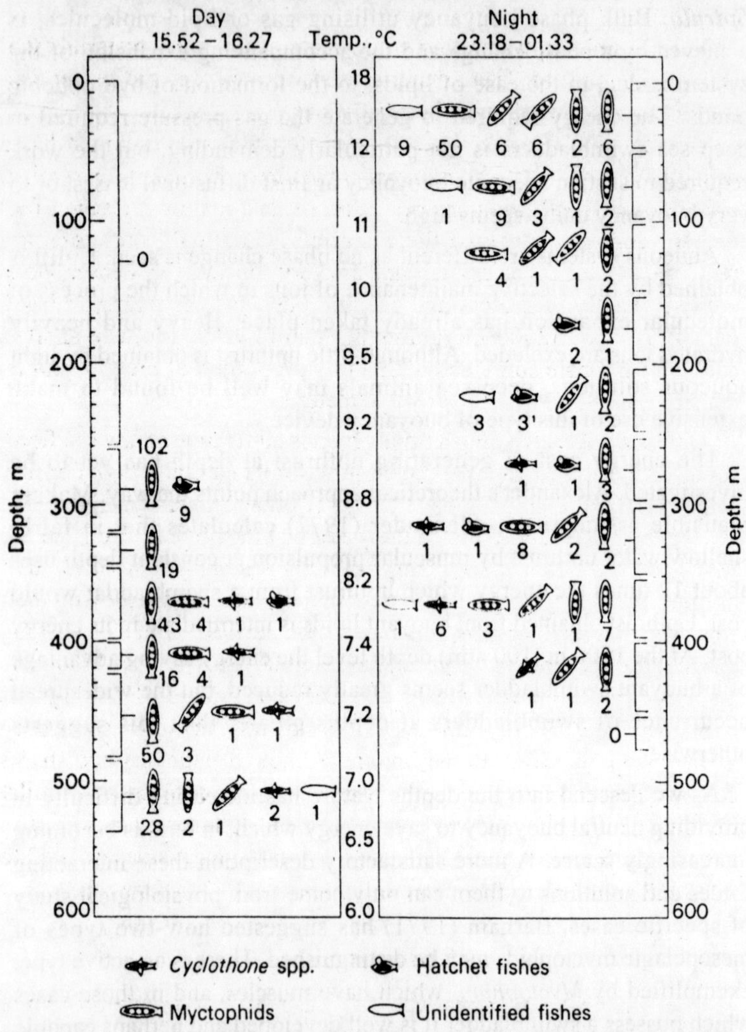

Fig. 4.8: A diagram showing numbers and posture of fishes observes from a deep submersible vehicle during two dives, 7 hours apart, in the San Diego Trough. Temperatures measured during the day–time dive are shown between the columns.

CHAPTER 5
BUOYANY IN ANIMALS

Muscle sinks in water, and so do most of the other materials that animals are made of. These materials sink because they are denser than the surrounding water, whether that water is fresh water, which has a density of 1000 kilograms per cubic meter or sea water which has a density of nearly 1030 kilogram per cubic meter. Muscle is denser than either (about 1060 kilogram per cubic meter), mainly because of its protein content. Many of the materials of animals skeletons are reinforced by inorganic crystals and are very considerably denser, Bone consists mainly of protein and calcium phosphate crystals and has a density of about 2000 kilograms per cubic meter. Molluse shell is calcium carbonate combined with a tiny proportion of protein, and its density is 2700 kilograms per cubic meter.

Unless a water-dwelling animals spends all its time on the sea floor, it must have some way of compensating for its naturally high density. Many swimming animals have floats or other buoyancy organs that reduce their densities to match the water that they live in. You can tell how good the match is, for many fishes, by watching them in aquaria: at times they hang almost motionless in the water, hardly moving their fins, yet they neither sink nor rise to the surface. Other swimming animals are denser than water but have no buoyancy organs. Flounders and most sharks, for example, have densities of 1060 to 1090 kilograms per cubic meter. These animals swim to stay high in the water; as soon as they stop, they sink to the bottom. Although flounders and sharks can be observed resting on the bottoms of aquaria, some fish of similar density, the large and vigorous tunnies, apparently never stop swimming. Squids without buoyancy organs have densities of 1055 to 1075 kilograms per cubic meter and also swim to avoid sinking.

Swim or Sink

The density of an animal can be measured by a simple procedure that depends solely on being able to weigh the animal, both in and out of the water. A little explanation is needed to show how the procedure works. The weight of animal is its mass multiplied by the gravitational acceleration, and mass is volume multiplied by density : 98

$$\begin{aligned} \text{weight} &= \text{mass} \times \text{gravity} \\ &= \text{volume} \times \text{density} \times \text{gravity} \end{aligned}$$

When an animal is submerged in water, an upthrust acts on it which, according to Archimedes Principle, equals the weight of volume of water equal to the volume of the animal:

$$\text{upthrust} = \text{volume} \times \text{water density} \times \text{gravity}$$

Dividing the first equation by the second gives us a formula for density:

$$\frac{\text{weight}}{\text{upthrust}} = \frac{\text{volume} \times \text{density} \times \text{gravity}}{\text{volume} \times \text{water density} \times \text{gravity}} = \frac{\text{density}}{\text{water density}}$$

$$\text{density} = \frac{\text{water density} \times \text{weight}}{\text{upthrust}}$$

So, to calculate the density of an animal we need to know its weight and the upthrust. To obtain its weight, we can weigh the animal i the ordinary way in air. Since the effect of the upthrust is to reduce the animal's weight in water, we can find the upthrust by weighin the animal again while it is suspended in water and subtracting it weight from its weight in air:

$$\text{upthrust} = \text{weight in air} - \text{weight in water}$$

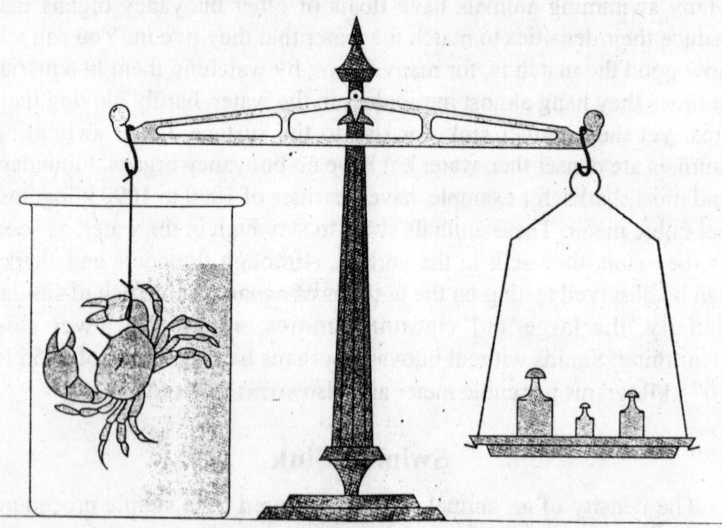

Fig. 5.1: The density of a crab can be measured by weighing the animal in water.

Once we have the two weights, a simple calculation provides the density:

$$\text{density} = \frac{\text{water density} \times \text{weight}}{\text{weight in air} - \text{weight in water}}$$

A simple comparison of the water density and the animal's density will tell us whether an animal will sink or float. Sometimes we want to know not merely whether an animal will sink, but how strong the force is that will make it sink. Just an on land an animal's weight is the force drawing the animal downward, so an animal's weight in water is the force making it sink.

weight in water = weight – upthrust

Using the equations given above for weight and upthrust,

$$\text{weight in water} = \text{volume} \times \text{density of animal} \times \text{gravity} - (\text{volume} \times \text{water density} \times \text{gravity})$$

$$\text{weight in water} = \text{volume} \times \text{excess density} \times \text{gravity}$$

(By excess density I mean the difference between the density of the animal and that of the water.) If we divide each side of the equation by the animal's weight, we get the simpler equation:

$$\frac{\text{weight in water}}{\text{weight}} = \frac{\text{excess density}}{\text{density}}$$

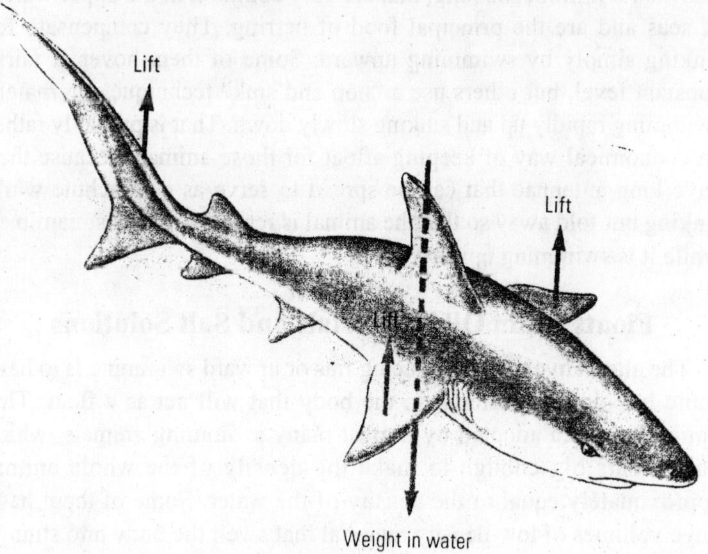

Fig. 5.2: The tail as well as the pectoral fins of a shark must provide lift to counteract the fish's submerged weight.

For example, a flounder with a density of 1080 kilograms per cubic meter has an excess density in seawater of 50 kilograms per cubic meter, and its weight in seawater is 50/1080 of its weight in air, or 4.6 percent of that weight. That is reasonably typical of animals that have no buoyancy organs: the sinking force that they have to overcome when they swim in the sea is usually about 5 percent of body weight.

Such animals must produce upward hydrodynamic forces to overcome this sinking force, just as birds much produce lift to fly. Sharks depend for upward lift on large pectoral fins (the pair of fins, one on each side of the body behind the head), which are held at an angle of attack. Shark's tails are asymmetrical, shaped to fan water downward as well as backward when the fish swims, and so provide an upward force as well as thrust. The lift on the fins and the upward force on the tail together balance the animal's weight in water. Because the fins are much closer to the shark's center of gravity than the tail is, most of the upward force has to come from them. Tunnies also depend mainly on their pectoral fins for lift to prevent them from sinking, and squids have fins that function in the same way.

These animals use their fins like airplane wings, to support themselves as they swim along. Some other dense swimmers behave more like helicopters. Copepods are small crustaceans, most of them less than 2 millimeters long, that are very common in the upper waters of seas and are the principal food of herring. They compensate for sinking simply by swimming upward. Some of them hover at fairly constant level, but others use a "hop and sink" technique, alternately swimming rapidly up and sinking slowly down. That is probably rather an economical way of keeping afloat for those animals because they have long antennae that can be spread to serve as a parachute while sinking but fold away so that the animal is reasonably well streamlined while it is swimming upward.

Floats From Oily Materials and Salt Solutions

The alternative to lift-producing fins or upward swimming is to have some low-density material in the body that will act as a float. This option has been adopted by a great many swimming animals, which have floats big enough to make the density of the whole animal approximately equal to the density of the water. Some of them have huge volumes of low-density material that swell the body into strange shapes.

The quantity of low density material required depends on how low its density is. The weight in water of an animal without a float is, as we have already seen,

weight in water without float = volume × excess density × gravity

Instead of having a downward weight in water, a float is buoyed with an upward force that can be calculated from its density shortfall (water density minus float density):

buoyancy of float = float volume × density shortfall × gravity

To give an animal exactly the same density as the water, so that it neither floats nor sinks, the buoyancy of the float must counteract the animal's weight in water:

buoyancy of float = weight in water without float

By replacing the terms in this formula with their equivalents, we can discover how the volume of the float is related to the density of the material:

float volume × density shortfall × gravity

= initial animal volume × excess density × gravity

$$\frac{\text{float volume}}{\text{initial animal volume}} = \frac{\text{excess density of animal without float}}{\text{density shortfall of float}}$$

If the float has a low density (a large density shortfall), a small float will suffice, but if it is only slightly less dense than the water (having a small density shortfall), it needs to be large.

Ideally, floats should be as small as possible, and therefore of low density. Fats are low-density materials that are quite plentiful in many animals, but they are only a little less dense than water, as can be shown by putting a piece of fat in water: it will float, but with only a very small fraction of its volume above the water surface. Typical fats have densities of about 930 kilograms per cubic meter, giving a density shortfall in seawater of about 100 kilograms per cubic meter. To achieve the same density as seawater, a flounder with an excess density of 50 kilograms per cubic meter would need enough fat to increase its volume by 50 percent. If the lounder lived in fresh water, it would have to be even more obese. Its excess density would then be 80 kilograms per cubic meter and the density shortfall of the fat only 70 kilograms per cubic meter, so the fat needed to make it float would more than double its volume.

Some other oily materials are less dense than fat, and smaller volumes of them suffice to give animals buoyancy. Squalene is a hydrocarbon (that is, a member of the same group of chemical compounds as fuel oils) with a density of only 860 kilograms per cubic meter, found in large quantities in the livers of some sharks. Its density shortfall in seawater is 170 kilograms per cubic meter, so the volume needed to float a fish with an initial excess density of 50 kilograms per cubic meter is only 50/170, or 29 percent, of the initial volume of the fish. To contain even this quantity of squalene, sharks that depend on it for buoyancy have enormously enlarged livers.

The best known of these sharks is the gigantic basking shark, which grows to lengths of at least 11 meters (some books say even longer) and masses of at least tonnes (18,000 pounds). Despite its size, it feeds on small plankton that it strains out of the water as it swims slowly along. Its squalene content makes the shark an attractive target for hunters, who sell the squalene for use in the leather industry.

Wax esters are another group of oily compounds, of about the same density as squalene. They serve to give buoyancy to lantern fishes (described later in this chapter) and also to the coelacanth *Latimeria*. Of all living fish, *Latimeria* is the closest to the evolutionary ancestry of the legged vertebrates, including ourselves. Although the coelacanths are well known as fossils, they were thought to have died out at the same time as the dinosaurs until a Miss Latimer saved one from a South African fisherman's catch in 1938.

As an alternative to organic compounds such as squalene and wax esters, animals can gain buoyancy from low-density solutions of salts in water. These solutions cannot be less dense than pure water (density 1000 kilograms per cubic meter) and thus are effective only in salt water, and even there they cannot have large density shortfalls: large volumes of these solutions are needed to compensate for even small excess densities in other parts of the animals. The jelly of jellyfish is slightly less dense than seawater because it contains a smaller proportion of heavy sulphate ions. It is a very dilute protein gel containing only about 1 percent organic matter. Jellyfish jelly also contains dissolved salts in the same osmotic concentration as the salts in seawater: if the jelly were more dilute, its water would be drawn out by osmosis and the jellyfish would shrivel up. Its density is slightly reduced because it contains less sulphate than seawater does, and so contains correspondingly more of other, lighter ions, but the effect is small because only 8 percent of the salt in seawater is sulphate. The density shortfall of jellyfish jelly in seawater is only about 1 kilogram per cubic meter; in consequence, a huge volume of jelly is needed to buoy up a tiny volume of living tissue. Indeed, the living tissues of jellyfish are confined to an extremely thin layer only one cell thick on the surface of the non-living jelly. Some jellyfishes are almost exactly the same density as the water they live in, but others are denser and must swim to keep afloat.

Solutions of salt water bring the densities of many squids very close to that of seawater. In these squids, the body cavities or some of the tissues are swollen by large volumes of a solution that resembles seawater but for the fact that most of the sodium has been replaced by ammonium ions. These solutions have densities of about 1010 kilograms per cubic meter, so their density shortfall in seawater is almost 20 kilograms per cubic meter—much more than the density shortfall of jellyfish jelly but much less than that of fats, squalene, and wax esters. Some squids look strangely bloated because they must hold large volumes of these solutions to achieve the same density as the water they live in.

These bloated squids live in the oceans, many of them at depths of several hundred meters. Although they are seldom seen alive, their remains are found more commonly than any other food in the stomachs of sperm whales. They must be very numerous, for it has been calculated that the quantity eaten annually by sperm whales exceeds the annual fish catch of the combined fishing fleets of the world.

Chambers of Gas

Floats filled with gas can be much smaller than other kinds because gases have very low densities. For example, at room temperature and atmospheric pressure, the density of air is only 1.3 kilograms per cubic meter, about $\frac{1}{800}$ the density of water.

One ingenious method for bringing gas into a float is used by the pearly nautilus, a primitive relative of the squids. It is found in coastal waters in the southwest Pacific, where it can be caught in lobster pots. Though related to the squids, it has a large shell, coiled like the shells of some snails. The soft parts of the animal's body occupy the quarter-turn of the shell nearest the opening, and the rest of the shell is divided into a series of gas-filled chambers. Even though the gas has a very low density, quite a large shell is needed to make the animal float, because the shell its self is so dense.

Nautiluses are caught at considerable depths, often at 100 meters or more. Hydrostatic pressure in the sea increases by 1 atmosphere for every 10 meters of depth, so nautiluses caught at 100 meters must have been living at high pressures. You might imagine that the gases in the shell chambers would also be at high pressures and that these gases would bubble out if a freshly caught animal was held under water while a hole was bored into its shell. Eric Denton and J.B Gilpin-Brown (then based in the marine laboratory in Plymouth, England) tried the experiment and got the opposite result: gas did not bubble out but water was sucked in, showing that there had been a partial vacuum in the intact shell. The two experimenters measured the volume of water sucked in by weighing the nautilus before and after the experiment, and with this information they were able to calculate the gas pressure: it was about 0.8 atmosphere in the older chambers, nearest the center of the spiral shell, and as little as 0.3 atmosphere in some of the newer chambers. The gas was mainly nitrogen.

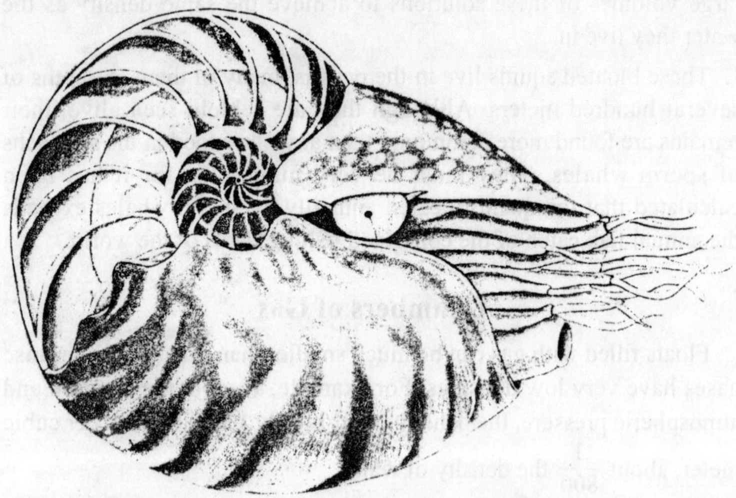

Fig. 5.3 : The pearly nautilus has gas filled buoyancy chambers in its shell.

Besides containing this low-pressure gas, the chambers contain a small amount of a solution of salts in water. Denton and Gilpin Brown measured the freezing point of this solution and showed that the solution was more dilute than the animal's blood. Strong solutions can suck water from weak ones, if the two solutions are separated by a membrane that allows water but not the dissolved salts to pass through. The process is called osmosis, and its effect is to make the strong solution weaker and the weak one stronger, so that their concentrations become more nearly equal. Water could be drawn by osmosis out of the dilute solution in the chambers and into the blood, but only under certain conditions of pressure. Osmosis can be prevented by a pressure difference sucking in the opposite direction—if the weak solution is at a lower pressure than the strong one. The pressure difference needed to prevent osmotic flow is called the difference in osmotic pressure between the solutions. Denton and Gilpin-Brown's measurements showed that the difference in osmotic pressure between nautilus blood and chamber fluid was 15 atmospheres. That means that water could be drawn out of the chambers by osmosis into the blood against pressure differences of up to 15 atmospheres.

These observations seem to show that the nautilus does not fill its shell chambers by pumping gas into them; rather, it sucks water out osmotically. The pressure in the sea is 1 atmosphere at the surface and increases by 1 atmosphere for every 10 meters of depth, so it is 15 atmospheres at a depth of 140 meters. Thus the observed osmotic pressure difference is just enough to suck water out of the chambers at that depth, leaving vacuum behind. After the chambers have been emptied in this way, gases will diffuse in from the animal's tissues. There is less gas in the newer chambers because there has been less time for gas to diffuse in.

The dissolved gases in the water at the surface of the sea are in equilibrium with the air, at a pressure of 1 atmosphere. Water deeper in the sea contains less dissolved gas, because the animals living in it use up oxygen. As it passes through the gills, the animal's blood exchanges gases with the water. Consequently, although these animals are living at depths where the pressure is many atmospheres, their blood and other tissues contain no more gas than would dissolve in them if they were kept in contact with air at 1 atmosphere pressure. That is why the pressure of the gas in the shell of a nautilus never rises above 1 atmosphere, however deep in the sea it is living.

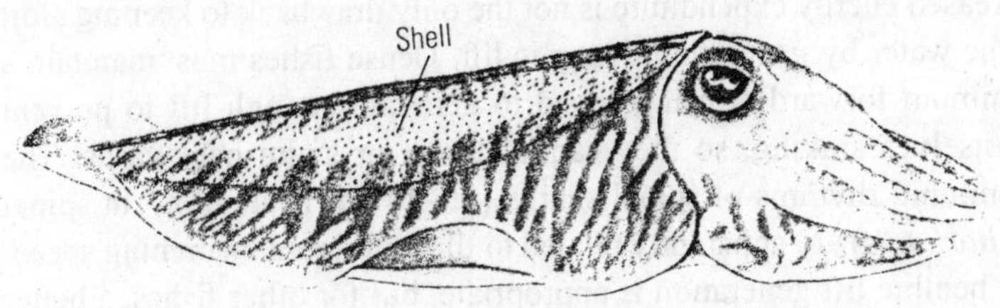

Fig. 5.4

Cuttlefish, other relatives of squid, have a more buoyant shell than the nautilus. This is the cuttlebone, which is often given to caged birds to peck. It has no protective function but is buried inside the body. It consists of gas spaces sandwiched between thin layers of shell, and its very light construction gives it a density of only 620 kilograms per cubic meter, less than the density of any of the buoyancy acids discussed so far. It occupies only 10 percent of the volume of the cuttlefish, but that is enough to make the density of the animal almost exactly the same as that of water. Cuttlefish can hang almost motionless in midwater or more slowly around by undulating their fins.

The rigid gas chambers of the nautilus and cuttlefish, filled with gas at low pressure, would collapse if the animals swam too deep. Denton and his colleagues put the shells of these animals into pressure chambers and increased the pressure until they collapsed. Nautilus shells collapsed at about 65 atmospheres (the pressure at a depth of 640 meters), but nautiluses seem not to swim deeper than about 500 meters. Cuttlebones collapsed at 24 atmospheres (the pressure at 230 meters), but cuttlefish seem to live no deeper than about 150 meters. These animals do not seem to go deep enough to put their shells at risk.

Buoyancy in Fishes

Dynamic lift

Some fishes are denser than the water in which they swim, and have to generate dynamic lift mainly by using their outspread pectorals as lifting foils. This process inevitably generates drag, which makes a significant contribution to the total drag and energy requirement of such fishes. For example a mackerel weighing around 25 g in sea water has to generate some 1.2×10^4 N of dynamic lift during level swimming—mackerel effectively are climbing a 1 in 15 hill all their lives! But increased energy expenditure is not the only drawback to keeping aloft in the water by generating dynamic lift. Dense fishes must maintain a minimum forward cruising speed to generate enough lift to prevent themselves sinking, so they cannot hover or swim backwards. The swimming rhythms of spinal sharks (35–40 tail beats min^{-1} in spinal *Scyliorhinus*) are apparently related to this minimum swimming speed. For benthic lift generation is appropriate, but for other fishes, a better solution would seem to be to store light materials to provide *static* lift (as do airships and submarines) and thus avoid the drawbacks of generating dynamic lift.

Static lift

Most of the materials making up a fish are denser than water. For example, much of the body consists of locomotor muscle (density 1050–1060 kg m^{-3} in common marine fishes). Skeletal tissues loaded with heavy mineral salts are correspondingly denser (2040 kg m^{-3} for typical teleost bones). Nevertheless, water is so much denser than air that even without special stores of low density materials, fishes are not so much denser than the water in which they swim. They have the option (denied to birds) of storing sufficient low-density material that they can make themselves the same density as the water; they can achieve neutral buoyancy and need expend no muscular energy to keep station in the water.

Fishes use two quite different materials to provide static lift. Gas is efficient in giving lift, since its density is very low, and most teleosts possess gas-filled swimbladders. Usually the swimbladder is around 5% of the volume of marine fishes, and around 7% in freshwater fishes, providing enough lift for neutral buoyancy. Fish swimbladders cannot significantly resist changing in volume as the fish swims up and down in the water and the ambient pressure changes, indeed, the

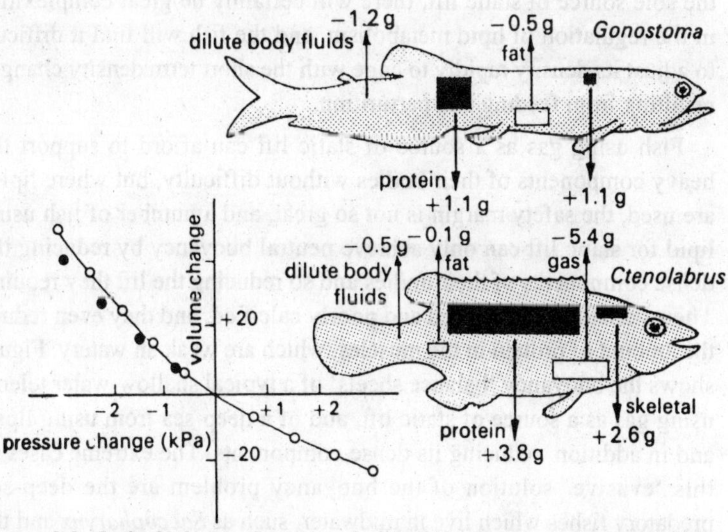

Figure 5.5: ***Left: Volume changes of airbubble (open circles) and fish swimbladder (filled circles) as ambient pressure changes. Right: buoyancy balance sheets for a mesopelagic fish (*****Gonostoma*****) and a shallow-water acanthopterygian (*****Ctenolabrus*****).***

swimbladders of almost all teleosts obey Boyle's law perfectly. So if a fish with a gas-filled swimbladder is to remain neutrally buoyant at different depths, it must secrete or absorb gas to keep the swimbladder at constant volume as the ambient pressure changes. To regulate the *mass* of gas within the swimbladder in this way requires complex mechanisms of great physiological interest.

Fats and oils are also used by fishes as sources of static lift. These are much less efficient and much bulkier for a given amount of lift. However, they have the great advantage that the lift provided varies little with depth, because changes in ambient pressure have relatively little effect on the volume of the fat or oil: if a fish using lipid as a source of static lift is neutrally buoyant at the surface of the sea, it will be close to neutral buoyancy at the sea bed, even at considerable depths.

In the short term then, lipid provides fewer problems of buoyancy regulation than does gas, but in the longer term, difficulties arise, because the lipids stored to provide lift may have other functions as well. For example, sharks which store oil in the liver and muscles may have to draw on this store as a fuel for continuous swimming, or as a food reserve for developing embryos or for the adult. Where lipid is the sole source of static lift, there will certainly be great complexities in the regulation of lipid metabolism, and the fish will find it difficult to adjust its density rapidly to cope with the short term density changes resulting from feeding and parturition.

Fish using gas as a source of static lift can afford to support the heavy components of their bodies without difficulty, but where lipids are used, the safety margin is not so great, and a number of fish using lipid for static lift can only achieve neutral buoyancy by reducing the dense components of their bodies and so reducing the lift they require. Their skeletons are reduced and poorly calcified, and they even reduce the amount of protein in the muscles, which are weak an watery. Figure shows the buoyancy 'balance sheets' of a typical shallow-water teleost using gas as a source of static lift, and of a deep-sea from using lipid, and in addition, reducing its dense components. The extreme cases of this 'evasive' solution of the buoyancy problem are the deep-sea predatory fishes which live in midwater, such as *Saccopharynx* and the deep-sea angler fishes. As Denton and Marshall (1958) pointed out, they are little more than floating traps, with weak skeletons and watery muscles; they attract their prey with luminous lures and need not pursue it actively like the fish of the upper ocean. To some extent reduction in density of body components has taken place during teleost evolution, from the thick-scaled 'ganoid' fishes to the thin-scaled clupeids of today. Many fishes also have planktonic larvae whose dense components are reduced, and which have spaces for low density dilute body fluids.

Lipid as a source of static lift

In four unrelated fish groups (and perhaps in others) enough lift is provided by stored lipids to achieve neutral buoyancy. Interestingly enough, the lipids stored are not biochemically similar, but they are all of particularly low density and so most efficient in providing lift. Fish lipids vary in density from around 930 kg^{-3} (cod liver oil, and the oils of dogfish livers) to the wax esters of myctophids, gempylids, and *Latimeria* (densities around 860 kg m^{-3}), and the hydrocarbons of some sharks (869 kg m^{-3}). The difference in specific gravity between cod liver oil, and the wax esters or shark hydrocarbons may not seem very striking, but 1 g of the less dense oil will provide 0.1675 g of lift in sea water whose density is 1027.5 kg m^{-3}, whereas a gram of the denser oil will only provide about half as much lift (0.0975 g). It is thus well worthwhile for the fish to store these lighter lipids rather than the more common triglyceride metabolic reserves, as all the fish that achieve neutral buoyancy using lipids actually do.

Squalene

The first fish found to use lipid to achieve neutral buoyancy were the deep-sea squaloid sharks studied by Corner *et al.* (1969) (Figure 4.2). Unlike the spur dog of the same family, these fish live near the bottom in deep water, and the habitat is one where the family has successfully diversified, for there are many genera known. All share two striking characteristics: very large livers (making them grossly corpulent) and very small pectoral fins. In most animals, including ourselves, the liver is around 4–6% of the total weight, but in these fish, it may be more than ¼ of the total weight, because it contains an enormous amount of pale yellow oil. On this oil the fish literally float—when their livers are removed they sink. In all species examined, the liver oil is of low density (870–880 kg m^{-3}) because it is mainly composed of the hydrocarbon squalene. Squalene, which was first isolated from such sharks, is formed by the condensation of isoprene units on the pathway leading to cholesterol. It has the low density of 860 kg m^{-3}, and so is admirably suited to provide static lift.

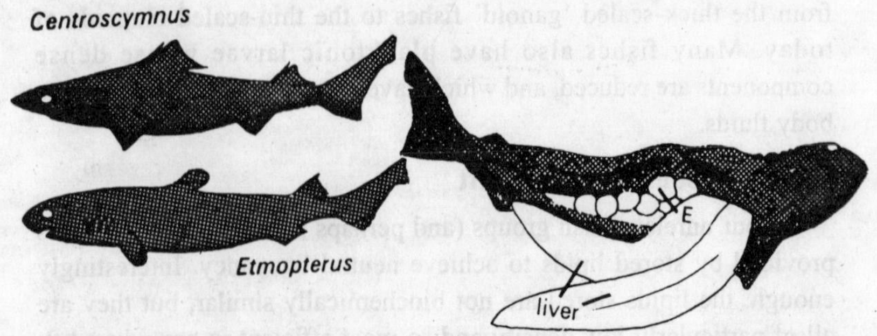

Figure 5.6 ***Neutrally buoyant deep-sea squaloid sharks. Right: female*** **Etmopterus** ***opened to show huge liver and large eggs (E). From sharks caught on a deep sea long line.***

We do not yet know how sharks regulate their liver lipids so as to balance their weight in water, but it appears from experiments upon *Squalus* that the fine adjustments required for neutral buoyancy may depend not on changing amounts of squalene, but rather on varying the other less abundant lipid constituents of the live oil. By attaching weights to *Squalus*, Maline and Barone (1970) found that the fish responded by increasing the amount of low-density alkoxydiglycerides in the liver oil, compared with the control fishes which had larger amounts of the denser triglycerides.

Deep-sea squaloids bear live young and have very large eggs (about the size of a billiard ball in a shark a meter long); ingeniously enough the eggs contain squalence and are neutrally buoyant themselves, so that pregnancy does not increase the density of the mother.

It is obvious that these fish need only have relatively small pectoral fins; they are used only during manoeuvring. Although the tails in all genera are markedly heterocercal, they evidently do not generate lift. Why do deep-sea squaloids need to be neutrally buoyant if they live on the sea bed? The answer is that they hover just off the bottom (as we know from cine films taken by deep cameras), unlike the dense bottom-dwelling elasmobranchs of shallow water, which rest *on* the bottom and are invariably very dense, like the dogfish.

Deepwater Holocephali evidently live in a similar way to the deep-sea squaloids, and like them, are close to neutral buoyancy, although they only manage this by virtue of reduction of dense components, and have poorly calcified skeletons. Their liver oil consists largely of squalene and this is also the main source of static lift in some teleosts. Eulachon (*Thaleichthys*), for instance, contain 20% of lipid by weight. Eulachon do not have a gas-filled swimbladder, and during their spawning migration, squalene forms a higher proportion of the total lipid that at other times: it seems that the fish metabolizes reserve triglycerides during these migrations and becomes denser, so that in this case, lipid metabolism is not sufficiently well regulated to cope with buoyancy and metabolic demands upon the total lipid pool, and at the same time, maintain neutral buoyancy.

Wax esters

A little squalene is also found in the living coelacanth *Latimeria*, but most of the massive amounts of lipid stored are wax esters. These make up 30% of the wet weight of the ventral musculature and over 60% of the wet weight of the swimbladder (which contains only lipid, no gas), and in the pericardial and pericranial tissues, the percentage is even higher. The habits of *Latimeria*, have been very little known but its fins certainly seem unsuited to provide dynamic lift, and the only specimen examined alive floated at the surface, so that it seems safe to assume that it is neutrally buoyant in life.

Wax esters are probably stored to provide lift in many families of mesopelagic teleosts, but the only ones examined so far have been the gempylids and a few of the numerous kind of myctophids. The gempylid *Ruvettus* is loaded with low-density oil (density 870 kg m^{-3}) which has purgative properties, hence the common name of castor oil fish. The cranial bones have been modified as oil tanks, and are the least dense tissues of the body. This large (1 m) predatory fish ranges from depths of 15 m to over 500 m, and is very close to neutral buoyancy, feeding on smaller fishes which undertake diurnal migrations.

The most remarkable of these are the myctophids, some undertaking daily a double journey of 500 m up and down in the water column. Although all myctophids have gas-filled swimbladders as larvae, in many species the swimbladder shrinks and becomes invested with more and more lipid as the fish gets older, until no gas remains. As in *Latimeria*, lift from gas is replaced by lift from lipid, which may eventually make up 15% of the wet weight of the fish. As we should expect, the species with a high lipid content are neutrally buoyant, and store low-density wax esters. Why should many myctophids have abandoned gas as a source of static lift? Although the evidence is only circumstantial, it seems a reasonable guess that is because there are difficulties in regulating buoyancy over a wide depth range when gas is the source of lift. Those species which as adults have much low density lipid and are neutrally buoyant, have a greater depth range than the juveniles, which still have gas in the swimbladder, similarly, species which nave gas-filled swimbladders as adults undergo less extensive vertical migrations than those which rely on lipid only.

On the whole, the evidence is that lipid storage for static lift is a secondary phenomenon in myctophids, and the most 'advanced' species in the family are those which have abandoned gas altogether as adults; it is interesting that amongst the family there are fishes showing all stages in this change over adapted to different life-styles.

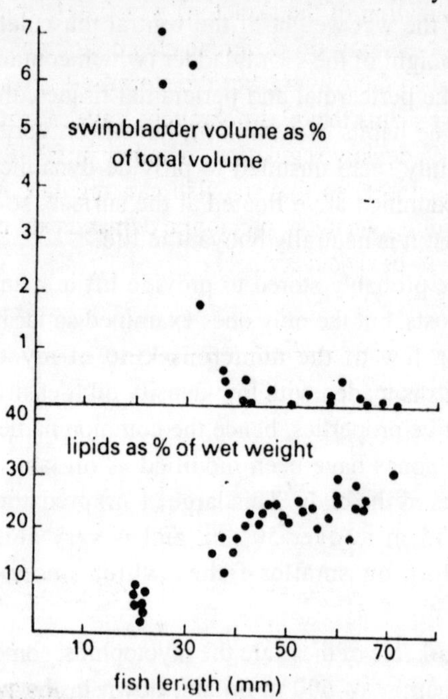

Figure 5.7: ***Changes in swimbladder gas content and lipid investment during growth of the myctophid* Diaphus theta. *After Butler and Pearcy (1972).***

Insufficient static lift for neutral buoyancy

So far, we have considered fishes of different groups which store low-density lipids to attain neutral buoyancy. What of fish which use lipid for static lift, but are not neutrally buoyant? Many teleosts are certainly in this category. Mackerel vary in density at different times of year because they have more or less lipid stored in the muscles; such a system is evidently a primitive one, for lipid is not set aside, as it were, to maintain a constant density. Rather, the lipid store is at the mercy of metabolic demand, and reduction in density, valuables as it must be, is simply a side effect of the storage of lipid for metabolic purpose.

We know more about elasmobranchs—Baldridge (1969) and Bone and Roberts (1969) showed that a wide range of elasmobranchs, including large pelagic sharks, are each of a characteristic density, some being fairly close to neutral buoyancy, others being very dense. Liver lipid is an important factor in determining the density of some species, but in most, it is the density of other tissues (amount of fat in the white muscle, mineralization of the skeleton) which determines the density of the species. Not only do different elasmobranchs store different amounts of lipid in their livers, but the stored oil is least dense in the dense species. It seems that elasmobranchs have managed to set apart the lipid used for density regulation, whether in the muscles or liver, from metabolic stores, so that the fish can regulate its density to a characteristic figure, whatever metabolic demands are made. How this is done remains to be discovered.

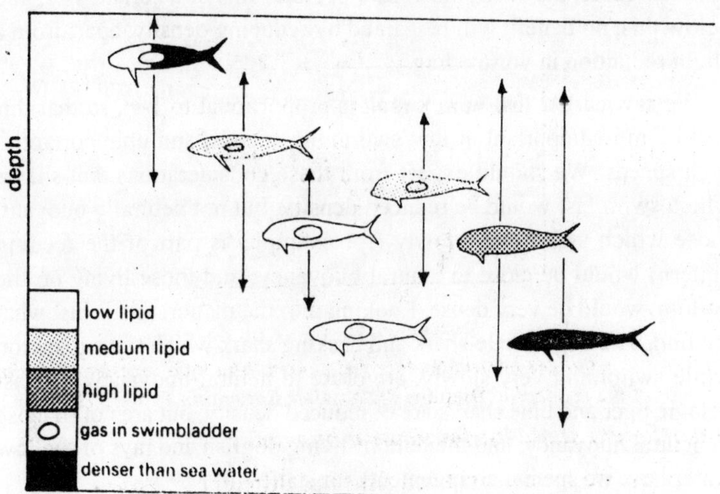

Figure 5.8: ***Functional types of myctophids. Vertical arrows indicate known or assumed extent of vertical migration.***

At the beginning of this section, the advantages of neutral buoyancy were extolled, and it seems odd to find fishes using static lift to maintain a fixed density below that of neutral buoyancy. The regulatory mechanisms are there, and at first sight, it seems a harder problem to maintain a constant density (say around 1.03) than to regulate to neutral buoyancy, and what is more, dynamic lift must still be generated, with the penalties that this incurs. However, it is not hard to see why many pelagic sharks use lipid to reduce their density, but not so far as to become neutrally buoyant.

Shark fins vary a good deal in flexibility, and in their stiffening by skeletal elements, but the basic design, unlike that of teleosts, makes them impossible to retract: they are not of varying geometry. If we bear this in mind, we can see how considerations of dynamic lift generation lead to different densities in different species, depending on their mode of life. Reduction in density by lipid storage means that less lift need be generated in level swimming, so that the shark can either reduce the size of its pectoral fins, or it can cruise more slowly without stalling. Since sharks use their pectoral fins for manoeuvring (to turn, change in pitch, and even sometimes, as does *Heterodontus*, to creep backwards along the sea bed), there is a limit to reduction in fin size. Reduction in fin areas is wholly beneficial (since it reduces drag), but to remain manoeuvrable, the shark must have pectoral fins of a certain size, and below this, no benefit will be gained by reducing density, apart from a slight reduction in vortex drag.

We saw earlier that vortex drag is proportional to $1/v^2$, so that this will be most important at low swimming speeds, and unimportant at high speeds. We should expect from these considerations that sharks which swim fast would be reduced density, but not neutrally buoyant; those which swim very slowly (for example as part of the feeding pattern) would be close to neutral buoyancy; and those living on the bottom, would be very dense. Looking into the matter, this is just what we find. The huge whale-shark and basking shark which sieve plankton while swimming very slowly, are close to neutral buoyancy; the fast pelagic tiger and blue sharks are of reduced density, but are not so close to neutral buoyancy; and the bottom-living dogfish and rays of shallow water have no special arrangements for static lift.

Since teleosts have differently designed fins, which *can* vary their geometry, the arguments above do not apply, and we should not expect to find any teleosts using lipid to reduce their density to a constant figure not close to neutral buoyancy; so far, none have been found.

Gas as a source of static lift

Because fish swimbladders obey Boyle's law, to obtain constant lift as they swim up and down in the water, fish are obliged not only to store gas, but also to secrete and absorb it. In the sea, the ambient pressure increases by 1 atmosphere (101.3 k Pa) for every 10 m increase in depth, so that a fish living at the surface (with the swimbladder gas at atmospheric pressure) will have the volume (and lift) of the swimbladder halved if it descends to 10m, and vice versa. It is hardly surprising that the regulatory mechanisms cannot cope effectively with such large changes, and that many surface-dwelling pelagic fish have either lost the swimbladder altogether (tunas and *Scomber*) or have a reduced swimbladder (*Scomberomorus*). Gas is not a suitable source of static lift for this mode of life, unless the fish remains strictly at the surface, as do hemirhamphids and flying fishes.

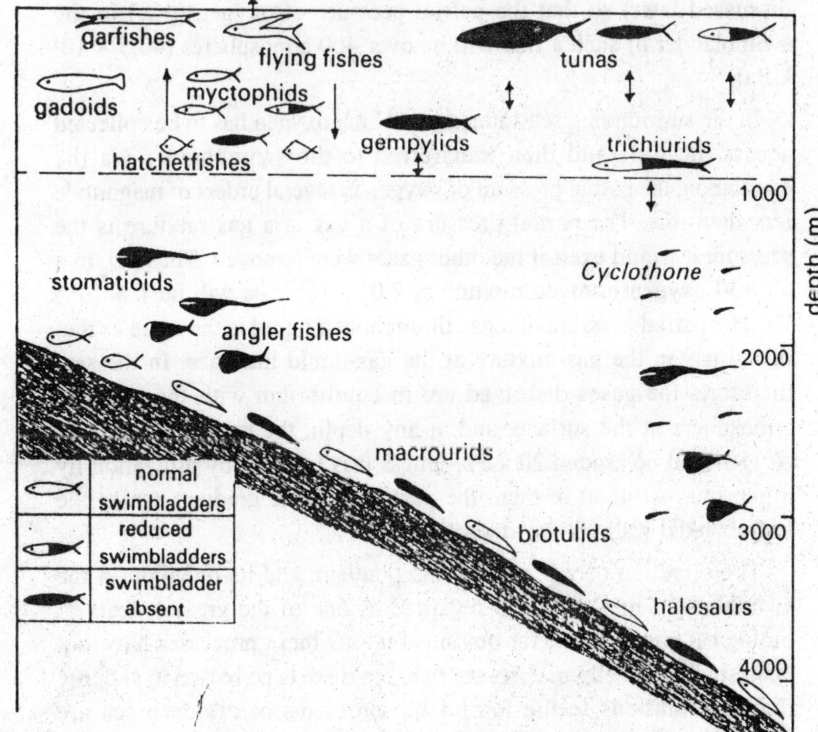

Fig. 5.9: ***Distribution of fishes with and without swimbladders in the oceans.***

At greater depths, however, changes in ambient pressure have relatively less severe effects; a fish changing its depth 10 m, at say, 400 m, will only alter the volume of its swimbladder by 1/40th, and the same applies with greater force, the greater the depth at which the fish lives.

The distribution of fish with gas-filled swimbladders in the sea has been examined by Marshall (1960) who has found that they occur down to the greatest depths (Figure 4.5), but are absent from the upper layers unless they live just at the surface. The deeper the fish lives, the less significant are changes in volume with depth, but the more formidable are the problems of gas secretion and retention. Yet fish with gas-filled swimbladders have been caught at great depths. The macrourid *Nematonurus armatus* has several times been caught below 4000 m, on one occasion at 4500m. At this depth the ambient pressure is 450 atmospheres (45.6×10^3 k Pa).

The swimbladder gas in deep-sea fish is mainly oxygen (for reasons discussed later) so that the partial pressure of oxygen (pO_2) in the swimbladder of such a fish will be over 400 atmospheres (40.5×10^3 k Pa).

In the surrounding seawater, from which oxygen has to be collected across the gills and then transferred to the swimbladder via the circulation, the partial pressure of oxygen is several orders of magnitude less than this. The partial pressure of a gas in a gas mixture is the pressure it would exert if the other gases were removed. Thus pO_2 in a 50 : 50 oxygen-nitrogen mixture at 2.0×10^3 k Pa will be 1×10^3 k Pa. The partial pressure of a gas throughout a liquid is the same as that of the gas in the gas mixture at the gas-liquid interface. In the sea, therefore, the gases dissolved are in equilibrium with those of the atmosphere at the surface, and at any depth, the partial pressure of oxygen will be around 20 k Pa, unless it is lowered by utilization by organisms. At 4500 m then, the partial pressure gradient across the swimbladder wall will be at least 2000 : 1.

The secretion of gas against this gradient, and its retention in the swimbladder, may justly be regarded as one of the greatest feats of biological engineering; for obvious reasons these processes have not been studied directly in deep-sea fish, but there is no reason to suppose that the solutions to the formidable problems of the deep sea are different in *kind* to those accessible for examination in shallow-water fishes. For one thing, the structural peculiarities of the swimbladder are essentially similar at whatever depth the fish studied normally lives.

Swimbladder Structure

Swimbladders vary in structure, partly because they perform other functions that hydrostatic ones; in this section we are concerned only with the structures that have evolved to enable the fish to cope with the problems of using gas as a source of static lift. In ontogeny, the swimbladder arises as a pouch from the roof of the foregut, and in many of the more primitive fish, such as elopomorphs or eels, the connection with the gut is retained as an open tube in the adult. These *physostomatous* fish are able to take in air at the surface, and to void it as they ascend: whether the swimbladder originally evolved as an hydrostatic organ or (more probably) as a respiratory organ, there is hardly any doubt that this is the primitive condition, as shown by its taxonomic distribution amongst teleosts.

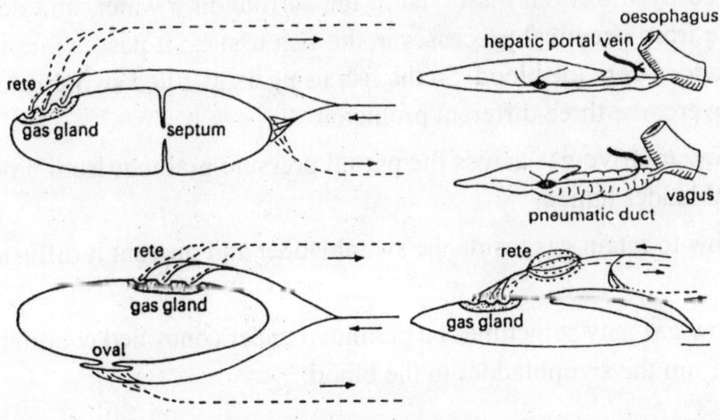

Figure 5.10: ***Structure of physoclistous (left) and physostomous swimbladders. Upper left: divided swimbladder (as in wrasse)—direct connection to systemic circulation on right. Lower left: gadoid swimbladder with direct connection to systemic circulation at oval. Lower right: eel swimbladder with direct connection to systemic circulation at pneumatic duct. Upper right: appearance of*** **Conger** ***swimbladder before (above) and after 5 min vagal stimulation, showing gas loss through enlarged pneumatic duct.***

The great majority of teleosts lose this open connection either at an early stage in larval development, or in post-larval life, and the adult swimbladder is entirely closed or *physoclistous*. The swimbladder is filled either before the connection is lost, or by secretion of gas from special cells in its wall. Gas cannot be lost very rapidly from such a closed system, and this is why deep-water physoclists come to the surface with the viscera protruding from the mouth, blown out by the enormously dilated swimbladder which is, often ruptured by the decrease in ambient pressure. As fish are not designed to be fished up, presumably the physoclistous arrangement is a better design than the physostomatous, since it is found in all advanced teleosts, though where the advantage lies is still not clear. The swimbladders of the great majority of teleosts share certain curious structural features. The connections with the blood system are complicated; there are sometimes diaphragms or sphincters isolating one part of the swimbladders from the gas inside; and parts of the living epithelium are specially modified.

Gas in the swimbladder

It is known from experiments with O_2^{18} that the swimbladder gas is derived from oxygen dissolved in the surrounding water, and does not arise from chemical processes in the fish tissues; it passes into the swimbladder from the blood. So the fish using a gas-filled swimbladder has to overcome three different problems:

1. How to drive gas across the partial pressure gradient from blood to swimbladder lumen.

2. How to retain gas inside the swimbladder and prevent it diffusing out.

3. How gas may sometimes be permitted under controlled conditions to pass from the swimbladder to the blood.

a. *Loss of gas*

Obviously enough, the last problem is the simplest. All that is required is a capillary system to bring systemic blood into contract with the swimbladder wall. Because of the partial pressure gradient, gases will rapidly diffuse from the swimbladder into the blood, and thence across the gills into the water. If it is possible to occlude this connection with the systemic circulation at will, then the problem of controlled loss of gas is solved, even in the physoclist swimbladder. There must be a second connection with the circulation via which gases may be made to enter the swimbladder, and it is not very hard to see how this second potential 'leak' could be closed. Provided that blood passing to the system can be arranged to flow close to the blood passing away, diffusion across the walls of in going and outgoing capillaries will minimize the loss of gas frcm the swimbladder. The rest of the swimbladder could be made relatively impermeable if it was made of materials in which gases were insoluble, or alternatively, by making the diffusion pathway from inside to outside very long. It is not so simple to infer what structural arrangements would be required to solve the first problem.

In physostomes, gas can be lost from the swimbladder via the pneumatic duct, which has a valve at its junction with the oesophagus. In some fish, such as *Conger*, the duct is flattened when the fish is in buoyancy equilibrium, but if the external pressure is reduced, the duct fills with gas. Since it is supplied with a capillary plexus leading to the systemic circulation, gas can now diffuse away via these capillaries; only if the external pressure if further reduced does the oesophageal valve open, and bubbles of gas are lost via the mouth. The appropriate muscles of the pneumatic duct wall and the blood vessels are under nervous control; vagal stimulation and injection of sympatheticomimetic drugs like adrenalin cause loss of gas. In physoclists, special regions of the swimleladder wall are supplied by the systemic circulation and there can be occluded either by an adjustable diaphragm across the swimbladder (wrasse) or by a sphincter muscle closing off an oval area of the wall (gadoids and perciformes).

b. Retention of gas

The problem of retaining gas in the swimbladder is a much more difficult design problem. The swimbladder wall must be impermeable to gas, yet it is thin and elastic. Krogh (1919) found that if the swimbladder wall consisted of normal connective tissue, it would be about 100 times as permeable as it actually is. Denton and his colleagues (1972) found by ingenious experiments with *Conger* swimbladders that their strikingly low permeability resulted from the investment of the swimbladder by a layer of cells containing sheets of guanine crystals some 3, μ m thick. The silvery guanine layer could be removed without damaging the underlying layers, when it was found that permeab.'ity rore forty-fold. In deep-sea fish, there is more guanine in the swimbladder wall than in *Conger*, and the impermeability results from the long tortuous diffusion pathways around the sheets of overlapping guanine crystals. We must admire the ingenuity with which fish make their swimbladders impermeable in this way, whilst still allowing them to be elastic and to change in volume; using the same material that it employs elsewhere for an entirely different purpose.

c. Gas secretion

Gas enters the swimbladder via blood capillaries which run to a specially modified area of the inner wall, the gas gland. It is evident that this is where gas is secreted into the bladder, for in actively secreting bladders, the cells of the gas gland are covered with foamy mucus. The connection with the systemic circulation is a potential leak from the swimbladder, and it is here that the counter current capillary arrangement of the rate mirabile occurs. Anatomically, there are three sorts of retial arrangement; functionally they are the same. Either incoming and outgoing capillaries are apposed some way from the gas gland; or they may be in connection with the gas gland itself; or in some fish (salmonids) the retial system is diffuse, forming a so-called micro-rete.

In transverse section, the enormous extend and close apposition of the retial vessels is clear, and we can well understand how the system operates to prevent loss of gases from the swimbladder. The deeper a fish normally lives, the greater the partial pressure difference between the swimbladder gas and the gases in the surrounding seawater, and in the systemic circulation. It is not surprising that the length and complexity of the retial system increases as does this partial pressure gradient, and that the longest retia yet observed (25 mm long) were found in the abyssal *Bassozetus taenia* (caught between 4575 and 5610 m).

The retia mirabilia then, operate to reduce the loss of gas from swimbladders which are in equilibrium, and where gas secretion is not taking place. The way they work when the fish has descended in the water and requires actively to secrete gas long remained a most challenging mystery, and is still not yet completely understood. What is needed, evidently, is more mechanism which reduces the amount of gas in the venous capillaries leaving the rete for the heart, as compared with that in the arterial capillaries entering the rete. This mechanism must be a very ingenious one, for if the partial pressure of the gases in the venous capillaries of the rete are lower than those in the arterial, we should expect gas to diffuse across from the arterial to the venous capillaries and the whole secretory process would grind (or rather, diffuse) to a halt. The baffling problem seemed that the arrangement of the rete mirabilia was well adapted to *retain* gas within the swimbladder, but also to *prevent* any gas entering it ! What is needed is some change in blood properties as it passes through the gas gland which will raise the partial pressure, although the actual gas *content* of the blood is decreased, so that gas will pass into the swimbladder, and at the same time, diffuse from venous to arterial capillaries within the rete.

In most fishes, we need only seek for some changes in the blood which will account for a higher partial pressure of oxygen (pO_2) in the venous capillaries, for the accumulation of gases like argon and nitrogen can be explained by purely physical effects linked to the secretion of molecular oxygen. That the secretion of oxygen is the important process is shown by the facts that newly secreted gas contains a much higher % of oxygen than does gas from equilibrated swimbladders, and that is deep-sea fishes, the swimbladder gas is nearly pure oxygen (over 80%).

Elegant experiments on shallow-water fish, in particular upon eels (which have bipolar retia suitable for cannulation and investigation of blood parameters during secretion), have shown that in these fish the important change in the blood as it flows through the gas gland is *increase in acidity*. The cells of the gas gland are rich in glycogen, carbonic anhydrase, and lactate dehydrogenase, and produce sufficient lactic acid to lower the pH of the blood passing through the gland by nearly one pH unit. The red cell haemoglobin unloads oxygen when it is acidified, and this Roof effect is a very marked one: a change of one pH unit unloading almost 50% of the oxygen, even against considerable oxygen pressure. As arterial blood flows through the rete, therefore, it receives oxygen and acid from the venous capillaries, and pO_2 is thereby increased so that it exceeds that within the swimbladder, and gas diffuses across the wall at the gas gland into the swimbladder. The venous blood leaving the gas gland has a higher pO_2 than the blood entering it, but a lower oxygen concentration (less oxyhaemoglobin) and it is also very acid. As it flows into the rete, oxygen, lactic acid and CO_2 will diffuse across into the arterial capillaries, the venous blood will become less acid and this increase in venous pH will load oxygen on to the haemoglobin, thus decreasing pO_2. But Berg and Steen (1968) found that the decrease of pO_2 following increase in pH was relatively slow, and took place outside the rete as the blood flowed back away from the rete towards the heart. The half-time of this Root-on shift, loading haemoglobin with oxygen as pH increases, was in the eel some

10-20 seconds, compared with the Roof-off unloading shift, which had a half-time of only 50 ms.

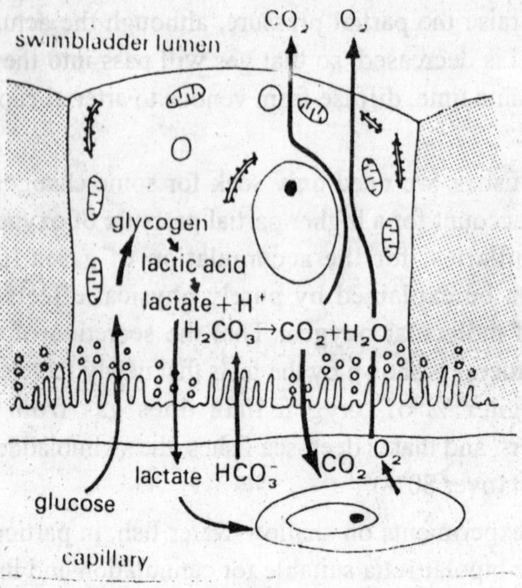

Fig. 5.11 ***Operation of gas gland cell.***

It is not yet clear that this scheme applies to *all* fish. Some store not oxygen, but almost pure nitrogen, and there is not as clear a correlation as we should expect between the magnitude of the Root effect and the depth at which a fish normally lives. One possibility is that osmotic changes in the blood within the gas gland alter the solubilities of gases and that gas is secreted in part (or in some fishes wholly) by a salting-out effect; so far, there is no direct evidence for this though it is simplest mechanism which would explain nitrogen secretion in some salmonids.

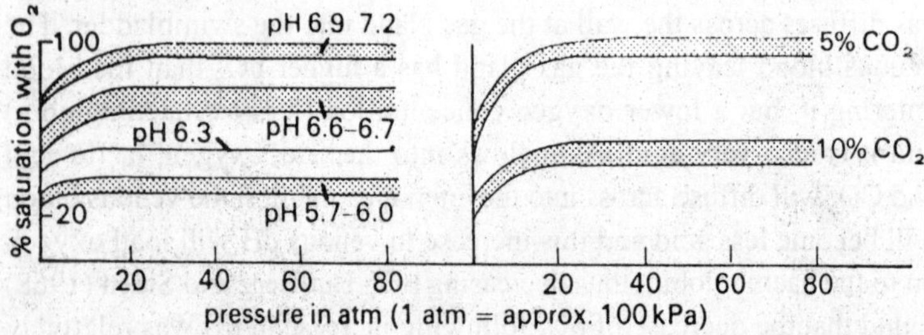

Fig. 5.12: ***Oxygen dissociation curves for blood of black grouper (*****Epinephalus mystacinus*****) at 10°C, showing large Root effect even at high O_2 pressure when blood is acidified either by adding lactic acid (left) or CO_2 (right).***

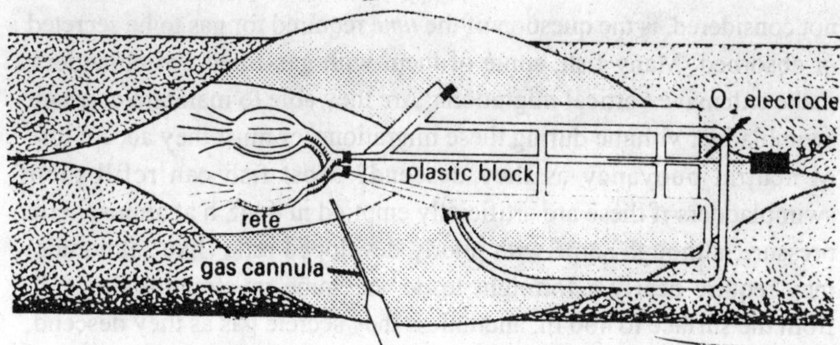

Fig. 5.13: ***Experimental arrangement to determine changes in blood properties during gas secretion by eel*** **Anguilla.**

The gradual unravelling of the mechanism of gas secretion in teleosts has occupied physiologists of the highest caliber, and is a fascinating story which we have only been able to touch on here. Denton (1961) gives a good historical account; but there are several aspects which are not yet understood and the subject is still being actively investigated.

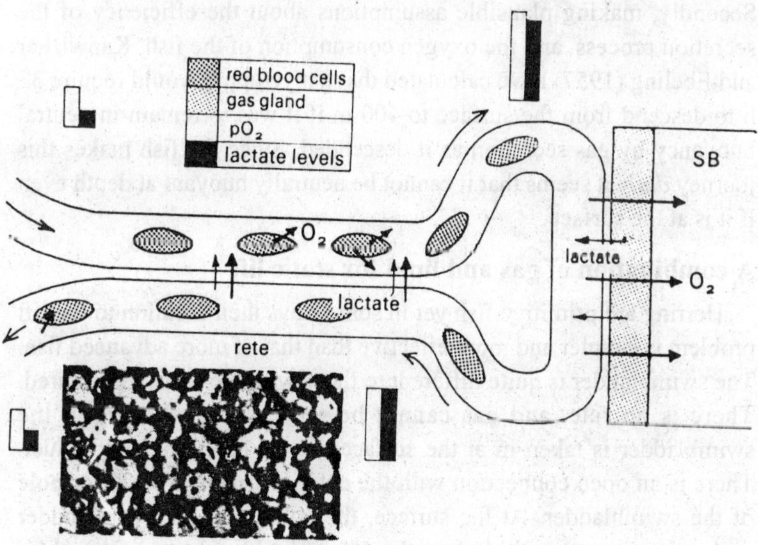

Fig. 5.14: ***Diagram showing how the rete operates during gas secretion into swimbladder (SB). Inset shows part of the rete of*** **Conger** ***in transverse section (terial capillaries smaller than venous).***

d. Rates of gas secretion

One difficulty about the use of gas for static lift, which we have not considered, is the question of the *time* required for gas to be secreted or absorbed. Many fish, some of them with gas-filled swimbladders, make extensive vertical migrations. Are they able to maintain constant swimbladder volume during these migrations, or must they accept loss of neutral buoyancy as they descend? Most fish can refill their swimbladders if these are artificially emptied in 4–48 h at atmospheric pressure, although some with poorly developed retia (like salmonids) may require nearly a fortnight to do so. Some myctophids descend, from the surface to 400 m, and unless they secrete gas as they descend, the volume of the swimbladder will decrease (there is thus a positive feed back on the rate of descent) until at 400 m the swimbaldder volume will only be 1/40th of that at the surface, and the fish will be denser than the ambient water. Several lines of approach suggest that such a fish could not secrete gas sufficiently rapidly to make itself neutrally buoyant both at the surface and at depth. First, secretion implies oxygen uptake from the water; so much oxygen would have to be secreted into the swimbladder that impossible rates of uptake at the gills are required. Secondly, making plausible assumptions about the efficiency of the secretion process, and the oxygen consumption of the fish, Kanwisher and Ebeling (1957) have calculated that a myctophid would require 33 h to descend from the surface to 400 m if it was to remain in neutral buoyancy by gas secretion as it descended. Since the fish makes this journey *daily* it seems that it cannot be neutrally buoyant at depth even if it is at the surface.

A combination of gas and lipid for static lift

Herring are primitive fish yet in some ways their solution to the lift problem is simpler and more effective than that of more advanced fish. The swimbladder is quite different to those we have so far considered. There is no rete, and gas cannot be secreted; all the gas in the swimbladder is taken in at the surface via the open pneumatic duct. There is an open connection with the exterior from the posterior pole of the swimbladder. At the surface, the volume of the swimbladder varies, but it is usually below the 5% of body volume required by ordinary marine teleosts. The tissues contain a good deal of lipid (up to 20%) and the fish is neutrally buoyant, swallowing less gas the more lipid there is in the tissues. As herring descend, swimbladder volume decreases, but lipid still contributes static lift, so that less dynamic lift has to be generated than if gas alone provided lift. Similarly, as the herring ascends in the water it can rise to the surface without pausing to equilibrate. Physoclist predators of the herring (like the cod) can only cope with changes in swimbladder volume of some 20 %, which means that a cod adjusted for neutral buoyancy at 50 m could only rise to 37 m before absorbing gas. So the herring, able to rise directly from any depth to the surface, has a positive advantage over those predators who rely on gas alone for static lift. Of course, this is no help when herring are hunted by sharks such as the mako (*Isurus*) or blue (*Prionace*) which use lipid for static lift!

The simple arrangement in the herring is a suitable one for fish which have a wide vertical range, and although the herring swimbladder is highly specialized for sound reception, it shows us today a primitive stage in the development of the swimbladder which is still a successful solution for fish which migrate diurnally. Probably many of the myctophids operate in the same way, although they need a rete since they have physoclistous swimbladders.

The Best Buoyancy Policy

This chapter has shown the variety of ways that aquatic animals stay afloat. Some aquatic animals are denser than the water they live in: they have to swim upward to keep off the bottom or rely on fins for hydrodynamic lift. The others have their densities reduced to match that of the water by floats that may be low-density organic compounds, solutions of unusual ionic composition, or gas-filled chambers. None of these possibilities is necessarily superior to the others. Each has its merits, and the best buoyancy policy for a particular animal depends on where it lives and what its habits are.

Fish and other animals that live in the upper waters of deep oceans do not have the option of resting on the bottom, because it is too far down. For those that live in shallower water, the bottom may be the best place to hide or to lie in ambush for prey, because it is easier to hide on the bottom than while floating in midwater. Many swimming animals such as dogfish and flounders have evolved camouflage patterns that make them very inconspicuous when resting in suitable places on the bottom. Many midwater fish such as herring have silvery sides, which make them reasonably inconspicuous in open water, and a few midwater fish are largely transparent, but it remains true that the bottom is the easiest place to hide. For the animals that rest there, being denser than water may be an advantage because the animals will rest firmly on the bottom and not be wafted along by the slightest current.

The penalty for being denser thar water is that they cannot rest in midwater but must keep swimming, so long as they want to stay off the bottom.

Bottom-living fish such as dogfish and flounders are not the only ones that are denser than water. Porbeagle sharks that swim perpetually are almost as dense as dogfish and, like them, depend on lift from fins. Tunnies seem never to rest on the bottom (at least in aquaria), yet they are as dense as flounders and also have no swimbladder. Many squids that are denser than water swim perpetually using fins and their jet-propulsion style of swimming to keep them from sinking. Can there be an advantage for animals like these in being denser than water?

Whether hydrodynamic lift or buoyancy organs are used, there is a cost to be paid for keeping off the bottom. To get hydrodynamic lift, work must be done against induced drag. Buoyancy organs do not increase induced drag, but by making the body larger they increase the drag that acts on it swimming. The options available to swimming animals as they evolve are to have a small, dense body and do work against induced drag, or to have a buoyancy organ that makes the body larger and do more work against drag on the body. The induced drag (for the same lift) decreases as speed increases, but the extra drag on a larger body increases as speed increases. This suggests that slow swimmers will do best with buoyancy organs and that fast swimmers will use less energy if they have no buoyancy organ, but get lift from fins.

Swimming animals seem to have evolved accordingly. The dense squids swim around quite fast. Whereas cuttlefish and the nautilus are relatively sluggish. The sharks that hunt for large prey are dense, but the basking shark, which swims very slowly straining out plankton, is buoyed up by squalence. Although most teleost fish have swimbladders, the fast-swimming tunnies have none.

The advantage changes from buoyancy organs to hydrodynamic lift at a particular swimming speed, and that speed must depend on the composition of the buoyancy organs. The buoyancy organs of lowest density can be smallest, so they increase drag least. In this respect, gas chambers are better than squalene and wax esters, which in turn are better than low-density solutions. Other considerations, however, may change this order of preference: we have seen how wax esters may be better than a swimbladder for lantern fishes that make large daily depth changes. The speed at which the advantage shifts from having a buoyancy organ to doing without one must be slow for the very bulky low-density solutions, faster for moderately bulky squalene, and faster still for compact swimbladders. It has been calculated that for 1-kilogram fish this critical speed should be about 0.5 meter per second for squalene and 0.8 meter per second for swimbladders; for 1-tonne fish it should be 3 and 4 meters per second, respectively. Although rather few data are available, it seems doubtful whether tunnies swim quite fast enough for the loss of the swimbladder to give a clear energy-saving advantage. Thus other possible advantage have been suggested instead. For example, the killer whales that feed on tunnies emit sounds and seem to use the echo to locate prey. Airfilled cavities such as swimbladders return strong echoes and would make their possessors conspicuous to echolocation.

Although most animals that live below the water's surface are as dense as or denser than water, a few occasional visitors have the opposite trait. These are the diving birds, which face the problem of being too buoyant.

Birds carry a good deal of air in their bodies, in their lungs and associated air saes. They also have air trapped between their feathers and skin that has an important heat insulating function, like the air trapped around human bodies by clothing. All this air gives birds low densities, which is why they ride so high when floating on water. When ducks dive, they have to swim strongly downward to overcome the buoyancy of the air that they carry with them.

The conclusion from this chapter must be that there is no one best buoyancy policy. It may be best for bottom-living fish to be denser than water, so that they sit firmly on the bottom. For fast swimming fish and squids it may be better to remain a little denser than the water than to be swollen by a large buoyancy organ; but slow-swimming fish and cuttlefish may do better with buoyancy organs that match their density to that of the water. For fish that remain at fairly constant depths, the swimbladder is likely to be the best type of buoyancy organ because it can be the least bulky, but because swimbladders are compressed by increased pressures, wax esters may be better for fish that make large vertical migrations.

CHAPTER 6
SWIMMING

Penguins use their tiny wings to "fly" but they often leap through the air picture. One theory is that leaping san enabling the animal to escape from the that resist its movement through water simply that the land leaps to breathe Adelie penguins.

Living Rowboats

Some animals row themselves along like tiny boats. In rowing, the blade of an oar is pushed backward through the water during the power stroke. Because the oar is moving backward, the drag on it acts forward, serving as thrust to propel the boat. As the boat moves forward, drag acts backward on it: the forward thrust from the oars must balance the backward drag on the boat. Water beetles swim like this, using their legs as oars.

To work efficiently, an oar needs a large blade. The same thrust can be obtained by pushing a lot of water backward at a low speed or a little water backward at a higher speed. Less power is needed if the oars push a lot of water at low speed, and for that the oars must be large. The same principle applies to flight, longer wings give the same lift with less induced drag because they push on more air.

The legs of water beetles have evolved to be good oars. The segments of each leg are flattened to form the central part of the oar blade, which is greatly enlarged by a fringe of long brestles. Because these bristles are closely spaced, they push on the water almost as effectively as would a blade made of a continuous sheet of material of the same length and breadth. A big advantage of the bristles is that they can be folded when required, making the blade much narrower.

After each power stroke, an oar must be brought forward again in a recovery stroke, ready for the next power stroke. During the recovery stroke, any drag on the oar must act backward and slow the animal down. A water beetle (or a person rowing a boat) wants as much drag as possible to act on the oar in the power stroke and as little as possible in the recovery stroke. People rowing boats lift the oars clear of the water for the recovery stroke, but animals do not do that, nor can they, if they are swimming far below the surface. Water beetles spread their leg bristles for the power stroke and fold them for the recovery stroke, when they also reduce drag by turning their flattened leg segments edge on to the water. The spreading and folding are done automatically by the pressure of the water on the bristles, which are hinged to be free to fold one way but not the other.

Ducks paddle with their feet. The principle is the same as rowing, but the feet move alternately like the paddles of a canoe instead of together like the oars of a rowboat. Like the legs of water beetles, the webbed feet are spread for the power stroke and folded for the recovery stroke. The feet paddle in the same way whether the bird is swimming on the surface or diving under water.

Swimming with Wings

Penguins also swim, but their method of swimming is strikingly different from that of ducks and (as we will see) much more economical of energy. Rather than using their feet, penguins swim with their wings and use them as hydrofoils rather than as paddles. The difference is that the thrust from paddles such as ducks' feet is drag, whereas the thrust from hydrofoils such as penguins' wings is lift.The blades of the propellers of ships are also hydrofoils (and those of airplanes are airfoils) that plrovide lift to serve as thrust. The penguin's method of swimming is also used by sea turtles, which beat their flippers up and down to propel themselves by lift.

Although the wings of penguins are shaped more of less like those of other birds, they are too small to fly with. The area of one wing of, say, a 4.2–kilogram penguin is 74 square centimeters, whereas that of an eagle of the same body mass would be 2600 square centimeters. The action of the wings as they beat up and down in the water does indeed look very much like that of flying birds, but there is an important difference. The downstroke is like the downstroke of flight: the wing has a positive angle of attack and provides upward, forward lift. The upstroke, however, is quite different. In flight, any lift during the upstroke acts upward and backward, but in the upstroke of penguin swimming the wing is given a negative angle of attack so that the lift acts downward and forward. The downstroke provides an upward, forward force and the upstroke a downward, forward one. The upward and downward components cancel out and the overall effect is the forward thrust that the penguin needs to propel it through the water. The reason for the difference between flight and the swimming of penguins is that the principle component of force required for flight is vertical, to counteract gravity, whereas the principal component of force that penguins need for swimming is horizontal. Penguins do not need a vertical force because they are about the same density as the water. This interpretation of penguin swimming is based mainly on films of captive penguins that had white lines painted on their wings, from the front edge to the rear, to make it easier to measure the wings's angle of attack.

When penguins and ducks swim at the surface of the water, they inevitably push a wave along in front of them, like the bow wave of a boat. The unfortunate effect of the bow wave is to increase the drag on the body. Experiments with streamlined (tropedo–shaped) bodies have shown that, because of the bow wave, considerably more drag acts on them whem they move just below the surface, or even half submerged, than when they teravel at the same speed well below the surface.

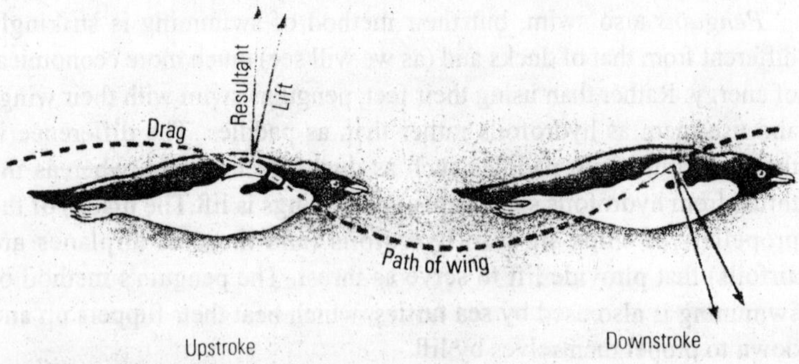

Fig: 6.1: As a penguin swims, the hydrodynamic lift on the wing acts forward and upward on the downstroke, forward and downward on the upstroke.

The height of the bow wave and the extra drag it brings depend on the body's size and speed. A rule that we have already encountered will help us to understand this.It has already been stated that if gravity is important, the motions of different–sized bodies cannot be dynamically similar unless their Froude numbers are equal. Gravity is important in the dynamics of water waves and so, for bow waves to be dynamically similar, the ships (or duck, or whatever) responsible for them must travel with equal Froude numbers. In this case,

$$\text{Froude number} = \frac{(\text{speed})^2}{\text{hull lenght} \times \text{gravity}}$$

The bow wave builds up higher as a ship goes faster. The drag it creates increases sharply at Froude numbers above 0.2 and reaches a maximum at a Froude numbers of 0.5. Ducks generally swim too slowly for bow wave resistance to be much of a problem. For example, Mallard ducks on a pond had "hulls" about 0.3 meter long and swam at about 0.5 meters per second, so their Froude numbers were only about $(0.5)^2$ $(0.3 \times 10) = 0.08$. (The gravitational acceleration is about 10 meters per second squared.) Mallards made to swin faster in laboratory tests seemed unable to sustain speeds above 0.7 meter per second (Froude number 0.16). Although ducks presumably could swim faster if they had larger leg muscles, a lot of extra muscle would be needed to give a little more speed, because drag increases so sharply with Froude number. Speedboats reach high Froude numbers by hydroplanning (skimming over the surface of the water), but no animals do that.

The history of boats suggests that the penguin style of swimming may be better than the duck style. Paddlesteamers propelled by drag, like rowboats and ducks, were once common, but they have long since been replaced by screw–driven ships propelled by lift, like penguins. This made it seem reasonable to suppose that the paddling of ducks would use more power than penguin swimming by hydrofoil action at the same speed. To find out whether it does, Bussell Baudinette and Peter Gill of Flinders University, Australia, measured the rates of oxygen consumption of swimming ducks and penguins. They trained the birds to swim in a flume, a channel through which water could be driven in a smooth, even stream. The bird was enclosed in a plexiglass chamber, which was suspended just above the surface of the water, so it had to breathe the air in this chamber. Air drawn through the chamber was analyzed to find out how much oxygen the bird was using, and the rate of use of energy was calculated from the oxygen consumption.

Baudinette and Gill compared birds of very similar masses, 1.1–kilogram Black ducks (*Anas superciliosa*) and 1.2–kilogram Little penguins (*Eudyptula minor*). (This is the Australian Black duck, a different species from the American bird of the same name.) When resting on the surface of still water, both metabolized at rates of about 6 watts per kilogram body mass. Both species could maintain speeds of up to 0.72 meter per second, and they could hold those speeds long enough for it to seem clear that their swimming was powered by aerobic metabolism. When both species swam on the surface at that speed the ducks used about 18 watts per kilogram and on the penguins only 13 watts per kilogram. To calculate the power actually used for swimming, we must subtract the 6 watts per kilogram used while resting on still water thus the energy cost of swimming is 12 watts per kilogram for the ducks and 7 watts per kilogram for the penguins, at the same speed. The difference may stem in part from the better streamlining of the penguins, but it also seems clear that the penguins' wings, beating up and down through a wide angle, must have pushed on more water than the ducks' feet. It is well known that the same thrust can be obtained at less energy cost by pushing a lot of water slowly than by pushing a little water fast.

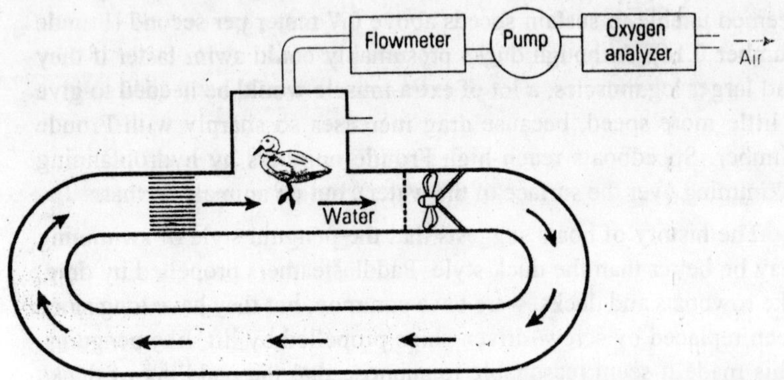

Fig. 6.2 : The oxygen consumption of swimming birds can be measured by analyzing the loss of oxygen in a small plexiglass chamber fitted over the bird.

The wing movements of the penguins were fast compared to the paddling movement of the ducks. Whereas the penguins beat their wings at 3 to 5 cycles per second, the ducks paddled at only 1.5 to 3 cycles per second. It may be interesting to see if thses rates of movement are similar to those of equivalent birds flying or walking. When Black ducks walk, they take about 2 strides per second, and when Mallard ducks of similar mass fly (I do not have the data for Black ducks), they beat their wings at about 5 cycles per second. Thus swimming penguins beat their wings at about the same frequency as flying birds of equal mass, and swimming ducks paddle at about the frequency of their walking strides.

The ducks always swam on the surface, but in some experiments the penguins were persuaded to swim well below the surface, visiting it only to breathe. The then produced no bow wave, and the energy cost of swimming was less. At 0.72 meter per second (the maximum speed for sustained surface swimming), they used only about 9 watts per kilogram, considerably less than the 13 watts per kilogram required for surface swimming. After subtracting the 6 watts per kilogram used when the penguins were resting on still water, we find that the energy cost of their swimming movements must have been about 3 watts per kilogram under water and 7 watts per kilogram on the surface. It is much more economical for penguins to swim well below the surface. They can also swim a little faster under water they could sustain speeds up to 0.84 meter per second, instead of 0.72. For ducks, however, surface swimming may be the more economical option because the air trapped under their feathers makes them so buoyant that they have to paddle hard to keep themselves submerged.

The Hydrofoil Tail

Dolphins also swim by up–and–down movement of hydrofoils, but their hydrofoils are tail flukes, not wings. The tail beats up and down as the dolphin swims, taking a wavy path through the water. Its angle of attack is presumably adjusted to give upward, forward lift in the downstroke and downward, forward lift in the upstroke, just like penguin wings.

Dolphins often swim close to the bows of ships, staying with a ship for hours even when it is traveling quite fast, at up to at least 10 meters per second (20 knots). This won them a great reputation for speed and stamina, but it has since been realized that they get help from the ship and could not by themselves keep swimming for long at the same speed. Their trick is so swim in the bow wave that is being pushed along by the ship.

Fig: 6.3 : Dolphins are propelled by the lift on their flukes, much as penguins are propelled by lift on their wings.

The speeds at which dolphins and porpoises can swim unaided have been measured in experiments with trained animals. The highest speed was recorded by a spotted porpoise (*Stenella attenuata*) that had been trained to chase a yellow wooden lure that was towed by a winch across a lagoon in Hawaii. The porpoise reached 11 meters per second, the speed of an Olympic class sprinter at the fastest stage of a 100 meter race. However, it could maintain that speed for only a few seconds.

A lot of power is needed to propel so large an animal, at such a high speed. The porpoise was 1.9 meters long, with a mass of 53 kilograms. It was calculated that to propel a rigid body of the same size and shape at the same speed, 2900 watts would be needed. This is a minimum estimate of the power required, because the calculation ignored the kinetic energy given to water in the wake and the likely effects on drag of the beating of the tail. The energy needed to propel a fish is more than would be needed for a rigid body of the same size and shape, traveling at the same speed.

One can compare the estimate of propoise power output with the maximum power output of a human athlete, The power produced by athletes has been measured by means of a bicycle ergometers (an exercise bicycle with instruments for measuring work and speed) fitted with hand cranks so that the athlete's arm muscles could do work, as well as the leg muscles. The greatest power that a 70–kilogram athlete could maintain for a few seconds was only 1400 watts, about half the estimated power output of the 53 kilogram porpoise.

Poroising is a well–known habit of dolphins and porpoises. They swim along, alternately swimming below the surface and leaping clear of it. This looks strenuous and playful, but it has been suggested that porpoising may reduce the energy cost of fast swimming. The idea is that because air is so much less dense than water, the drag on a body moving through air is much less than the drag on a body moving at the same speed through water. Leaping costs energy, of course, but the suggestion is that if the animal is traveling fast enough, the energy cost of the leap is outweighed by the energy saved by escaping from the drag of the water.

Scientists argue about the energy cost of the leap. One opinion is that the principal energy cost is the kinetic energy associated with the upward component of the animal's velocity as it leaves the water. Against this view, it has been pointed out that although the dolphin is traveling downward when it reenters the water, its kinetic energy then is almost the same as when it left the water traveling upward, so no energy is needed to give kinetic energy to the animal itself. However, like any body moving through water, a dolphin pulls along with itself a thin coating of water, called a boundary layer when the animal leaves the water, this boundary layer falls off (photographs show water falling away from leaping dolphins) and its kinetic energy is lost. Whether the work of the leap is the upward kinetic energy of the dolphin itself or the kinetic energy of the lost boundary layer or both, it is proportional to the square of the speed. From the theory of projectiles, we know that the length of the leap also is proportional to the square of the speed if dolphins traveling at different speeds leave the water at the same angle, they will leap four times the distance for twice the speed. Since both the length of the leap and its likely energy cost are proportional to the square of speed, the energy cost *per unit distance* of leaps is the same for all speeds. The higher the speed, however, the higher the drag in water; thus the energy cost per unit distance swum in water increases with increasing speed. This argument seems to tell us that there must be a speed above which leaping saves energy. The doubts about the energy cost of leaping make it difficult to estimate what speed may be, and few measurements have been made of the speeds at which dolphins and propoises do or do not leap. Therefore it is difficult to test the hypothesis that porpoising saves energy in fast swimming.

Penguins also porpoise, and their porpoising habits have been studied more systematically by Clifford Hui of the University of California, Los Angeles. He found that penguins porpoised at speeds of 3 to 4.5 meters per second, but seldom at lower speeds, which seems to fit the energy–saving hypothesis. The leaps were short, however, never measuring more than 22 percent of the total distance traveled, and so any energy savings must have been small. Hui observed that the penguins were out of the water for about 0.3 second in each leap, just a little longer than a walking penguin needed for each intake of breath. He suggested that porpoising saves energy by giving the bird opportunities to breathe without having to swim at the surface where drag would be increased by the bow wave.

Tunnies and similarly shaped fishes use their tail fins as hydrofoils in essentially the same way as dolphins use their flukes. The principal difference is that dolphins have horizontal flukes, which beat up and down in swimming, whereas tunnies have vertical tail fins that beat from side to side: the swimming action of a tunny is like that of a dolphin lying on its side.

Tunnies and their relatives have a reputation for speed, and attempts have been made to find out how fast they can swim by letting them run while attached to an instrumented fishing line. A fish was hooked, then allowed to pull the line out freely, and the instruments recorded the rate at which the line unwound. This method gave and astonishing speed of 21 meters per second for a wahoo (*Acanthocybium solanderi*) and for another species of tuna. This speed is so much greater than the highest speeds recorded by any other method, for any swimming animal. As the line was drawn out, the instruments sensed magnetic markers spaced at intervals along it, and each sensing produced a blip on a strip of paper in a recording instrument. From the number of blips per unit time, it was a simple matter to calculate speed. It seems possible that the record was corrupted by electrical "noise" and that some of the blips that were counted as showing the passage of markers were merely noise. The highest speed recorded for a fish by any other means seems to be 10.5 metrers per second for a 0.5–meter skipjack tuna (*Katsuwonus pelamis*).

The lift on a tunny's tail (or a dolphin's fluke or a penguin's wing) propels the animals, but the drag on the tail (or fluke or wing) slows it down. A tail that supplies lift needs a special shape to minimize drag. It has already been seen for the highest ratio of lift to drag, wings should have high aspect ratios–they should be long and narrow. Similarly, we expect tunny tail fins to be tall and narrow, and in fact they have much higher aspect ratios than the tail fins of most other fish.

Jet Propulsion

We will return soon to the swimming of fishes, but first we will look at jet propulsion, a swimming technique used by squids and a few other animals. To drive themselves along, these animals squirt water out of cavities in their bodies. A jetting squid does not move steadily like a jet aircraft but in a series of jerks, speeding up as it squirts water out and slowing down as it takes in the water that will be squirted out in the next jet. Medium–sized squids of about 0.4–kg body mass make about one jet per second when they swim fast.

The mechanism for squirting water is a variant of the mechanism squids use to pass water through the gills. The gills of squid are hidden inside the body in a cavity enclosed by a muscular walls. The wall is called the mantle, and the cavity it encloses is called the mantle cavity. Fish breathe by drawing water in through the mouth and blowing it out through the gill covers; on the way it passes over the gills. Similarly, squid draw water into the mantle cavity through one opening and blow it out through another, so that it passes over the gills on the way. The water enters through a slit across the width of the belly and leaves by way of a round opening, the funnel. (Valves prevent flow in the reverse direction.) Gentle rhythmic contractions of the mantle muscles are enough to keep water flowing over the gills for respiration, but more forceful contractions eject a rapid jet through the funnel and can propel the animal at high speed. If the funnel is in its resting position, pointing toward the animal's head, the jet propels the animal backward, but the funnel can be curled around to point backward and drive the animal forward.

Oceanic "flying squids" swim fast enough to leap quite high out of the water, and sometimes they even land on the decks of ships. If the heights or lengths of these leaps were known, that information could be used to calculate the speeds, but no one seems to have taken the right measurements.

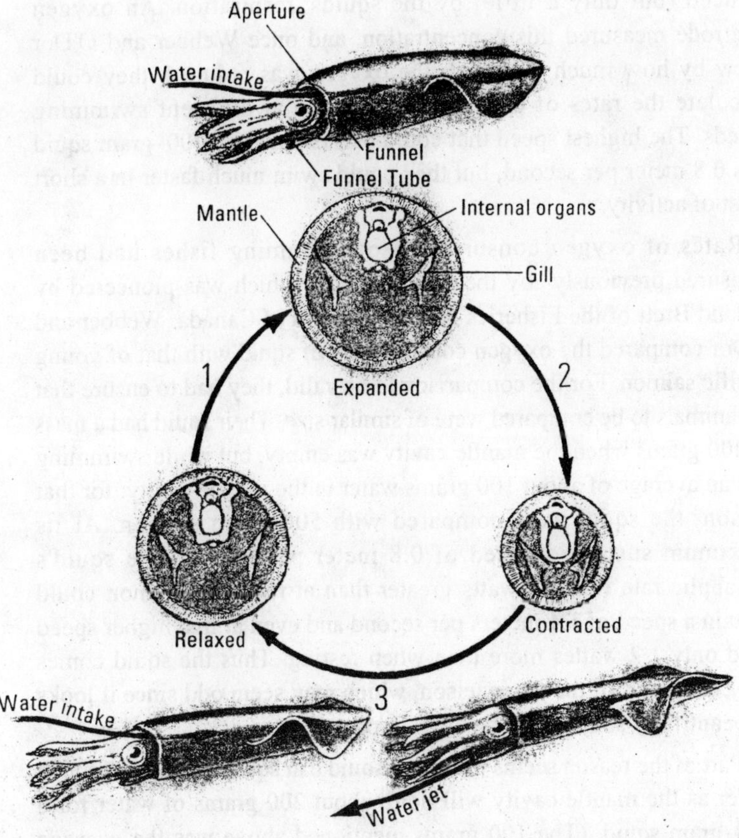

Fig: 6.4 : Squid propel themselves by squirting jets of water out of the mantle cavity, which contains the gills. Water is drawn in through a wide slit in the belly and squirted out through the round funnel.

Dale Webber and Ron O'Dor at Dalhousie University, Nova Scotia, measured the rates of oxygen consumption of swimming squids. They let squids propel themselves in a water tunnel consisting of a loop of large pipes through which water was kept circulating by a powerful pump. The tunnel was designed like a wind tunnel so that the flow was as smooth as possible, and as even as possible across the width of the pipe, in the section where the squid swam. Provided the speed of the current was not too great, trained squid swam, steadily against the current for was not too great, trained squis swam steadily against the current for several hours. As the same water circulated repeatedly around the loop, the concentration of dissolved oxygen was gradually reduced (but only a little) by the squids' respiration. An oxygen electrode measured this concentration, and once Webber and O'Dor knew by how much the much the oxygen was reduced, they could calculate the rates of oxygen consumption at different swimming speeds. The highest speed that could be sustained by 400–gram squid was 0.8 meter per second, but they could swim much faster in a short burst of activity.

Rates of oxygen consumption of swimming fishes had been measured previously by the same method, which was pioneered by Roland Brett of the Fisheries Research Board of Canada. Webber and O'Dor compared the oxygen consumption of squid with that of young Pacific salmon. For the comparison to be valid, they had to ensure that the animals to be compared were of similar size. Their squid had a mass of 400 grams when the mantle cavity was empty, but while swimming had an average of about 100 grams water in the mantle cavity: for that reason, the squid were compared with 500–gram salmon. At its maximum sustained speed of 0.8 meter per second, the squid's metabplic rate was 1.6 watts greater than at rest. The salmon could sustain a speed of 1.4 meters per second and even at that higher speed used only 1.2 wattes more than when resting. Thus the squid comes very badly out of the compaeison, which may seem odd since it looks as beautifully streamlined as the salmon.

Part of the reason seems to be that squid can squirt out only as much water as the mantle cavity will hold, about 200 grams of water for a 400–gram squid. (The 100 grams mentioned above was the *average* content of the cavity over a cycle of expansion and contraction.) In contrast, fish can push with their tails on very large masses of water. It has repeatedly been made the point that less energy is needed to supply the same thrust when large masses are accelerated to low speeds than when smaller masses are accelerated to higher speeds.

In an ingenious experiment, Charles McCutchen of the National Institutes od Health, Bethesda, Maryland, measured the masses of water on which fish push. He half–filled an aquarium with cold water, then topped it up very carefully with warmer water so that the warm floated on top of the cold, mixing with it very little. When a fish swam at the interface between the warm and cold water, it stirred some of the cold up into the warm and some of the warm down into the cold. Because the refractive index of water (the extent to which it bends light rays) depends on the temperature. McCutchen was able to make the disturbance in the water visible by suitable lighting. Thus he was able to film the movements of the water, as well as those of the fish.

A typical film sequence from his experiments shows a small fish coasting along, then making two strong tail beats that accelerate it and send it off in a new direction. The tail beats set two masses of water moving, one to the left and one to the right. It is impossible to measure the masses of moving water directly from the film, but here is how McCutchen calculated them. He knew the mass of the fish. The film gave the initial and final velocities of the fish and the velocities of the masses of water moved by the two tail beats. (Velocity means speed *and direction* of movement.) A fundamental physical principle, the Principle of Conservation of Momentum, told him that the change in the monentum (mass multiplied by velocity) of the fish must be equalled by the change in the momentum of the water. By applying this principle, he was able to calculate that the two masses of water were each about three times the mass of the fish. Each tail beat of the fish pushes on water amounting to three times its body mass. In comparison, each contraction of the squid's mantle squirts out water amounting to only about half the body mass. This difference makes the swimming of fish much more efficient than the jet propulsion of squids.

Undulating Swimmers

Rowing, hydrofoil swimming, and jet propulsion have already been described. The fourth and last major swimming technique is undulation, the technique used by eels. The movements are like the serpentine crawling of snakes: the body is thrown into waves that are made to travel backward along it. In serpentine crawling the waves remain stationary *relative to the ground* as the body moves forward, but in the eel's, style of swimming the waves travel backward through the water as the body moves forward.

The mechanism was first explained clearly in 1933 ny James Gray of the University of Cambridge, England, one of the pioneers of studies of animal locomotion. He took a film of an eel swimming and superimposed outlines traced from successive frames. Then he selected one short section of the body and highlighted its position in successive outlines. His analysis showed that, whether the segment was moving toward the left or toward the right, it was always angled to push water backward.

Not only eels but also some worms swim by undulating their bodies. Cuttlefish swim fast by jet propulsion and most true fish swim fast by tail movements, but to make slow swimming movements cuttlefish and many fish undulate their fins. A fish swimming in this way keeps its body straight and its main swimming muscles inactive: only the fins undulate, powered by small muscles at their bases.

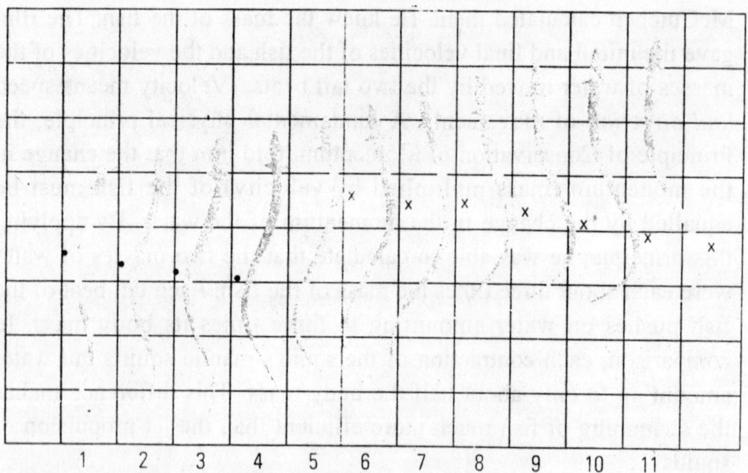

Fig. 6.6 : An cel (Anguilla anguilla) swims by undulating its body.

This style of swimming makes some teleost fishes wonderfully maneuverable. In addition to tail fins, they have fins on their backs, on their bellies, and on their sides. They can send waves backward along their fins to drive themselves forward, or they can send waves forward to drive themselves backward. They can send waves upward or downward in the tail fin and the fins on their sides to make themselves move vertically up or down. A fish swimming by beating its t, is like an airplane; it can only travel forward. But a fish swimming undulating its fins can move forward, backward, up, or down, like helicopter. This maneuverability is especially usuful to fishes that live on coral reefs and feed on coral polyps or on small animals that live in crevices. They have to move very precisely among the clutt of the reef to reach their food.

The Energy Cost of Swimming

Eels and tunnies represent two contrasting style of movement. Eels have only small tail fins and swim by undulating their entire bodies. In contrast, tunnies have tall tails act as hydrofoils; they bent their tails from side to side by bending the rear part of the body keeping the rest almost straight. The swimming of most other fish is a combination of the undulatory swimming of eels and the hydrofoil swimming of tunnies. These other fishes have large, hydrofoil tail fins like tunnies, but they also undulate the whole body, like eels.

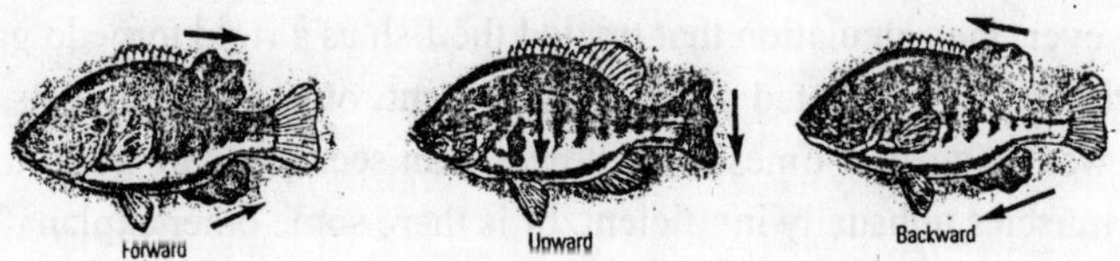

Fig. 6.7 : Telecost fishes can swim forward, vertically, or backward by undulating appropriate fins.

A fish starting from rest drives water backward: to give itself forward momentum, it must give the water backward momentum. When it swims at constant speed, however, its momentum is constant and so the total momentum of the water must also remain unchanged. Alternate strokes of the tail drive water to left and right, giving it momentum alternately to either side, but the leftward momentum given by one stroke is balance by the rightward momentum given by the next.

That view of fish swimming is the basis of an influential theory that shows how to calculate the power output of a swimming fish. Its author, James Lighthill, is a remarkable hydrodynamicist who has been at differernt times director of the Royal Aeronautical Establishment, Farnborough, England; a professor of mathematics at Cambridge

University; and provost of University College, London. Lighthill showed how the mass and velocity of the water pushed by each tail beat could be calculated if the movements of the fish were known. The force needed to give sideways momentum to the water could also be calculated, and the power output of the muscles moving the tail was this force multiplied by the tail's sideways velocity. A great deal could be calculated if the fish's movement were known.

There is another, more obvious way to calculate the power needed for swimming. A typical fish is a well–streamlined body like a torpedo, and it is easy to calculate the power needed to propel a tropedo of any size at any speed, as hydrodynamics books explain. Foer example, an 87–gram goldfish had used oxygen at a rate corresponding to an energy consumption of about 10 milliwatts when resting and 60 milliwatts when swimming in a water tunnel at 0.64 meter per second. (A milliwatt is a thousandth part of a watt.) The difference of 50 milliwatts was presumably needed for swimming. Muscles commonly work with efficiencies of about 0.2: about 5 joules of metabolic energy are needed to do 1 joule of work. Thus the 50–milliwatt metabolic energy cost of swimming suggests a mechanical power output of about 10 milliwatts. However, the calculation that treated the fish as a rigid torpedo gave a much smaller estimated power requirement, only 2.4 milliwatts. The fish was using four times more power than seemed to be needed. Are fish muscles unusually inefficient, or is there some other explanation?.

About this time, Paul Webb–was experimenting with trout in a water tunnel, measuring their oxygen consumption and also filming them so that their power output could be calculated according to Lighthill's theory. Webb was also interested in the efficiency of the swimming muscles and devised an ingenious method to measure it. He attached baffles to the backs of some of his fish so that when they swam they had to do work against the drag on the baffle (which could be calculated easily) in addition to the work needed to propel their own bodies. He measured the oxygen consumption of fish swimming at the same speed, with and without baffles. The difference told him how much metabolic power was needed to do the work against the drag on the baffle, making it possible to calculate the efficiency of the muscles. His results give efficiencies of about 0.14, just a little lower than might have been expected.

Webb now had two methods for calculating how much mechanical power was required for swimming without baffles. He could use Lighthill's theory, or he could multiply the metabolic energy cost by the efficiency. The two estimates agreed well, but both were about three times higher than the result of the torpedo calculation. Webb concluded that the Lighthill theory gives a much better estimate of power requirements than the simpler and more obvious one.

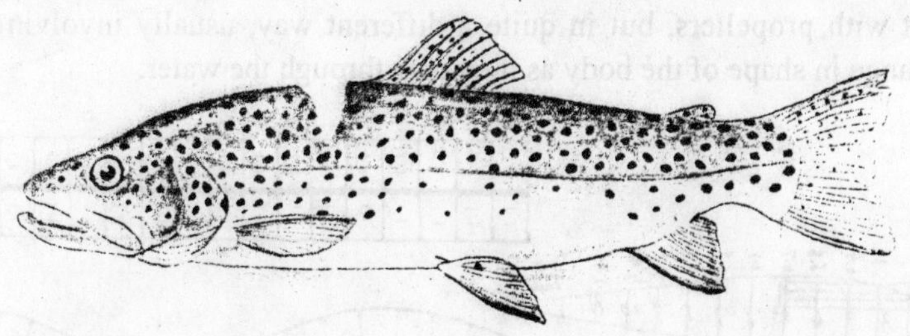

Fig. 6.8 : A trout with a baffle attached, for an experiment by Paul Webb.

There seem to be two reasons why the torpedo calculation underestimates the energy cost of swimming. First, it ignores the kinetic energy given to the water, which is pushed to either side as the tail beats. This is not very important, however, because the large masses *of water that are pushed move with only low sideways velocities.* Second, and much more important, the calculation treats the fish as a rigid torpedo, but it is not rigid. The undulating movements of swimming fish probably thin the boundary layer, increasing drag on the fish's body.

Swimming in Fishes

The problem of analysis

Fishes swim with such apparent ease that it is hard for us to realize just how difficult it is move rapidly through water, and how well adapted fishes are for locomotion in a dense medium. It is not surprising that our own underwater designs are often very similar in shape to fishes, for the density and viscosity of the medium dictate the same streamlined from: the solid of least resistance, which the great pioneer aerodynamicist Cayley took from the body of a trout, looks just like a modern submarine hull. Yet there is an essential difference between the fish design and our own which makes the analysis of fish swimming a difficult and intriguing problem.

Submarines are rigid vehicles which use rotating propellers to provide the thrust required to overcome the drag on their hulls. It is possible. therefore, to make rigid models and to test them in water tunnels to collect data which can be used to estimate the drag they incur and the power needed to drive them along. Data from such tests and from experience with actual submarines and aircraft from the basis for theoretical analyses of the hydrodynamic of rigid bodies. Unfortunately for biologists, fishes are usually far from rigid, and they generate thrust not with propellers, but in quite a different way, usually involving change in shape of the body as they pass through the water.

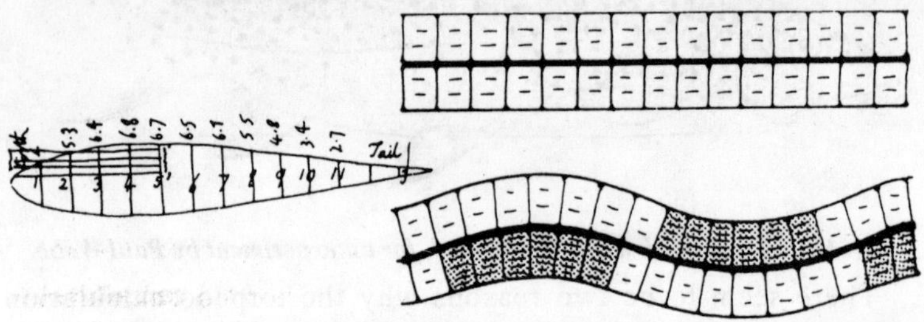

Fig 6.9: ***Left: Cayley's drawing of a trout; right: schematic diagram illustrating bending of a body with flexible backbone (thick black line) by segmented muscle blocks (contracted, stippled).***

Various fishes have adopted special methods of locomotion in which they use paired and unpared fins to row, flap, or undulate themselves along, but the great majority of fishes propel themselves by a combination of two processes—the backward passage of transverse waves along the body, and lateral movements of the caudal fin. In such fish as scombroids, thrust is generated only by the caudal fin, and the body flexes rather little, whereas in eels, or in dogfish, there are large amplitude body waves passing towards the tail. Most fishes fall somewhere between these two extremes, sending smaller amplitude waves down the body which terminates in a substantial caudal fin, as in gadoids or salmonids. In general, the fastest swimming fishes rely on the oscillation of a high aspect ratio caudal fin, attached to the body by a narrow caudal peduncle; slowly-swimming fish undulate through the water like eels, and as we shall see, this is a less efficient process.

This flexible method of thrust generation makes it impossible to apply directly hydrodynamic data from the performance of rigid bodies to the analysis of fish swimming. We could, for example, make a model of an eel, or stiffen an actual eel in some way, and measure its drag in a water tunnel, but we should not expect that the drag we measure would be the same as that which must be overcome by the swimming fish, nor that the surface flow patterns around the model would be even remotely like those around the living fish. Although the hydrodynamics of fish swimming are thus more complex, and less well understood than those of rigid bodies, good progress has been made in the past decade, both in obtaining kinematic and respirometric data from swimming fishes set up in water tunnels, and in working out suitable mathematical models of fish swimming. We can now make reasonable estimates of the power needed for swimming, and of the efficiency of the process.

In this chapter we shall look first at the machinery which fishes use to move their tails and bodies; next see how these movements generate propulsive thrust; and finally, look into the efficiency of swimming, and at the adaptations which some fishes have devised to reduce the cost of swimming—some of these remarkably resemble devices found on modern aircraft.

The muscular machinery

Myotomal structure

The backward passage of body waves and the lateral oscillation of the tail both result from the operation of the segmented axial musculature, divided up into myotomes by the myoseptal connective-tissue partitions on to which the muscle fibres insert. The myotomal muscle fibres run roughly parallel to the long axis of the fish, and because the notochord or back bone is incompressible, contraction of some or all of the myotomes on one side of the body will result in lateral movements. Stated simply like this, it seems easy to understand how the myotomal apparatus works, and the main problem would seem to be the organization of the control system in the spinal cord ensuring that the appropriate link in the chain of successive myotomes along the body contracted at the correct point in the swimming cycle.

Certainly, this is an interesting problem but a closer look at the myotomes of any real fish raises other questions; there are mechanical subtleties in the myotomal layout which are not yet fully understood.

Firstly, no fish myotomes are simple rectangluar blocks. Not only do the myocommata slope obliquely forwards and backwards from the myotomal surface just under the skin, to their insertion on to the backbone, so that transverse sections always cut across several myotomes, but the outer borders of the myotomes are folded. In amphioxus they are simply folded once, into a series of V-shapes, but in lampreys and higher fishes the folding is rather more complex, and gives a shallow W-shape. The phylogenetic increase in complexity from amphioxus to teleosts is paralleled in teleost ontogeny, where the myotomes from first as V-shaped blocks before folding further to give the adult W-shape. It is not easy to make 2-dimensional structures. Probably the best approach is to lightly boil a fair-sized fish so that the myotomes can be separated, and then to try and model myotomes for oneself out of plasticine. Within these complicated myotomes, only the most superficial muscle fibres are actually arranged parallel to the long axis of the fish. Those deeper in the myotomes are arranged in a kind of spiral pattern so that each lies at a different angle to the long axis, even up to 30°. At first sight this arrangement seems very peculiar, for a completely parallel arrangement might seem more what is wanted, but as Alexander (1969) showed, the spiral design allows each muscle fibre within the myotome to contract at the same speed, in this way operating most economically. Alexander and Kashin's analyses were notable, for they offered the first functional explanation for a puzzling feature of myotomal design, and what is more, hint at the reason for the complex folding of the myotomes, probably arranged to allow optimum muscle fibre packing in a wide myotome.

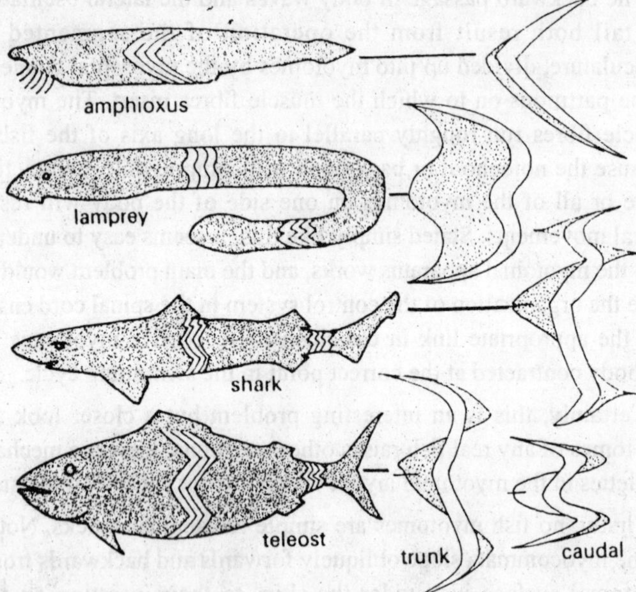

Fig. 6.11: Myotomal form in different fishes Single myotomes are seen at right, their outer surface unstippled.

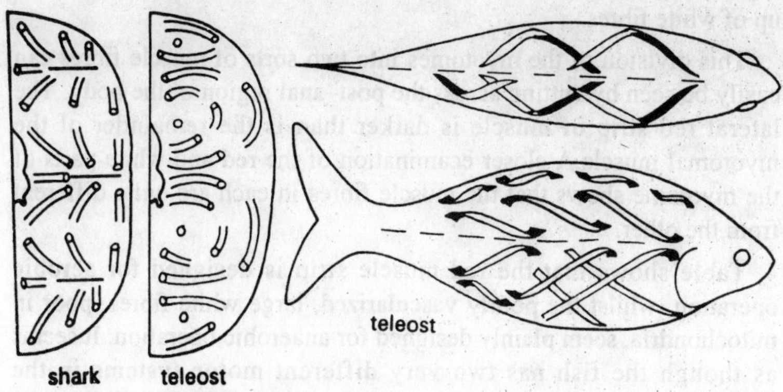

Fig. 6.11: Left: thick transverse slices across a shark and a teleost showing orientation of myotomal muscle fibres. Note tendone in shark (horizontal lines). Right: dorsal and lateral views of typical teleost showing course of muscle fibres in successive myotomes along the body. The helices shown were obtained by taking the origin of one muscle fibre from the point at which the muscle fibre in the myotome next anterior insert on to the common myoseptum, and so on along the fish.

The myosepta are composed of a meshwork of inextensible collagen fibres, which can be deformed, but not stretched (like a sheet of paper), and they are in teleosts often stiffened to limit their deformation by vertebral ribs and by intermuscular bones. It is not yet clear exactly how these stiffening elements work (they are absent from all sharks) but it seems probable that they allow only laterlal movements when the myotomal musculature contracts, and serve to reduce the general flexibility of the body. In oceanic scombroids, and other high–speed teleosts, the posterior myotomes are elongate in their lateral folding, thus giving rise to tendons which operate the caudal peduncle.

The advantage of this arrangement is clear: it is the same as that which determines that the leg muscles of fast–running terrestrial animals lie within the body, rather than across the joints in the limbs. The mass and inertia of the caudal peduncle is kept as low as possible by the development of these tendones so that the rate of tail beat may be increased and the width of the peduncle be reduced for hydrodynamic reasons. An additional benefit is that the inertia of the anterior part of the body is increased, so that the tendency for the tail to oscillate the body is reduced.

Myotomal muscle fibres

The muscle fibres of the myotomes in almost all fishs are of two quite different kinds, easily distinguished by their colour. A thin red superficial layer covers the major part of the myotome which is made up of white fibres.

This division of the myotomes into two sorts of muscle fibres can easily be seen by cutting across the post–anal region of the body. The lateral red strip of muscle is darker than is the remainder of the myotomal muscle.A closer ecamination of the red and white parts of the myotome shows that the muscle fibres in each are quite different from the other.

Table shows that the red muscle strip is designed for aerobic operation, whilst the poorly vascularized, large white fibres, poor in mitochondria, seem plainly designed for anaerobic operation. It seems as though the fish has two very different motor systems in the myotomes, operating on different fuels (or at least utilizing different routes for ATP production from the fules), and we might guess that the two regions of the myotome are used by the fish for different kinds of swimming.

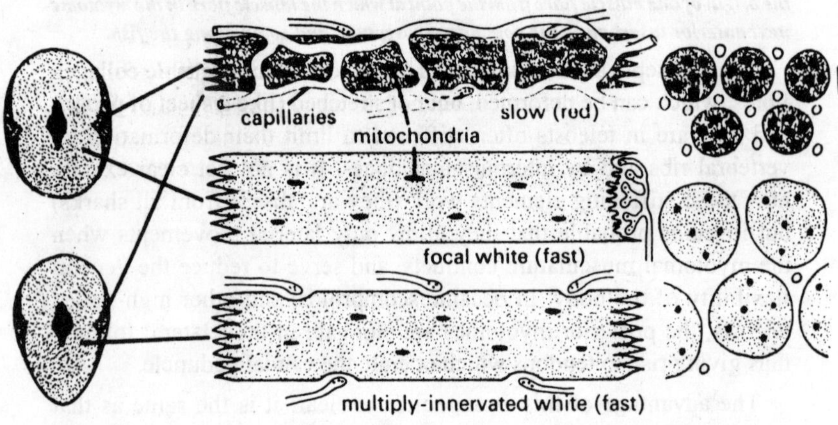

Fig 6.12: Summary diagram showing structure and innervation pattern of red and white myotomal muscle fibres in fishes. Note that multiply–innervated white fibres are found only in higher teleosts.

Several lines of approach support this guess, but the most direct is simply to insert wire or needle electrodes into muscles of swimming fish, and record electrical activity under different conditions of swimming.

Electromyograms (EMG's) recorded in this way from herring swimming in a tunnel respirometer, or from 'spinal' dogfish (which are particularly conveninent since they will continue to swim although

Table 6.1: A comparison between the fast and slow muscle fibres in fish myotomes.

Slow	*Fast*
smaller diameter (20–50% of fast fibers)	Large diameter (may be more than 300*)
Well vascularized	Poorly vascularized
Usually abundant myoglobin, red	No myoglobin, usually white
Abundant large mitochondria	Few smaller mithochondria
Oxidative enzyme systems	Enzymes of anaerobic glycolysis
Lower activity of Ca^{2+}–activated myosin ATPase	High activity of enzyme
Little low molecular wt. Ca^{2+}–binding protein	Rich in low molecular wt. Ca^{2+}–binding protein
Lipid and glycogen stores	Glycogen store, usually little lipid
Sarcotubular system lower volume than in fast fibers	Relatively larger sarcotubular system
Distributed cholinergic innervation	Focal or distributed cholinergic innervation
No propagated muscle action potentials	Propagated action potentials; may not always occur in multiply–innervated fibres
Long–lasting contractions evoked by depolarizing agents	Brief contractions evoked by depolarizing agents

the brain has been removed) show that the red portion of the myotome is active when the fish swims slowly, and that under these cruising conditions, no activity is found in the white portion of the myotome. Only when the fish is swimming in rapid bursts are muscle action potentials visible from the white muscle. Although the fish can continue to use the red part of the myotome for long periods and some fishes (such as mackerel or pelagic sharks) cruise continually the whole of their lives using red muscle fibres, the operation of the white part of the myotome is severely limited in dogfish and herring, being exhausted after 1–2 min of continuous operation. This is not such a drawback as might appear, for it is very rare for any fish to swim rapidly for more than a few strokes of the tail–it then glides into slower swimming. In fact, 2 min continuous operation in dogfish would allow the fish to travel about 600 m. Most of the drag which the fish must overcome as it swims along results from the viscosity of the water (water particles tend to become attached to the surface of the fish, forming a boundary layer), and is called skin friction drag (D_{sf}). We shall look at the different sorts of drag in the next section, but for the moment, all we need to know is that skin friction drag is proportional to the *square* of the speed at which the fish is swimming. At constant speed, drag must be equal to the thrust the fish produces (or it would slow down or speed up), and so the power required from the muscles to generate this thrust will be proportional to the *cube* of its swimming speed.

This means that speed in the water is extremely costly in terms of the power needed from the muscles. Doubling the speed from, say, the 25cm s^{-1} cruising speed of a dogfish, to 50cm^{-1} requires not twice, but eight times the cruising power (other things being equal). It is small wonder that in most fish the white part of the myotome is much bigger than the red portion, and that the white part of the myotome is mich bigger than the red portion, and that the white muscle fibres are highly specialized for maximum power production. Ultimately, the power produced by a given amount of muscle depends on the actin-myosin interaction, so that any modifications which will allow more myofilaments in a given volume will increase the power output. Thus the white muscle fibres are large (up to 300 μm diameter), have very few mitochondria interrupting the myofilament array, and have few muscle capillaries occupying space which could be filled by muscle fibres. These specializations for maximum power output naturally imply that the white fibres operate anaerobically, and as we find from their enzyme profiles and from studies on metabolite depletion, they operate by anaerobic glycolysis. This is a relatively inefficient way of gaining ATP for driving the actin–myosin interaction, and the ion pumps of the fibres, since this route provides only 3 mole of ATP per glycosan unit. Fast swimming is indeed an expensive process. But all designs are a compromise between conflicting requirements, and the fish accepts a relatively low chemical efficiency for its high–speed swimming machinery in return for a highly–efficient system that is used for continuous operation.

One difficulty arising from the use of anaerobic glycolysis for short bursts of speed is that large amounts of lactate are produced. Castellani and Somero (1981) have shown that a good correlation exists (as it should) between muscle buffering capacity and activity in different fish.

The red muscle fibres make up in most fish only 5–15% of the total myotome, and operate by aerobic glycolysis (yielding 38 mols of ATP per glycosan unit) or by aerobic lipolysis (yielding over 100 mole of ATP per mole); plainly a much more efficient use of the fuel. These processes require oxygen and so red fibres have a rich capillary bed, contain myoglobin for internal oxygen transport, and contain vast numbers of large mitochondria–in some scombroids, half the volume of the red fibres consists of mitochondria. Of course, this means that the power produced per unit of muscle volume will be much less than that for the white fibres, but at cruising speeds what is needed is efficiency and economy of operation, not maximum power output.

The two motor systems and terrestrial animals

The manifold defferences between the red and white fibres, emphasize the difference in design of the two for their two different roles, and reinforce the idea that fishes have two separate, independent motor systems in their myotomes, used under different swimming conditions. The arrangement is analogous to that of modern military jet aircraft, where economy in the cruise condition contrasts with the vastly increased power obtained by the occasional emergency use of reheat; extravagant in fuel, but acceptable for short bursts.

The dual motor system of the fish myotome is very different to the arrangement of muscle fibres in higher terrestrial animals. In water there is little penalty for carrying around a large mass of white muscle that is only of occasional use, but on land the animal is not buoyed up, and such a system would incur a severe weight penalty. So our muscles, and those of other terrestrial animals, are arranged in a different way, and our muscle fibres are all more similar to each other than are those of the red and white portions of the fish myotome.

Cruising speed and red muscle

The proportion of red fibres in the myotomes gives a good estimate of the importance of sustained cruising in the life–style of the fish, so that, for example. salmon and spur dogs(*Squalus*) have a relatively larger amount of red muscle than do dogfish or similar sluggish fishcs. But it is not possible simply to increase the mass of red miscle (to increase cruising speed) beyond a certain limit, for as we have seen, it requires abundant oxygen to operate, and this has to be acquired by the gills. Probably 15% of the total muscle is the greatest mass of red muscle that can be adequately supplied by the gills, unless (as in scombroids) there are special adaptations to increase oxygen acquisition and transport.

Warm red muscle

In some fishes, such as the more advanced scombroids and the big isurid sharks, body temperature is above ambient and the red muscles are run at relatively high temperatures. There is no problem in generating heat (it is an inevitable byproduct of muscular contraction, which is why we shiver when we are chilled), but at the gills of fish the blood is so intimately in contact with the surrounding water that it must be at ambient temperature; special arrangements have to be devised to maintain high muscle temperatures. Warm fishes use countercurrent heat exchangers, by making the warm blood leaving the red muscles pass through a capillary rete mirabile where it runs close to the cool oxygenated blood, and so heat is exchanged to the incoming blood and the muscle is able to maintain its elevated temperature.

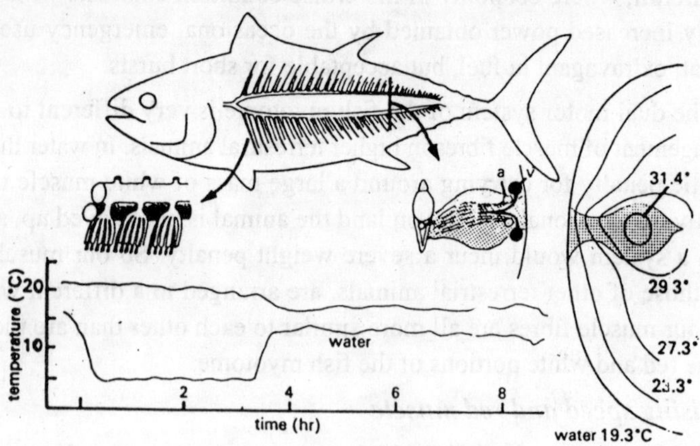

*Fig: 6.13: Organization and function of the retial thermoregulation system in the bluefin tuna (***Thunnus thynnus***). Upper: bluefin showing lateral vessels (details of branches to left, position of lateral vessels and retial system in red muscle to right. a, artery; v, vein). Right: thermal profile obtained with thermistor probes, red muscle stippled. Bottom: records of water temperature (lower line) and stomach temperature (upper) obtained by telemetry from free-swimming bluefin, showing independence of body temperature from changes in water temperature.*

Similar countercurrent rete mirabile are found in the flippers of cetaceans, and in the swimbbladder gas generating system both places where the added capillary resistance is outweighed by the advantages of heat and gas retention. Although warmblooded fishes have long been knowm, it is still not clear *why* it should be advantageous to be warmblooded. One possibility is that more power could be extracted from a given volume of muscle if it is 'run' at a higher temperature but the increase in white muscle volume that this would allow (remembering that the red muscle volume is peobably limited by its oxygen requirement) seems but a small fain for the complexity of the retial system. Perhapa more important is the possibility that the fish is able to operate its cruising musculature at a more or less constant temperature, in waters of different ambient temperature. That is, like mammals, the advanced warmblooded fish can maintain a more or less constant 'milieu interieure', and can adapt its enzyme systems to operate at a single most efficient temperature.

Functional overlap between the two motor systems

This brief resume of the dual motor system in the fish myotome has suggested that the two systems are indeed quite separate and employed during different kinds of swimming. No biologist will be surprised to learn that although such a generalization is broadly true, there are some

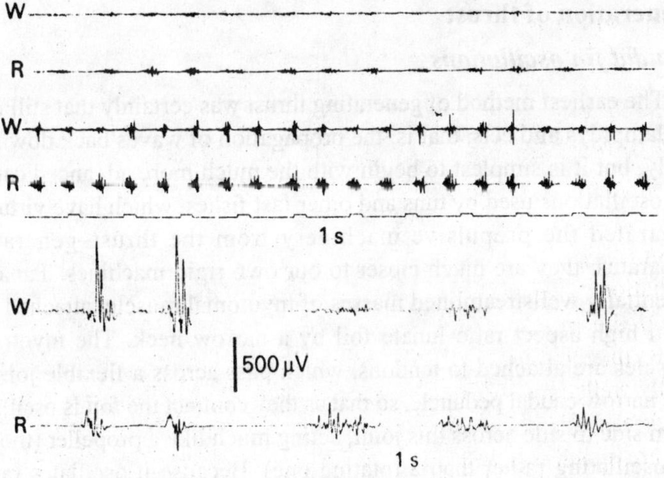

Fig. 6.14: Electromyogram from red (R) and white (W) regions of myotome in carp swum at different speeds in a tunnel respirator respirometer. Note that even at the slowest speed at which the fish could be persuaded to swim (0.75 body lengths^{-1}, top) low–level electrical activity is seen in the white zone, increasing as swimming speed increases to 1.76 b.i.s^{-1}. Bottom record shows spike–like potentials recorded from white zone at 2.0 b.l. s^{-1}. Time marks seconds.

exceptions. All the available experimental evidence suggests that the two motor system are separate in sharks, lungfish, lampreys, amphioxus, *Amia,* and in some primitive teleosts like clupeids or eels. But in the more advanced teleosts (that is in most fish) the fast motor system is probably used not only during bursts of fast swimming, but also during slow sustained cruising.

In carp, for example, if we record from their red and white muscles whilst they are swimming at different speeds in a tunnel repirometer we find low–level electrical activity in the white muscle even at the lowest speeds at which we can persuade the fish to swim. At higher speeds, bursts of action potentials of the usual kind are seen from the white muscle just as in dogfish or trout. The reason for this difference is not yet fully understood, but seems to be related to the unusual (distributed) innervation pattern of the fast fibres in these teleosts. It does means that some use can be made of the otherwise inactive white muscle during sustained cruising.. Just how white fibres whose whole design seems to be organized for anaerobic operation can be used during sustained swimming remains a mystery, for an oxygen debt is not accumulated by these fish during long periods of sustained cruising.

Generation of thrust

Caudal fin oscillations

The earliest method of generating thrust was certainly that still used by lampreys and eels, that is, the propagation of waves back down the body, but it is simplest to begin with the much more advanced caudal fin oscillations used by tuns and other fast fishes, which have virtually separated the propulsive machinery from the thrust–generating apparatus: they are much closer to our own rigid machines. Tuna are essentially wellstreamlined masses of myotomal muscle, attached to a rigid high aspect ratio lunate foil by a narrow neck. The myotomal muscles are attached to tendons, which pass across a flexible joint in this narrow caudal peduncle, so that as they contract the foil is oscillated from side to side across this joint, acting much like a propeller (though an oscillating rather than a rotating one). Because it oscillates rather than rotates, the caudal foil operates alternatively with the same surface forming the upper and lower surface of the foil, and since it comes to a halt, as it were, at the end of each stroke, the thrust produces is oscillatory, though smoothed by operation at high frequency, up to the remarkable frequency of 10 tail beats per s. How does this oscillatory foil generate thrust, and how is it specially adapted to do so? Here we are on firm ground, for the process is the same as that by which aeeroplane wings generate lift, or propellers thrust.

Circulation, lift and thrust

The forces which act on a foil as it passes through a fluid arise from the displacement of fluid by the foil; they are of two kinds. Frictional forces (resulting from viscous effects) can for the moment be neglected, since they are only indirectly concerned. Lift or thrust forces can only result from a net difference in the pressure of the fluid on either side of the foil. As an aeroplane wing passes through the air, it is both sucked upwards from above, and pushed up from below, for there are higher pressures below the wing and lower pressures above.

How do these pressure difference arise? The relationship between pressure and velocity in the flow of an incompressible fluid is given by

$$P+\frac{1}{2}\rho V^2 = \text{Constant}$$

where P = static pressure; ρ = density, and V = velocity. The quanitity $\frac{1}{2}\rho V^2$ (often abbreviated to q) is the *dynamic pressure,* the force experienced if one puts one's hand out of the window of a moving car, and it is this which is the fundamental source of aerodynamic or hydrodynamic forces. As we shall see, the dynamic pressure appears in formulae for the calculation of thrust, lift and drag.

Since the density of the fluid does not change at the speeds at which fish swim, clearly the pressure distribution which one can measure over a foil in a fluid must result from differences in the velocity of flow over the two surfaces, and the velocity is higher above a lifting foil than below it. If small particles like polystyrene beads are distributed in water, and a foil is moved through the water, the displacement of the water caused by the passage of the foil can be observed by taking high–speed cine–films. Let us follow the displacement of two particles A and B from their initial positions at rest before the foil arrives, until their final positions after it has passed. We find that they follow curved paths of opposite sense, that is , they are forced to *circulate* by the passage of the foil. Since the displacement of particle A above the foil is greater in the same time than is that of B below it, the algebraic sum of the two results in a net colckwise rotation of *circulation,* and because it is in this sense, the relative velocity of fluid above the foil will be greater than that below it, and lift will be generated.

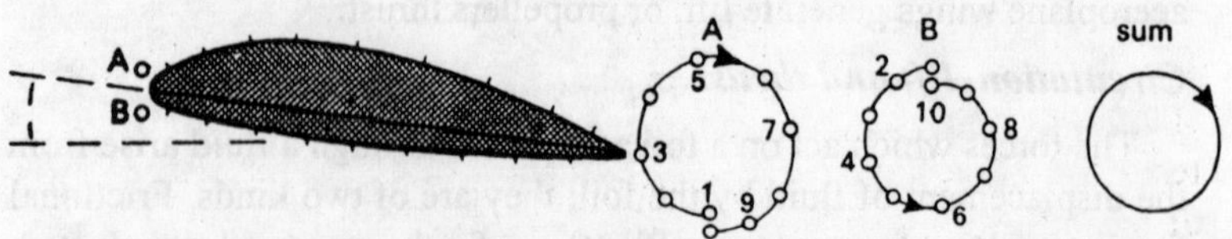

Fig. 6.15 : Circulation and lift. Two particles A and B are caused to change position as the foil passes through the fluid. Right: the successive positions of the two particles (numbering as on left). The sum of their movements is a net clockwise circulation producing lift.

The reduction of drag associated with lift and thrust

The circulation over the main part of the foil continues downstream in a series of tip vortices at the tips of the foil as it moves through the fluid, giving rise to an upwash just behind the back of the tip of the foil, and a downwash across the span, decreasing to the root of the foil. This process requires energy, and the energy that has to be expended to maintain the tip vortices is a component of the drag the lifting foil incurs as it passes along, called vortex drag (D_v).

It is not hard to guess that as the tip vortices depend on the circulation, the greater the circulation, the greater the amount of vortex drag which is incurred. Like aeroplanes. fish take advantage of two possibilities to reduce this drag. First, since a given lift or thrust can be produced by a weak circulation over a long span, or by a strong circulation over a short span, doubling the span can halve the vortex drag incurred, and for this reason, the caudal fins of fast–swimming fishes are long and thin, as are the lifting pectoral fins of fast fishes like tuna, which are denser than water. Their aspect ratrio, AR (span^2/ fin area) is as high as 6; as high as can be arranged without sacrificing structural strength. Secondly, it can be shown that for a given aspect ratio, vortex drag is least if the spanwise lift distribution is elliptical; this can be conveniently achieved by making the plan form of the foil elliptical, and this is the shape both of the caudal foils of scombroids, and of the lifting pectoral fins of fast sharks and many large fishes like marlin or swordfish. High AR tail fins of this kind, and the similar lifting pectoral fins, are designed to minimize vortex drag, and are suitable for high–speed cruising. But they are not very suitable for rapid manoeuvring, or for rapid accelerations, and this is why fishes which need these capabilities have caudal fins that are much broader and of lower aspect ratio. To provide thrust or (in the case of an aeroplane or

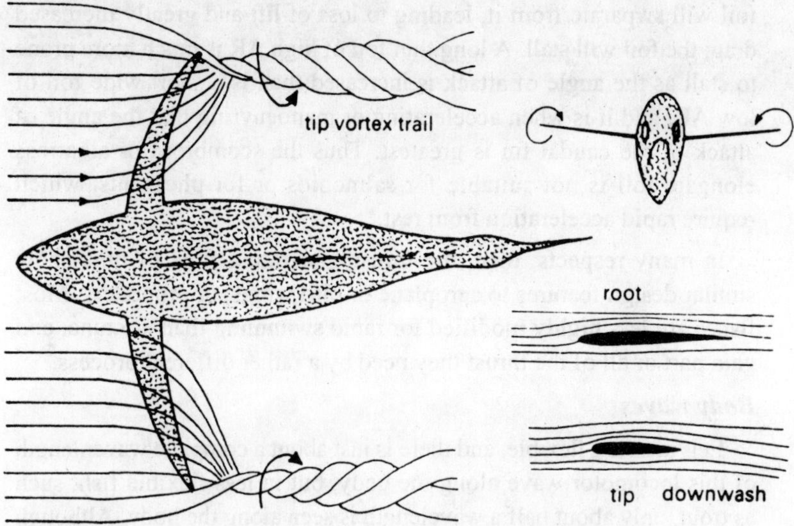

Fig. 6.16: Flow patterns around a gliding skipjack (*Katsuwonus*). Tip vortices cause vortex of lift–associated drag. Lower tight; flow over root and near tip of pectoral fins showing downwash near tips.

dense fish) lift, the foil has to be operated at an angle of attack to the fluid, or the circulation of fluid above and below the foil will be equal, and no lift or thrust will result. At small angles of attack, fluid will flow around the foil in an ordered way, but if the angle of attack is increased above a certain limit, the flow around the upper part of the

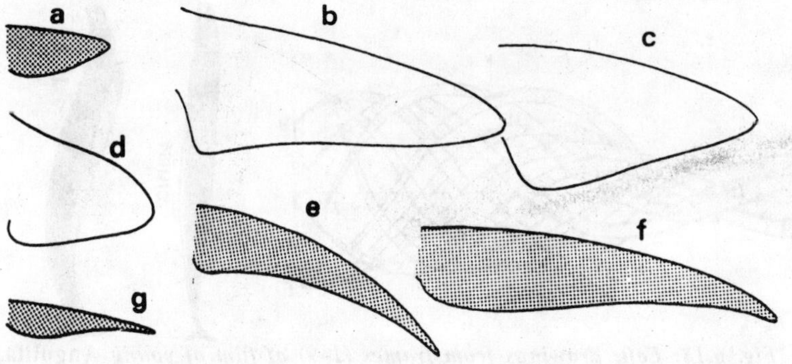

Fig. 6.17: The lifting pectoral foils of different fishes (teleosts stippled). (a) Mackerek (*Scomber*); (b)* Carcharhinus longimanus*; (c) basking shark (*Cetorhinus*); (d) dogfish (*Scyliorhinus*); (e) carangid (*Trachurus*); (f) longfin tuna (*Thunnus alalunga*; (g) swordfish (*Xiphias*).

foil will swparate from it, leading to loss of lift and greatly increased drag; the foil will stall. A long thin foil of high AR is much more prone to stall as the angle of attack is increased than is a short wide foil of low AR, and it is when accelerating or manoeuvring that the angle of attack of the caudal fin is greatest. Thus the scombroid or albatross elongate foil is not suitable for salmonids or for pheasants, which require rapid acceleration from rest.

In many respects, then, the thrust–generating foils of tuna show similar design features to aeroplane or bird wings, or propellers. Most fishes are less highly modified for rapid swimming than are tuna, and gain part or all of the thrust they need by a rather different process.

Body waves

Eels are very flexible, and there is just about a complete wavenlength of this locomotor wave along the body, but in less flexible fish, such as trout, only about half a wavelength is seen along the body. Although we could consider this process in a similar way to the operation of the tuna tail, that is, in terms of curculation theory, it is more appropriate, as Lighthill (1971) showed, to think of it in terms of the forces'felt' by the water masses next to the body surface. These are the forces which result from the inertia of the water next to the fish, and are proportional to the rate of change of relative velocity of the surface of the animal, with respect to the water next to it. The result of the passage of the locomotor wave bacl wards along the fish is to increase the momentum of water passing backwards, and it is the rate of shedding of this momentum into the wake which is proportional to the thrust.

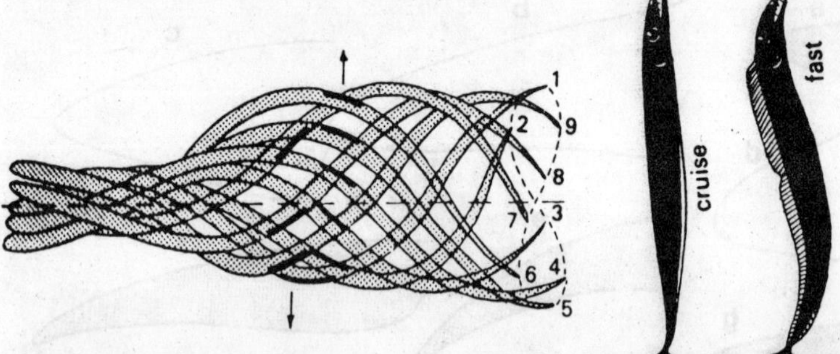

***Fig. 6.18: Left: drawings from frames (1–9) of film of young* Anguilla, *superimposed to show short sections of the body (thick lines) as planes inclined to axis of progression when the eel moves forwards. The oblique dashed lines is the axis of progression, the tail tip describes a figure–eight pattern. Right: the two swimming patterns of the trichiurid* Aphanopus.**

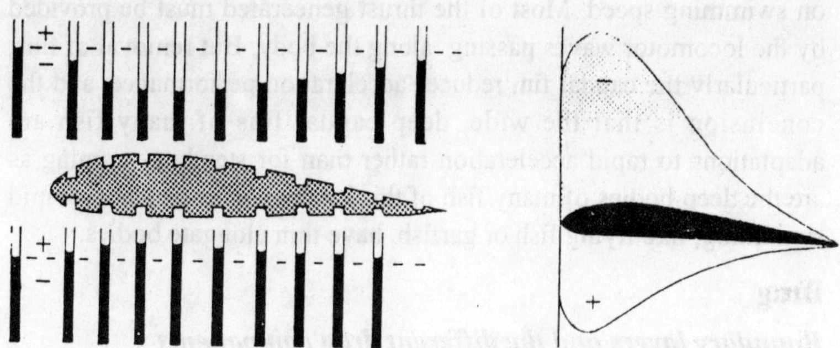

Fig. 6.19: Pressure distribution around a foil at a positive angle of attack in a wind or water tunnel. Left: manometers connected to openings in upper and lower surfaces show pressures above and below ambient (dotted line). Right: typical pressure distribution around such a lifting foil.

The eel is a relatively inefficient swimmer (although it travels enormously long distances to spawn) because the body is more or less round in section, and so the water affected by its passage through it is of relatively low mass. By deepening the body to give a flat cross–section like many carangids, or the leptocephalus larva the virtual mass of water which is given momentum by the swimming movements is much increased, and greater thrust can be produced. Many fish which swim in this way deepen the section of the body with median fins of different kinds (retractable in teleosts, but not in sharks) and although these median fins play an important role in stabilizing the fish in roll, they are equally important in thrust generation. It is not necessaty to have a continuous fin. An interrupted series of fins (like those of many gadoids) works equally well, provided their spacing is appropriate, and this gives rise to less drag.

Steady swimming with a caudal foil, or by the passage of locomotor waves along the body, or by a combination of the two, has been most studied, and reasonable estimates of the effeciency of the process have been obtained. But less is known about acceleration, and as Webb (1977) has shown, the optimum body shapes for steady performance are different from those required for fast–starts performance. Some fishes actually adopt different shapes for the two requirements. The deep–water scabbard fish, for example, has an elongate body terminating in a small caudal fin of fairly high aspect ratio. It swims slowly by keeping the body rigid and oscillating this foil. To accelerate rapidly, it unfurls its long median dorsal fin, and passes locomotor waves along the body. Amputation experiments on salmonids have shown that for steady swimming, caudal fin removal had little effect on swimming speed. Most of the thrust geneerated must be provided by the locomotor waves passing along the body. But removal of fins, particularly the caudal fin, reduces acceleration performance, and the conclusion is that the wide, deep caudal fins of many fish are adaptations to rapid acceleration rather than for steady swimming as are the deep bodies of many fish of this kind (such as carp). For rapid swimming, like flying fish or garfish, have thin elongate bodies.

Drag

Boundary layers and the different drag components

The thrust that the fish generates is at constant speed, equal to the drag forces resisting forward motion: how do these arise, and how can they be minimized by fish? Obviously, drag–reduction is greatly advantageous for fish, since it will mean that for a given power output, they can travel faster, or can use less power at a given speed. Since water is viscous, it tends to sticks to the surface of the fish, and a thin boundary layer forms, in which there is a steep velocity gradient from the still water layer next to the fish, carried along with it, to the water outside the boundary layer, This velocity gradient means that there is a large shear stress which results in drag.

The properties of the boundary layer depend upon the ratio between viscous and inertial forces action on the fish, given by the dimensionless Reynolds number (R_e). This is defined as

$$\frac{\text{length} \times \text{velocity} \times \text{density of medium}}{\text{viscosity of medium}}.$$

or LV/v, where v is the kinematic viscosity of the medium, viscosity / ρ, approximately 0.01 in water.

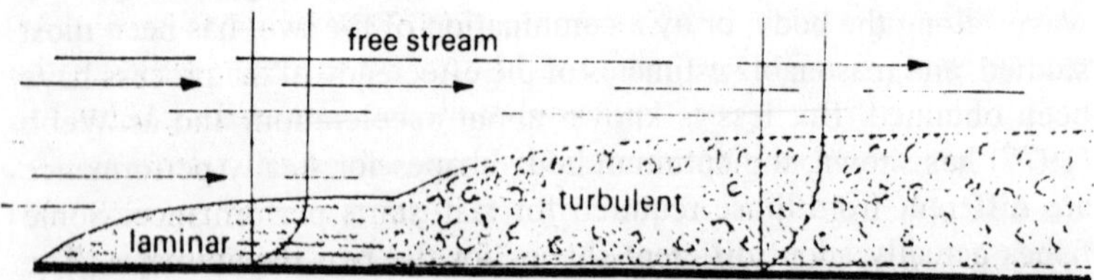

Fig: 6.20: Development of boundary layer on a flat plate in a fluid flowing left to right. The anterior thinner laminar layer has a steeper velocity gradient than the posterior turbulent layer.

The Reynolds number gives us an immediate idea of the relative imortance of viscous and inertial forces, and in practice, at Reynolds numbers below 10^5 viscous forces predominate and the boundary layer on a flat plate will be laminar. At higher Reynolds nimbers than this, say over 10^6, inertial forces are more important, and flow within the boundary layer will normally be turbulent.

In a flat plate of sufficient length, we shall find that near its front upstream edge, the boundary layer will be laminar, but as we pass downstream it becomes thicker, and as L (and hence Re) increase, it changes into a turbulent boundary layer. The point of transition from laminar to turbulent depends much on surface irregularities, for since energy to maintain the laminar layer can only enter by diffusion across the layers within it, it is rather sensitive to any adverse pressure gradients which may be produced by surface irregularities.

The skin friction drag on our flat plate is given by $D_{sf} = \frac{1}{2}\rho V^2 ACf$, where ρ = density; V = velocity of flow; A = watted area; and Cf is a drag coefficient.

The value for Cf is much lower for laminar than for turbulent boundary layers. But because laminar boundary layers are relatively unstable, as we have seen, they tend to separate from the surface, and give rise to a turbulent wake behind the object in the flow. A large wake means a large pressure difference between nose and tail, and thus a large pressure drag, so that separation is to be avoided as far as possible, Abrupt change of contour and awkward excrescences will cause the flow to separate; fishes, however, are smoothly shaped and well streamlined–many have elaborate transparent eyes fairings, for example, or fold their fins into slots when they are not required. Fish can very greatly reduce the form drag component of pressure drag. Indeed, a mackerel, say, tested in a water tunnel as a dead rigid object is so well streamlined that its drag is very similar to that of a flat plate of the same wetted area: there is almost no pressure drag. In the swimming fish, however, some part of the vortices trailing behind in the wake result from the feneration of thrust (and of lift in dense fishes), so that although fishes minimize vortex drag as far as possible, this second component of pressure drag is an inevitable consequence of thrust and lift generation. The total drag that a fish or an aeroplane has to overcome as it goes along thus has two different components.

total drag; skin friction drag + pressure drag (form drag + vortex drag).

Streamlining and attention to increasing the aspect ratio of the lifting and thrust–generating foils will minimize pressure drag, which is usually of much less consequece than skin friction drag. What can be done to minimize skin friction drag?

Drag–reducing mechanisms

It is important to realize first, as we saw at the outset, that skin friction drag on most fishes is not liklely to be the same as that for a flat plate of equivalent wetted area. There are good theoretical and experimental grounds for supposing that it will be considerably greater, by a factor of 2–5 times, for those fishes which make substantial lateral body movements. For example, the power output esrimated from oxygen consumption measurements on goldfish swimming at constant speed was found to be 3.6 times that needed to overcome the drag on a rigid model of the same wetted area (Smit *et al.,* 1971). It is obviously important for fast swimming fishes to reduce the lateral movements of the body, and hence this augmented skin friction drag, and this is just what they do; only the slower swimmers have large amplitude body movements. The great majority of fishes are sufficiently small that even at burst speeds, they do not exceed a Reynolds number of 10^6, so that they do not need to have special devices for reducing skin friction drag, since boundary layer flow will be laminar. But larger and faster fishes swim at Reynolds numbers above this, where inertial forces are more important, and the boundary layer will be turbulent, except in the anterior region. Since the drag coefficient for a turbulent boundary layer is much higher than that for a laminar layer, anything the fish can do to maintain a laminar layer, or to delay transition to a turbulent layer, will reduce its drag. Furthermore, separation of the boundary layer from the surface (with the penalty of pressure drag added to skin friction drag) is more likely at higher Reynolds numbers when the boundary layer is thicker.

Trachypterid and stromateid fishes have a remarkable 'porous' integument which has a canal system filled with sea water just under the smooth outer surface; and it seems likely that this curious arrangrment is designed to maintain a laminar boundary layer, by a process of distributed dynamic damping. Walters (1963), who first examined trachypterid fished such as *Desmoderma,* suggested that small adverse pressures in the boundary layer which might otherwise lead to transition or to separation, would be damped by fluid sinking into the pores and being transmitted to regions of lower pressure. This mechanism has not been tested experimentally.

Other fishes have a different kind of skin, which seems to be specialized for maintaining a turbulent boundary layer, and preventing separation. In this case, the system is better understood. The castor oil fish, *Ruvettus,* has a regular pattern of very sharp pointed scales all over its body, which project about 1.0 mm above the body surface. These will project through the boundary layer, and entrain vortices from the free stream. This input into the boundary layer will help to stabilize it and prevent separation. Similar devices are found on some aeroplane wings. One problem is that they will *increase* drag at low speeds, and so they are only found on fish which can accept reduced efficiency at cruising speeds in order to obtain a high burst speed. Perhaps the ctenoid scales of most advanced teleosts work as vortex generators, but experimental data are still needed here.

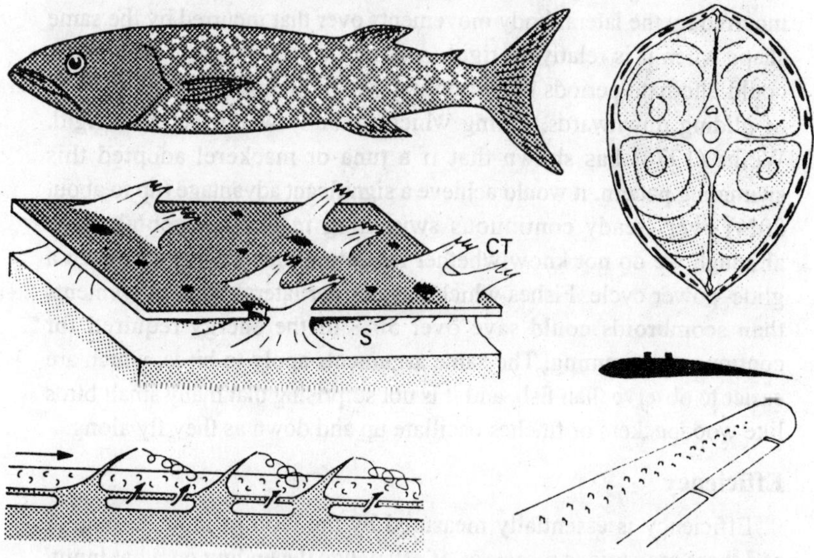

Fig. 6.21: The specialized integument of the castoroil fish* Ruvettus. *Upper left: ctenoid scales (white dots) form a regular pattern over the surface. Upper right: transverse section of body showing sub–dermal spaces (black). Middle; stereogram of integument showing ctenoid scales (CT) and sub–dermal spaces (S). Anterior to left. Lower left: assumed operation of integumenty injecting momentum into boundary layer (see text). Lower right: vortex generators on aircraft wing.

As well as vortex generators, *Ruvettus* also has another system for drage reduction by injecting fluid into the boundary layer. The bases of the ctenoid scales form pillars suporting the outer skin away from the connective tissue layer over the muscle, and so the fish has a system of spaces under its skin. These are full of sea water, and communicate with the boundary layer by openings facing obliquely backwards. As the body oscillates, water will alternately be sucked into, and ejected from, these sub–cutaneous spaces, and so *Ruvettus* presumably once again stabilizes its boundary layer. *Ruvettus* certainly has the most complex integument of any fish yet examined, although other fishes, like the salp–eating stromateoid *Tetragonurus*, have analogous injection systems. Similalr fluid injection into the boundary layer is common in aircraft practice, as for instance in 'blown' flaps and control surfaces, fed from the jet engines.

Quite a different approach to the problem of decreasing drag and thus increasing range (or saving fuel) arises from the increased drag incurred by the lateral body movements over that incurred by the same shape when it is relatively rigid. Any fish which is denser than water could alternate periods of swimming obliquely upwards with periods of gliding downwards, during which its body would be more rigid. Weihs (1974) has shown that if a tuna or mackerel adopted this swimming pattern, it would achieve a significant advantage (up to about 20%) over steady continuous swimming in one horizontal plane although we do not know whether migrating tuna do swim in such a glide–power cycle. Fishes which have larger lateral body movements than scombroids could save over 50% of the energy required for continuous swimming. The same arguments apply to birds, which are easier to observe than fish, and it is not surprising that many small birds like woodpeckers or finches oscillate up and down as they fly along.

Efficiency

Efficiency is essentially measured by the ratio of input to output, and there are various measures of efficiency depending on what input/output properties to be considered. One can compare the swimming fish to his own machines, such as motor cars, which gain power (as does the swimming musculature) by converting chemical to mechanical energy, and then have to transmit this power to the road to drive themselves along. Engineers define two principal kinds of efficiency for such machines.

1. Mechanical efficiency (η_m): $\dfrac{\text{brake horsepower}}{\text{indicated horsepower}} \times 100.$

This is the ratio of the power developed at the pistons vs, the power available at the output shaft after frictional losses in bearings, gear trains, etc. In practice, the mechanical efficiency of a motor car is usually around 80–85%.

2. Thermal efficiency (η_t): $\frac{\text{work done}}{\text{heat supplied}} \times 100$.

This is the ratio of one kind of energy, heat energy supplied during the combusrion of fuel in the cylinders, to the useful energy obtained in the form of work done by the engine driving the wheels. Diesel engines have thermal efficiencies around 40%, petrol engines rather less.

Overall efficiency, *H,* will be the product of these two, so that the overall effeciency of a well–designed diesel lorry will be something like 30%. How do fish compare with this? In the case of the fish, *N,* may be considered as the product of the efficiency of the caudal propeller (usually called η_p), and that of the conversion of chemical to mechanical energy in the muscles (η_m). Values for *H* during sustained aerobic swimmong by salmonids work out between 5% at low speed to around 20–22% at the maximum speeds sustainable. Calculatons for a different biological machine, man, during sustained activity, give similar values.

Obviously, if we can measure *H* from oxygen consumption during swimming, and if we knew either η_m and η_p, we could obtain both. But unfortunately η_m has not been measured for any fish owing to difficulties with the way in which myotomal muscles insert into compliant myocommata. Still, fish muscles are fairly similar to those of other animals in which the measurements enabling η_m to be determined have been made, and by substituting these values, Webb (1971) calculated that propeller efficiency in trout rose to a maximum around 75% at a swimming speed of 2 body lengths per s. It seems very probable that scombroids and other high–speed fish, swimming by oscillating lunate tails foils, will have a higher η_m than this, and thus that overall efficiency *H* will be higher than for trout. So fish compare quite favourably with our transport designs, and indeed, with ourselves!

Speed

Fish vary widely in overall size (from larvae only a few millimetres long to the12 m whale–shark), and the speed of only a very few within this range has been studied, but it seems generally true that fish smaller than half a metre long are capable of maximum speeds of about 10 times their body length per second. It does not seem likely that larger fish are capable of such a performance (which would give the whale–shark a maximum speed of 422 kph!), although they can travel at relatively high speeds. Accurate measurements of the speed of large fish are hard to come by, but measurements have been made up to 42 kph of line run–out on hooked tuna about 1 m long. A novel approach to the estimate of maximum swimming speeds has recently been made by Wardle (1975) who reasoned that the maximum contraction velocity of the white muscle could be related to swimming speeds, since swimming speed is closely related to tail beat frequency. His observations have only so far been made on the muscles of fish under a metre long, but they give maximum swimming speeds in good agreement with those observed. In the case of tuna where maximum contraction speed of the white muscle is similar to that of other fish, but whose maximum swimming speeds seem well established as much higher, it is possible that this is achieved by reducing body flexure and maintaining a constant angle of attack of the caudal tail as it sweeps across. Wardle's concept of stride length is an interesting one, but the case of tuna requires further investigation.

CHAPTER 7
CLIMBING, JUMPING AND CRAWLING

Climbing Adaptations

The animals that inhabit trees are said to lead an arboreal or scansorial mode of life. Climbing on the part of the arboreal animals is because of the necessity to procure food and safety. Relatively feeble and defenceless creatures may take to the trees for safety and retreat. There, they get abundant and easily obtainable food. From the stand-point of view of adaptation of arboreal forms, three different types of adaptations can be distinguished in them, as follows:

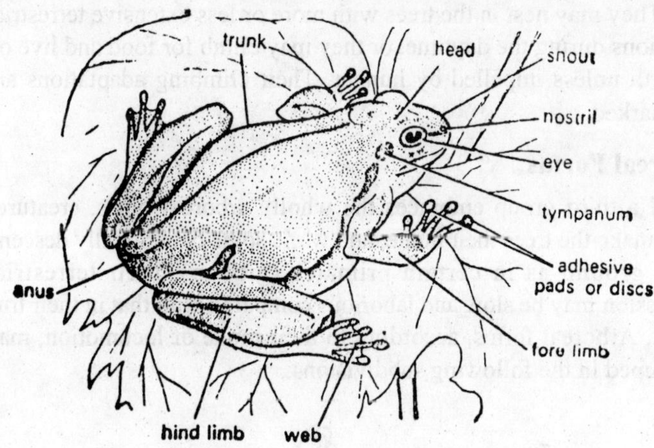

Fig. 7.1: Hyla

Categories of Scansorial Animals

On the basis of bionomic classification (*i.e*, habitat) scansorial animal may be of three types:

Wall and Rock Climbers

The classification of scansorial animals from the stand point of their adaptation groups them into three subdivisions, of which the first are the wall and rock climbers. These are not necessarily tree-inhabiting at all, but are, like the gecko lizards, well suited for climbing on the walls of buildings as well as on similar surfaces in nature. The geckoes are, however, a very old and widely distributed group, and the range of their individual adaptation is great, hence it may well be that their scansorial adaptation is after all a response to arboreal life, and that the peculiar structure of their climbing organs rendered subsequent rock-climbing possible. Among mamals there is a genus of flying squirrels limited to high altitudes at Gilgit and perhaps in Thibet, and thought to live on rocks, perhaps among precipices. Here, again, we have a form whose ancestry may have been aboreal, but if not, it would afford an interesting instance of volant adaptation without an intermediate arboreal habitat.

Terrestrio-Arboreal Forms

The second category, the terrestrio-arboreal, embraces a number of carnivores, rodents, and insectivores which while capable of climbing, nevertheless are still perfectly at home upon the ground beneath the trees. They may nest in the trees with more or less extensive terrestrial excursions during the daytime, or they may climb for food and live on the earth unless impelled by hunger. Their climbing adaptations are very marked.

Arboreal Forms

Still a third group embraces the wholly arboreal types, creatures which make the trees their home, and while some occasionally descend to the ground as in certain primates (gibbon), their terrestrial progression may be slow and laborious compared with that in their true habitat. Arboreal forms, according to their mode of locomotion, may be grouped in the following subdivisions:

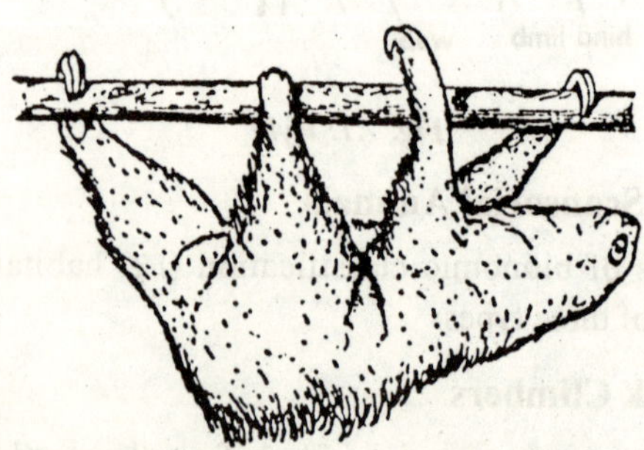

Fig. 7.2: Sloth, Cholaepus, walking suspended from a branch.

1. *Branch runners*, like the squirrels, marsupials, lemurs, and chameleons, which live and progress on all fours on the upper surface of the branches. The group embraces, nevertheless, some instances of very perfect arboreal adaptation, as the great majority of tree-dwellers are here included

2. *Forms suspended beneath branches* : The sloths for instance, are so constituted that they cannot walk upon the branches but rest and move suspended from them by the powerful recurved claws of all four limbs. Sometimes when quiescent, if a convenient branch lie sufficiently near, the sloth may rest his back on it and relax the hold of one or more of his feet, but the inverted position is rarely reversed. On the ground the animal progresses with the utmost difficulty. The bats should perhaps also be included under this head, as they rest suspended by the hind limbs, head down. The same position of rest is assumed by the so-called flying lemur, *Galeopithecus*, really not a lemur at all but an insectivore.

3. *Forms swinging by the fore limbs* (brachiators, Gr. *βραχlων*,, arm). These forms show a very remarkable method of progression by means of the fore limbs, swinging with great speed and accuracy from limb and from tree to tree. The hind limbs are comparable to those of the tree-dwelling marsupials and the creatures rest and progress on the tops of the branches at times, although the fore limbs are almost the sole organs of more rapid locomotion. Here long many of the primates, more especially the great or manlike apes, and our prc-human ancestors.

Modifications

Body : Climbing adaptation, as in the other lines of adaptive radiation, implies certain body modifications as well as those of limbs. Body contour is of little moment in climbing, but strengthening of chest and ribs and of shoulder and hip girdles is of importance. Nevertheless, in thoroughly arboreal types the section of the thorax anteriorly is subcircular, and the ribs are much curved, in contrast with the compressed thorax and flat anterior ribs of quadrupedal running types. The ribs, especially in the sloths, are numerous and afford ample support to the contained viscera in their inverted position. The dorsolumbar series of vertebrae is often elongated, especially in the tree-sloths of the genus *Cholaepus*, where the number has apparently been increased from about nineteen (normal for the order) to from twenty-five to twenty-seven as a response to arboreal need. The same an arboreal rodent, has twenty-three as compared with the normal nineteen, and *Dendrohyrax*, the only arboreal ungulate now alive, has six more than its terrestrial, hoofed relatives.

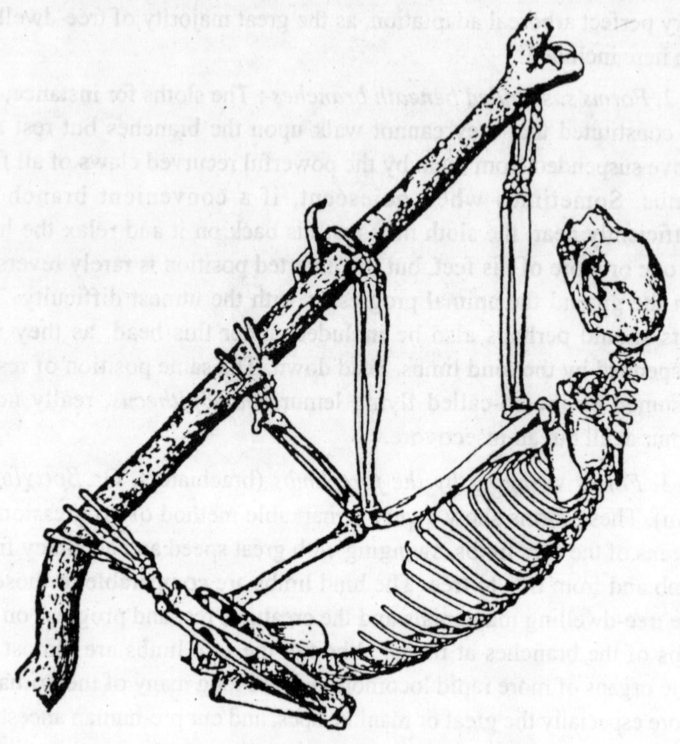

Fig. 7.3: Skeleton of sloth, Cholaepus didactylus.

Limb Girdles : The shoulder girdle especially is strong in that both elements, the clavicles and scapulae, are well developed although in terrestrial types the clavicles tend to diminish, even in closely related forms, and may entirely disappear as in cursorial quadrupedal forms. The fore-and- aft swing of the limb of a deer or horse would be distinctly limited by a clavicle, but in a climbing type whose arms are subjected to much more varied and violent strain the clavicle is very essential, as it withstands the compression of the powerful breast muscles. The scapula is also well developed, but not exceptionally so.

Pelvic Girdle : The ilium or hip-bone especially shows modification in such types as the sloths and primates as it is broadened out as a support for the viscera. This is markedly true of the sloths, whose inverted posture necessitates additional support, since the mesenteries or membrances which sling the intestine to the dorsal wall lose much of their efficiency when the body is erect or inverted.

Limbs : In contrast with the cursorial types, it is the proximal limb segments which now elongate, especially in the suspended and brachiating forms, those of the sloths again being very long, while in the great apes the relative length bears a direct ratio to the creature's climbing powers, reaching the extreme in the gibbons (*Hylobates*) whose arms are so long that the knuckles of the hand touch the ground when the animal stands erect. The progress of the gibbon from branch to branch and tree to tree is remarkable. Climbing forms are generally plantigrade, some of the raccoons secondarily so. In certain lemurs (*Tarsius; Galago*) the tarsus may be elongate, but this is probably due to the fact that the creatures leap as well as climb and the elongation of this segment is a response to the former need rather than to the latter.

Feet : The feet of arboreal animals may be either prehensile, that is, grasping, with more or less opposable digits, or nonprehensile. In the *non-prehensile type* the claws may to be well developed as in the squirrels or the cats, giving a fairly tenacious hold. In the Canada tree-porcupine (*Erethizon*) the plantigrade feet are armed with long curved claws, in addition to which the soles bear spines and tubercles which aid in climbing.

Adhesive pads either on the tips of the digits or on the soles of the feet occur several isolated instances, such as the tree-frogs, geckoes and *Dendrohyrax* among mammals. The frogs are aided by a stickily secretion of their pads. "Tree frogs, when hopping on to a vertical plane of clean glass, slide down a little probably until the secretion stiffens, or dries into greater consistency..........Wet leaves or moist glass-walls afford no hold. The adhesion of these frogs is assisted in most cases by their soft and moist bellies, just as a dead frog will stick to a pane of glass".

The geckoes, by means of their adhesive digits, climb up absolutely smooth and vertical surfaces, or, back downward, along a whitewashed ceiling. The apparatus, Gadow says, is complicated in its minute detail, but very simple in principle. The adhesion is effected not by sticky matter, but by small and numerous vacua.

Dendrohyrax, the tree-hyrax, is allied to the coney of Scripture. The tree-hyraxes frequent the trunk and larger branches of trees, sleeping in holes high up in the big trees, especially, according to Roosevelt's observations, the cedars. The adhesive organs have been described by G.E. Dobson, who says that these animals are enabled to climb perpendicular walls and trees without the use of claws. The thickly padded, tuberculated soles are drawn up by certain flexor muscles, thus leaving a partial vacuum by means of which the animal retains its hold.

The primitive type of prehensile foot has been developed in the two great mammalian groups, that of the marsupial being represented by the opossum (*Marmosa*), that of the early placentals by the creodonts, the archaic flesh-eating mammals, the foot of which has been shown to be a terrestrial modification of a grasping type.

Feet of the *prehensile type* are found today in the marsupials and primates. In the former group it is the hallux or great toe which is offset so as to oppose the fourth digit, the second and third being bound together in a common integument (syndactyly) and so slender that their combined strength about equals that of the outermost of *fifth* digit. In marsupials which have become terrestrial the offset great toe has become vestigia or may entirely have disappeared, as in the kangaroos. In the primates, while the foot is perhaps most apt to show this opposable first digit, it also exists in the hand, although it is nowhere developed to the degree shown in mankind, wherein the final perfection of the hand as a organs of prehension has developed since its release from the necessity of arboreal locomotion.

Syndactyly (Gr. συν , together, and δά*KTU*λ digit) has already been referred to as occurring in the marsupials, and even such as are no longer tree-inhabiting, like the kangaroo, still exhibit this feature in unreduced condition. It doubtless arose primarily, however, as an arboreal adaptation. The koala shows a rather remarkable modification for climbing, for the foot has a long, widely offset, clawless great toe, syndactylous second and third toes, which are clawed, and powerful clawed fourth and fifth toes, the former being the longer. The hand, on the contrary, has five subsequal digit, all of which bear sharp claws; but two digits, numbers 1 and 2, oppose the other three. Its clinging powers are so great that event death will not dislodge the creature from the tree in which it is shoot.

Among reptiles, the true African chameleons (*Chamaeleon*) exhibit remarkable syndactyly, as it extends to both fore and hind feet. On the hand the first three fingers form the inner bundle and are opposed to the outer two which are likewise syndactylously bound. The foot is similar but reversed, in that the inner bundle contains two, the outer one three digits. These very admirable grasping organs are supplemented by a prehensile tail, so that the creature is very firmly anchored in position, which is rendered necessary perhaps in part by its method of securing insect prey by the unerring aim of the enormously extensile tongue.

In the so-called scansorial birds such as the parrots, woodpeckers, and the like, the outermost toe has been rotated backward in such a way that it and the hallux oppose the second and third toes, the fifth, as in all birds, being absent. This gives a very firm grasp for the actual grip of a branch as in the parrots, or, reinforced by strong claws, enables the animal to cling to the roughened bark of a tree trunk. In the parrots, woodpeckers, and cuckoos the rotation of the outer toe is permanent and the foot is called zygodactylous (Gr.ξνγὸν, yolk); certain other owls, etc, may turn it backward or not at will.

Digital Reduction

While arboreal forms usually have need of all of their digits, occasionally one sees *digital reduction*. The foot of the koala, with syndactylous second and third toes, functions as fourtoed, even though consisting structurally of five; certain of the primates (lemurs), on the other hand, some of which resemble the koala superficially very much, have actually lost the second digit so that the opposability of the first in grasping a limb is unimpeded. In the lemur (potto, etc.), the fourth digit is the largest as in the koala. Digital reduction is also seen in the tree-sloths, the two-toed sloth *Cholaepus* having but two in the hand and three in the foot, while in the three toed sloth *Bradypus* there are three in each, and the hand and foot are both somewhat elongated, especially in the powerful hook-like claws which, like the feet of the koala, retain their grip on the length even after the animal has been shot.

Fig.7.4: Foot of a woodpecker, **Picus viridicanus,** ***showing fourth toe reversed for grasping.***

Tail : The tail may be prehensile or not as in the case of the feet. If non-prehensile, there are ectodermal spines or scales on the under side, as in the flying squirrel *Anomalurus*, which prevent the animal from slipping down. The same effect is produced in the woodpecker by stiff spiny feathers which are braced against the tree trunk to which the creature clings. The posture is familiar, and enables the bird to drill into the wood for the grubs upon which it feeds, or to excavate cavities for its nest or for the storage of food.

Prehensile tails are found in a number of unrelated instances, as, for example, the chameleon lizards which have been mentioned, the opossums, the tamandua which is one of the anteaters, and certain of the New World monkeys (Cebidae) such as the spider monkey, the howlers, and the capuchins. Where the prehensile powers are well developed the tail is naked on the under surface near the tip. One of the most perfectly adapted of these forms is the spider monkey, *Ateles*, in which the tail is highly prehensile and functions as a "fifth hand". Perhaps as a correlation with this excellent grasping organ the real hands have lost the thumb, but the four long digits which remain form a splendid hooklike device for suspending the body. Not all South American monkeys have a prehensile tail; on the other hand, none of the Old World forms do, so that its presence is diagnostic of a New World ape.

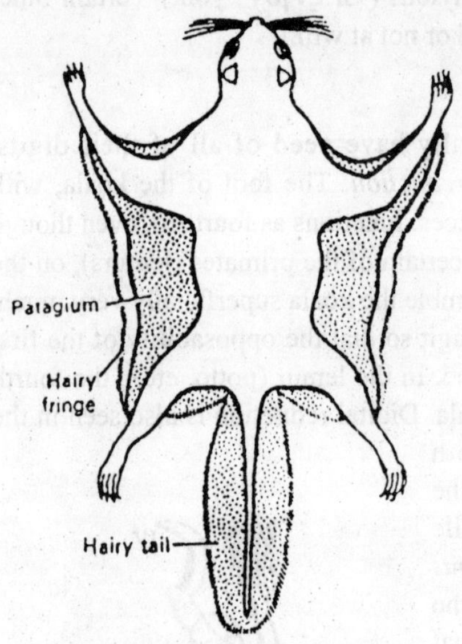

*Fig. 7.5: Flying squirrel (*Scruropterus*)*

Development of Accessory Organs

Due to scansorial habits various accessory climbing organs might be developed. These include spines and tubercles on the forearm in some lemurs (*Hapalemur griseus*). In *Lemur catta* there is a specific climbing organ, which is composed of a patch of hardened skin on the forearm, which projects to a large extent. Both these organs have glands connected with them.

Jumping

Animals jump for many different reasons. Lions jump onto prey to

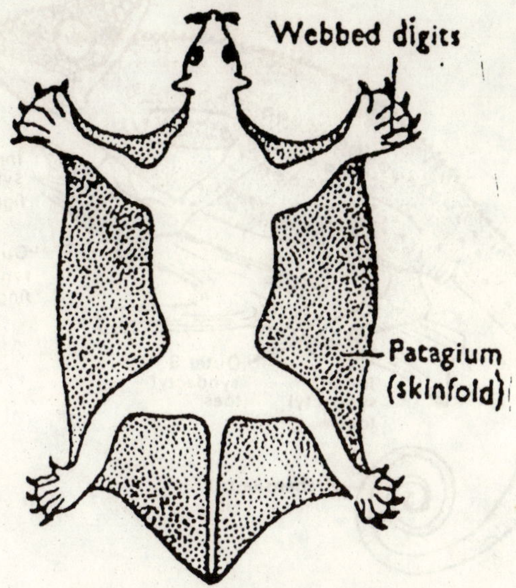

Fig. 7.6: Flying lemur (*Galeopithecus*).

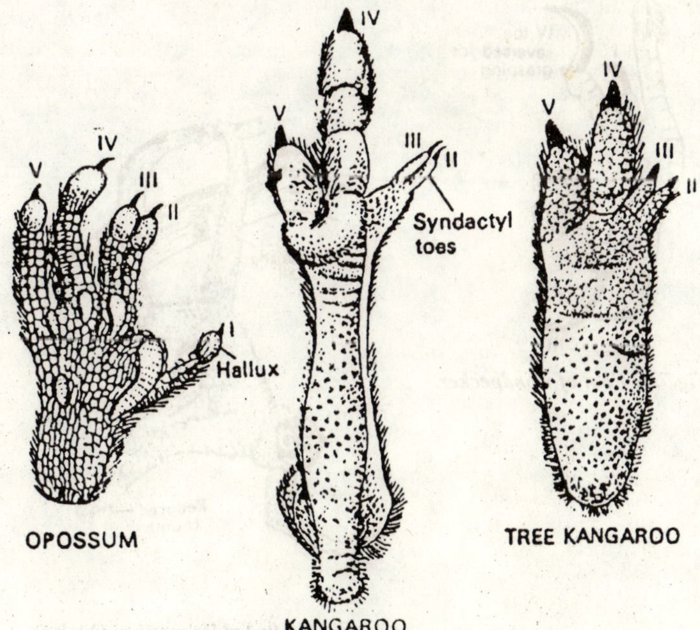

Fig. 7.7: Hind feet of marsupials

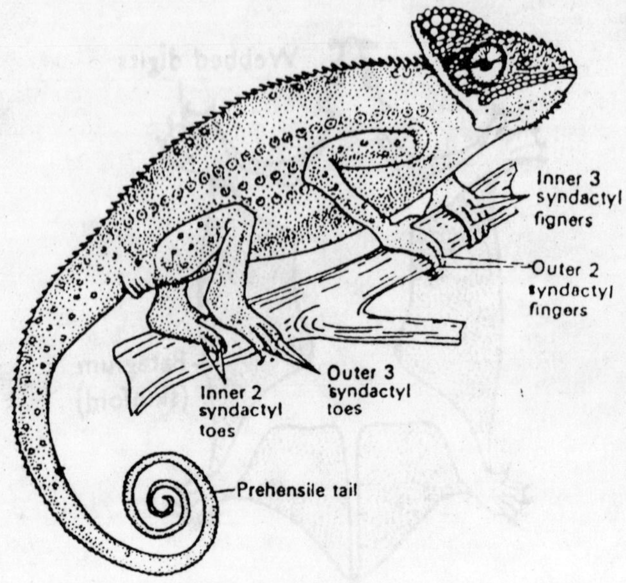

***Fig.7.8: Syndactyly of* Chamaeleon**

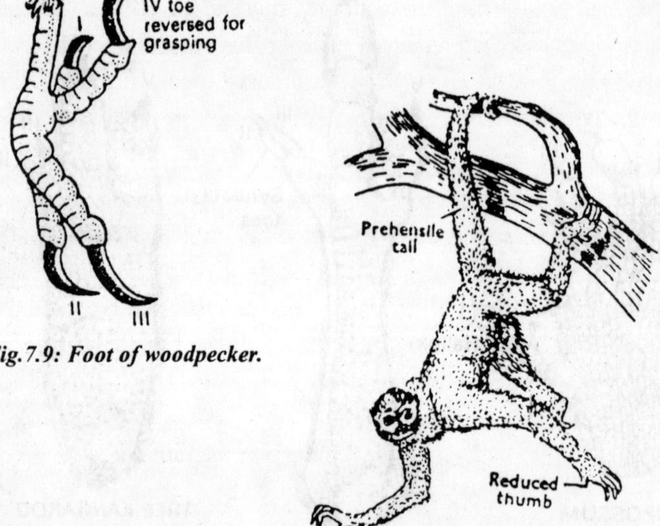

Fig.7.9: Foot of woodpecker.

Fig. 7.10: Spider monkey (*Ateles*) pentadactylous.

bring them down. Frogs and grasshoppers jump in directions that are hard to predict, to escape from danger, Squirrels and bushbabies jump between the branches of trees. Fleas resting on leaves jump when they senses the warmth of the body of a passing mammal of bird: with luck, they may be able to attach themselves to the potential host.

Fig. 7.11 : Bushbabies (this one is Galago senegalensis) leap between branches in their forest habitat.

The secret of a good jump is speed. A fast takeoff will carry an animal closer to its target or farther from its enemy. As we will see, it turns out that the smaller the animal, the more difficult it is to reach a good jumping speed by the action of muscle alone. The muscles of smaller creatures need an assist, and for this reason we will find the most varied, and the most bizarre, jumping mechanisms in insects. Indeed, there is one type of beetle that manages to jump without even using its legs.

What ever the function of the jump, there is a simple relationship between the speed and angle at which the animal takes off and the distance it covers or the height it clears. It is particularly easy to calculatc thc height of a vertical jump. The kinetic energy of a body of mass *m* travelling at speed *v* is $\frac{1}{2}mv^2$. If the body rises through a height *h*, it gains potential energy *mgh* (g is the gravitational acceleration). In a vertical jump all the kinetic energy at takeoff is converted into potential energy at the top of the jump.

$$mgh = \frac{1}{2}mv^2$$

The height is therefore

$$h = \frac{v^2}{2g}$$

In these equations, h is the increase in height to the body's center of gravity from the instant of takeoff to the top of the jump. For example, in a standing jump a human athlete might leave the ground with a vertical velocity of 3 meters per second, and the center of gravity would rise about $3^2/(2 \times 10) = 0.45$ meter. (The gravitational acceleration is about 10 meter per second squared.) The body's center of gravity is a little above the hips, so it would be about 1 meter from the ground at takeoff and would rise to 1.45 meters at the top of the jump. Senegal bushbabies are very much smaller than we are, weighing only 250 grams, but these tree-dwelling primates are very much better jumpers. A pet bushbaby was once observed jumping from the floor to the top of a 2.26-meter door. Its center of gravity could have been no more than 0.25 meter from the floor at takeoff, so the height gain h was 2 meters, which requires a takeoff speed of more than 6 meters per second.

People often rates fleas as the most remarkable of animal jumpers. Though they are so small, fleas can jump heights of at least 130 millimeters (5 inches). In absolute values their jumps are low in comparison to human jumps, but relative to body length they are much higher. Excluding the legs, flea's body is only about 2 millimeters long and a man's body (again excluding the legs) about 1 meter. Therefore it might be argued that a flea's 130-millimeter jump is equivalent to an impossible 65-meter human jump.

The fallacy of that argument was pointed out in 1950 in an article by the famous muscle phyiologist A.V. Hill. Hill wanted to figure out how the height of a jump depended on the size of a jumper. He would then be able to compare the expected jump heights for different-sized animals, including a flea and a man. He first argued that the work available for a jump should be proportional to body mass. The work done in a single muscle contraction is the force multiplied by the shortening distance. That force is proportional to the cross sectional areas of the muscles, and we can expect the muscles to be able to shorten by amounts proportional to their initial lengths. Therefore, the work done in a single contraction is proportional to the cross-sectional area multiplied by the length—that is, to the volume of the muscle. If the animals being compared have equal proportions of muscles in their bodies, we can expect the work available for a jump to be proportional to body mass.

The work required for a jump to height h equals the potential energy gain, *mgh*. If this work is proportional to body mass *m* in animals of different sizes, all should be able to jump to the same height *h*. The flea that raises its center of mass 0.13 meter in a jump is doing much less well than a human who rises 0.45 meter. We should not ask why fleas jump so high, but why their jumps are so feeble.

Hill's theory said that different-sized animals should jump to equal heights, but ignored a problem faced by very small jumpers. People, bushbabies, and fleas all prepare for a jump by bending their legs. To take off, they extend the legs rapidly, accelerating over a distance of about 0.4 meter for people, 0.16 meter for bushbabies, and only about 0.5 millimeter for fleas. Humans accelerate from rest to about 3 meters per second at takeoff, so the average speed, as the legs extend, is about 1.5 meters per second; thus the 0.4-meter acceleration distance is covered in about 0.27 second. Films confirm that this estimate is about right. The simple equation on page 59 tells us that a flea jumping to a height of 0.13 meter would have to take off at 1.6 meters per second. (It would actually have to take off rather faster because the equation ignores the effect of air resistance, which slows small jumpers much more than large ones). As the flea accelerates during takeoff from rest to over 1.6 meters per second, its *average* speed would be a little over 0.8 meter per second: it would cover the 0.5-millimeter takeoff distance in only 0.6 milliseconds.

No muscle can complete a single contraction in so short a time. Some midges beat their wings at 1000 cycles per second; a midge has only 0.5 millisecond for each up or down stroke. Such frequencies depend on the wings and their muscles going into a state or resonant oscillation. A jump requires a single contraction to extend the legs, and no known muscle could do that in as little as 0.6 milliseconds.

The flea's jump is possible only because the animal has a built-in catapult, a tiny block of rubber like protein at the base of each in leg. These blocks are too small for mechanical testing but seems to be made of a protein called resilin that is also found in the tendons of some dragonfly wing muscles and in the hinges that join locust wings to the body. Tests on these larger (but still very small) pieces of resilin show that its properties are very like those of soft rubber. It can be stretched to three times its initial length and returns nearly all the energy in its elastic recoil.

When children use catapults they first stretch the rubber, storing up elastic strain energy in it. If the catapult is a strong one the child may be unable to stretch it fast, but a slow stretch will store the elastic strain energy equally well. When the catapult is released, the rubber recoils very rapidly indeed, projecting the stone much faster than it could have been thrown. Work done slowly on a catapult is returned much faster. Fleas store up strain energy in their resilin by a relatively slow contraction of their leg muscles lasting, it seems, about 100 milliseconds. A trigger mechanism releases the resilin, which recoils, extending the legs in only 0.6 millisecond and throwing the insect into the air.

The strong hind legs of a locust launch it into the air at a speed of up to 3.2 meters per seconds, fast enough to propel the insect about 0.5 meter (20 inches) if it jumped vertically. In fact locusts jumps at an angle and rise less high but can clear horizontal distances of about 1 meter. Like fleas, they must rely on elastic recoil, but instead of using a tiny block of resilin as a catapult they store elastic strain energy by means of a tug-of-war between two muscles.

About halfway along the locust leg is a joint called the knee, by analogy with the human leg. Kept bent when the animal is standing, the joint extends very rapidly to throw the animal into the air. A large extensor muscle filling most of the locust's fat thigh extends the knee and does most of the work needed to power the jump. A much smaller flexor muscle bends the knee and prevents the joint from extending too soon. Like many other insect muscles, these are pennate; the fibres converge on a central tendon made up of material very like the exoskeleton that encloses the whole body. When the insect is preparing for a jump, both muscles contract. The extensor is much the stronger but, while the knee is bent, its tendon runs much closer to the axis of the joint than does the tendon of the flexor. This difference of lever arms enables the relatively weak flexor to hold the knee bent, against the stronger pull of the extensor. The large forces that are developed at this preparatory stage compress the semilunar processes (flexible parts of the leg skeleton, close above the knee) and stretch the extensor tendon, storing elastic strain energy in both. To jump, the animal suddenly relaxes the flexor muscle, allowing the knee to extend, which it does very rapidly, driven by the elastic recoil of the semilunar processes and the extensor tendon.

One of the most curious of jumping insects is the click beetle, which can jump twice as high as a flea without so much as bending a leg. The click beetle flexes its back instead. These beetles are often found crawling on blades of grass, but drop to the ground if disturbed. If they land upside down and cannot right themselves quickly, they jump by a sudden jackknife movement of the back that may throw them as high as 0.3 meter (12 inches) off the ground. The jump has been filmed using a special camera running at 3100 frames per second. (Ordinary films are taken at only 18 to 24 frames per second). Even at this exceedingly high rate, the films showed the beetles accelerating to takeoff in just two frames, or 0.6 millisecond. As in flea and locust jumping, the rapid acceleration of a click beetle jump is made possible by a catapult mechanism.

All these insects take off so quickly that there is scarcely time for muscles to start shortening during takeoff; the jumps are almost entirely powered by elastic recoil. Larger jumpers such as bushbabies and human beings have time to shorten their muscles during takeoff, but these more muscle-dependent jumpers also benefit from tendon elasticity. If you ask people to do a standing jump, they will generally bend their knees and then immediately extend them. If they pause with knees bent and then jump, they cannot jump so high. The reason is that muscles can exert more force while being stretched than while shortening or holding constant length. The muscles that extend the legs at takeoff also stop the downward preparatory movement; as they stretch to counteract the binding of the knee, they are able to develop large forces. These forces stretch the tendons in turn, storing strain energy that is returned in elastic recoil, increasing the height of the jump.

Swinging Through Trees

Squirrels and many monkeys travelling through the forest run along the tops of horizontal branches much as they run on the ground, but leap as necessary from one branch to another, often moving from tree to tree; Bushbabies and lemurs move through trees in a different way, clinging to a vertical branches and leaping from one to the next, But some tree dwellers have found an efficient way to move through the forest while barely using their legs at all—these animals propel themselves by the strength of their arms.

Apes and a few monkeys swing by their arms below branches, using the technique called brachiation. Particularly good at brachiation are gibbons, Southeast Asian apes with very long arms and relatively short bodies and legs. Gibbons often have to travel along slender branches to reach fruit. Their method of travel is especially effective for traversing a slender branch, because walking on top of a thin branch is a feat of balance like tightrope walking, but swinging below a branch presents no such problems.

Children make swings go higher by "pumping:—bending and extending their legs at appropriate stages of the swing. Gibbons make use of the same technique, but instead of using it to go higher, they speed up their swinging through the trees. As they swing under a branch they bend their legs, raising their potential energy. At the top of the swing, as they reach out to catch the next branch, they extend their legs again. If at this stage they are flying through the air between branches, the extending of the legs lowers the feet and raises the trunk a little but leaves the path of the center of gravity unaltered. And if the center of gravity's path is unchanged, the body's energy is unchanges as well. Bending the legs increases the body's energy and extending them leaves it unchanged, so the effect of the whole cycle of pumping movements is to increase the energy, making the animal travel faster.

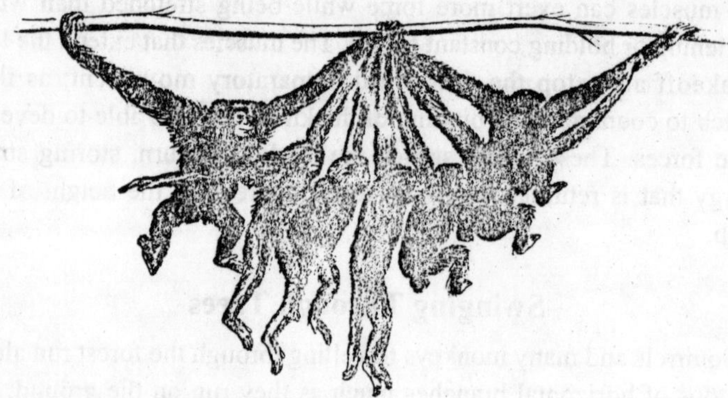

Fig. 7.12 : A branching siamang (hylobates syndactylus, a species of gibbon) builds up speed like a child "pumping" a swing.

A child pumping a swing stand up as it passes through the vertical and bends the knees at the top of the swing. This stand-and-bend sequence seems the reverse of the gibbon's action if you look only at the legs: the child is straightening its legs when the rope is vertical, but the gibbon is bending its legs when its arm is vertical. However, the effect on the center of mass is the same in both cases: it is being raised at the stage when rope or arm is vertical.

Gripping a Smooth Surface

Gibbons spend their lives in the trees, seldom or never descending. In contrast, squirrels and many monkeys often come down to the ground, where they find much of their food. To get back up a tree, these animals need to climb a vertical trunk. In accomplishing their climb, not only must they be correctly balanced to counteract the force of gravity, but they need a way to hang on to the surface. Indeed, for some animals, the difficulty of maintaining a good grip can make getting down trickier than getting up. A adroit as climbing mammals are at maneuvring up and down trees, however, the true masters are some insects and lizards that are able to hold on to smooth surfaces, even when upside down.

To climb a vertical trunk, an animal must pull with its fore limbs and push with the hind. The force exerted by the limbs must balance the animal's weight, so you might think that only vertical forces would be needed. But if the only forces on the animal were its weight acting downward and upward forces on its paws, it would not be in equilibrium: that combination of forces would make it rotate head over heels, for equilibrium, the forces on the feet must slope. The force on the fore feet may be more vertical and force on the hind more horizontal, or vice versa. The (downward) weight is then balanced by the (upward) components of the forces on the paws; the (forward) component of force on the fore paws is balanced by the (backward) component on the hind paws; and forces that would rotate the animal clockwise are balanced by others that would rotate it counter clockwise.

Squirrels have claws that dig into bark and help them climb. The hind claws are pushed into the bark, so the animal does not have to rely on friction to prevent it from sliding down the tree. The fore claws are hooked into the bark, enabling it to pull on them. Claws on the hind feet are not essential for their kind of climbing unless the surface is slippery, but claws on the fore feet are essential. The squirrel could not pull on its fore feet if its claws were not dug in.

The opposite is true for coming down a tree trunk headfirst: the hind claws must be hooked in. Not only does the animal need hind claws, but it also must be able to turn its feet back to front so that the claws can pull in the required direction. Squirrels can do this, and so can kinkajous and various other mammals that live in trees. Domestic cats cannot reverse their hind feet, though the South American tree-living cats called margays can. A domestic cat that climbs up a tree easily may have great difficulty coming down.

Fig. 7.13 : This kinkajou's (Potos flavus) ankles can be reversed to enable it to hang by its hind feet.

Reversing the hind foot as squirrels and kinkajous can do requires a very mobile ankle. To understand what is required, sit on the floor with the soles of your feet flat on the floor in front of you. You will be able to turn your feet inward so that the soles of the feet are vertical, each facing the other. To do as squirrels and kinkajous do, you would have to be able to turn them through a further 90 degrees until the soles were horizontal, facing upward.

Coming down tree trunks is relatively easy for birds because their gripping feet have one or more backward-pointing toes, so there are claws curving in the right direction to hook into the bark whether the bird is running up or down the tree. Creepers (known as tree creepers in Britain) work their way up tree trunks, searching for insects in the crevices in the bark, then fly to get down again, but nuthatches run down tree trunks as well as up. Creepers and wood peckers have stiff tail feathers, which they press against the tree trunk, using them to supply the force that climbing squirrels get from their hind feet.

Only one group of small monkeys, the marmosets, have claws, and even they lack a claw on the big toe. Other monkeys have finger nails and toenails, like humans, and rely on friction for their grip. To be able to pull with their arms (which they must do to climb vertical trunks), they must be able to reach well around toward the back of the tree. They need long arms to be able to climb even moderately large tree trunks, and they cannot climb very thick ones.

The accomplishments of monkeys pale beside those of creatures that seem able to climb the smoothest surfaces and even defy gravity by walking upside down. Flies and other insects can walk up vertical walls and even windowpanes that give no chance for claws to hook in. They can also walk upside down on the ceiling. Small lizards called geckoes are often seen hanging upside down on the ceilings of houses in the tropics, and tree frogs can cling to vertical surfaces by means of the soft pads on their feet. All these animals depend on adhesive feet.

Aphids (greenfly and similar bugs) suck out the juices from plants through mouthparts that resemble a hypodermic needle. They have to walk on smooth plant surfaces, including vertical stems and the undersides of leaves, and they have to adhere strongly enough not to be blown off or shaken off as the leaf waves in the wind. The force of adhesion has been measured by putting an aphid on a clean piece of glass on the pan of a sensitive scientific balance. The balance is adjusted to read zero with the glass and aphid in place, then a thread that has been dipped in glue is dangled over the aphid so as to stick to it. The investigator pulls gently upward on the thread, making the balance register a negative load. The balance reading at the instant when the aphid detaches is about –6 milligrams (thousandths of a gram) for aphids weighing 0.3 milligram. These aphids can hold on to glass with a force 20 times their own weight.

That experiment measured the force at right angles to the glass needed to detach the aphid, but aphids can resist forces at least as large when they act parallel to the glass surface, as a gust of wind might. This has been demonstrated by letting aphids crawl up the sides of glass centrifuge tubes and then spinning them at various speeds and observing whether they were thrown to the bottom of the tube.

Aphids have tiny claws on their feet, but you would not expect these to be any use for attaching to smooth glass surfaces. Thus it is not surprising that aphids can still adhere after the claws have been cut off with a fine scalpel. The attachment organ is a spongy pad on the foot. Various experiments have been performed to find out how it works.

One possibility is that the pads attach by suction. If there were muscles that could lift the center of the pad, the pressure under the pad would be reduced and it would hold on like a sucker. However, no likely muscle has been found, and the force of adhesion is affected very little by anesthetizing the aphid with carbon dioxide, which makes muscles relax. Furthermore, and this is the conclusive evidence, aphids remain firmly attached in a vaccum.

An important clue comes from the observation that aphids lose the ability to adhere to smooth surfaces after walking for a while on silica gel, which absorbs water strongly. Aphids that had walked on silica gel for 15 minutes took about 30 minutes to recover their power of adhesion—unless they were placed on water-soaked filter paper. Allowed to walk on this moist surface, they recovered within 2 minutes. Adhesion seems to depend on water.

The diagram at right shows how the mechanism probably works. There is a thin film of water between the foot and the surface to which it is attached. Pulling the foot away from the surface, widening the gap, would draw the water surface in around the edges of the foot. But because this inward pull is resisted by surface tension, a large negative pressure can develop in the water. The negative pressure holds the aphid in place.

The feet of beetles and some other wall-climbing insects are very different from those of aphids. Instead of having spongy pads, their feet are covered by a dense pole of very fine, flexible setae (bristles) with wider ends. Such setae can make exceedingly close contact with and solid surface, whether it be smooth or rough, and it has been suggested that they attach by van der Walls forces, which are forces of attraction between molecules. These forces are very weak unless the adhering surfaces are very close together; the force is inversely proportional to the cube of the distance, so if you move twice as close you get eight times the force. For insect adhesion to work this way, adhering surfaces would have to be no more than about 10 nanometers, or about the diameter of a large protein molecule, apart. (A nanometer is one millionth of a millimeter.) There seems to be no direct evidence that van der Walls forces are involved, but all the other obvious possibilities have been fairly convincingly eliminated: no liquid or adhesive has been observed, beetles remain attached in a vaccum, and they are not detached by an antistatic gun, which would release them if they depended on electrostatic forces.

Geckoes have setae on their feet like those of beetles, only finer. Even the ends, where they widen, are only about 200 nanometers across, and at this width the setae are too narrow to be visible by light microscopy. Electron microscopy is convenient for looking at the setae on beetle feet and is essential for observing the setae of geckoes.

Traveling Waves

Smooth gliding of snakes, without legs, may seem mysterious until we realize what is happening. Seeing a snake for the first time, you might almost imagine that tiny wheel must be hidden under its belly, but the actual mechanisms need neither legs nor wheels. In the most usual method of crawling used by snakes, the body forms an S, or a more complicated wavy shape, and then slides foreward along the wavy path of its own curves. The head slides forward forming new waves, and the tail slides forward losing waves from the rear. It is as if a wavy line were drawn on the ground and then the animal slid forward along it, every part of the body following all the curves of the line.

The mechanism is simple. The snake forms its body into waves, winding between stones, tussocks of grass, or even slight irregularities on the ground. It then makes the waves travel backward along its body, from head to tail. The body can slide past stones or other obstacles more easily than it can rise over them, so the waves remain stationary on the ground and the snake moves forward. This method of crawling works well on rough ground but not on smooth surfaces that have no obstacles for the snake to push against. A snake on a slippery floor gets nowhere.

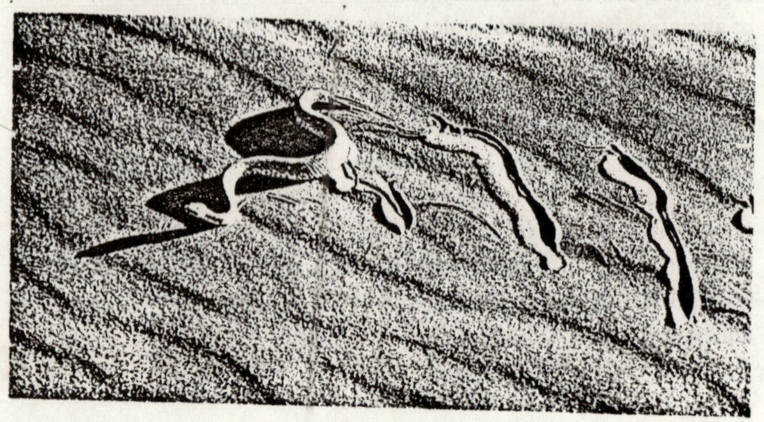

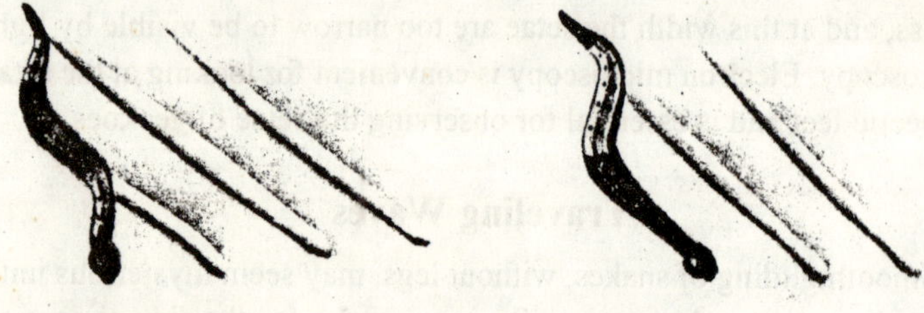

Fig. 7.14

This same mechanism is also effective for burrowing through loose sand: less force is needed to push the snout forward into the sand than to push all the waves of the body backward through the sand. The movements are like those of a swimming eel, and this method of burrowing is sometimes described as swimming through sand.

Only a few snakes burrow, but it seems likely that snakes evolved form burrowing lizards. Many lizards live in sandy deserts, where they are in a danger of overheating in the sun. They can escape this danger by burrowing in the heat of the day and coming to the surface only when it is cooler. Even quite shallow burrowing is enough to make a big difference in temperature. For example, 2.5 centimeters below the surface of bare soil in a hot part of Australia, the temperature fluctuates at midsummer between a daily minimum of about 25°C and a maximum of 53°C, but at a depth of 30 centimeters in the same soil the fluctuation is only between 35 and 38°C.

A lizard with normal-sized legs could fold them against its sides and swim through sand, but they would be rather in the way. Burrowing is easier without such appendages, and probably for this reason many desert lizards have reduced legs or no legs at all. The legless ones are easily mistaken for snakes, but there is more to being a snake than just having no legs. The easiest way to tell a lizard from a snake is to look at the scales along the underside. A lizard has irregular scales there, but a snake has a single row of large rectangular scales.

The wavy motion of snakes and legless lizards may remind you of the side-to-side bending movements that legged lizards make as they run, but there is a fundamental difference. The legless animals form traveling waves: each wave crest starts at the head and travels back along the whole length of the animal. The legged animals, however, form standing waves: crests form at one or two particular points on the body and do not move. Standing waves are fine for running but would be ineffective for burrowing.

The travelling-wave style of crawling that has been described so far is the most usual gait of snakes and is known appropriately as serpentine crawling. It works well on firm, rough ground, but less well on loose sand, which offers no fixed points for pushing on, and this method would be no use at all for climbing vertical tree trunks. For travel on these surfaces, two other snake gaits have evolved, both also formed by traveling waves. Sidewinding is the gait that rattlesnakes use to travel over loose sand. Bends travel backward along the body as in serpentine crawling, but there is no sliding of the belly over the ground. Instead, each part of the body is stationary while on the ground but forms new curves in the air as it is lifted periodically to a new position. Marks left in the sand behind the snake show where the body has lain.

A quite different technique (again using bends) enables snakes to climb up grooves in the trunks of trees or fissures in rock. Parts of the body are folded up like the pleats of an accordion to wedge them tightly in the groove or fissure, an arrangement that has inspired the name "concertina locomotion" for this style of travel. At the front of a wedged region the folds are opening out, pushing the snakes's head forward. At the back of a wedged region new folds are forming, drawing the tail forward. Thus groups of folds seem to travel backward along the length of the snake.

More energy is needed to drag a box along a road than to pull a wheeled cart of the same weight because the sliding of the box is resisted by friction. Using similar reasoning, you might assume that the sliding motion of a snake would require more energy than walking at the same speed, but you would be wrong. Michael Walton Jayne, and Al Bennett, have measured the rates of oxygen consumption of black racer snakes crawling on a moving belt. When the snakes traveled by sepenatine crawling, the team of scientists found that the energy cost per unit distance was about the same for the snakes as for lizards and mammals of equal mass (100 grams). The snakes were slow (they could manage short bursts at 1.4 meters per second, but the highest speed they could sustain was only one tenth of that), but they were not uneconomical. Concertina locomotion was much slower and also used more energy per unit distance.

Stretching and Squeezing

All the movements that have discusses so far have been powered by muscles that pull on jointed skeletons. Our legs and those of other mammals and of insects have skeletons of stiff rods or tubes, jointed together. Even backbones (including those of snakes) consist of rigid vertebrae connected by movable joints. Without our skeletons we would be flabby and ineffective, but worms, and molluses such as slugs, move very effectively without any such stiffening.

Earthworm burrowing works on the same principle as the concertina locomotion of snakes: some parts of the body are jammed tightly in the borrow while others move forward. The earthworm, however, is jammed into the available space by making the body swell, not by throwing it into folds. An earthworm's body consists of a line of about 150 segments, which appear as rings on the outside of the body. Inside, the fluid filled body cavity is divided into more or less water tight compartments, one for each segment. The body wall includes two layers of muscle, one of longitudinal fibers running lengthwise along the body and one of circular fibers running circumferentially. When the longitudinal muscle of a segment contracts, that segment gets shorter, but because its fluid contents cannot escape it also gets fatter. When the circular muscle contracts, the segment gets thinner but also longer. The segment becomes short and fat or long and thin as the two sets of muscle contract in turn, but its volume remains constant.

In the diagram, segments 7 to 16 and 24 to 29 are short and fat, jammed in the burrow. Segment 7 is about to lengthen and push segments 1 to 6 forward, driving the worm's head onward through the soil. Segment 17 is about to shorten and jam itself tightly in the burrow, and as it shortens it will pull the segments behind it forward. Thus the segments that are fat at any particular instant are stationary and those that are thin are moving forward. Each segment moves forward intermittently.

For this method of burrowing to work as described, the segments at the front of a fat region must always be getting thinner, becoming part of the thin region in front, and the segments at the front of a thin region must be getting fatter, joining the fat region on front. Thus waves of thickening must travel backward along the body.

Earthworms spend most of their time underground, but movements that serve for burrowing also enable them to crawl on the surface. The reason is partly that the thick parts of the body rest on the ground and are held in place by friction, while the thin parts are raised or rest more lightly on the ground. In addition, the worm is prevented from slipping backward by bristles that protrude slightly from the underside of the body, tilted at an angle that lets the worm slide more easily forward than backward. You can feel these bristles if you stoke the underside of a worm with your fingertip. The skin feels relatively smooth as your finger moves backward, but rougher as it moves forward, catching on the bristles. These bristles work like the scales on cross-country skis, which allow the skis to slide forward freely but prevent backward sliding as you climb a slope. The same principle is applied by ratchets that enable machinery to rotate in one direction but not the other.

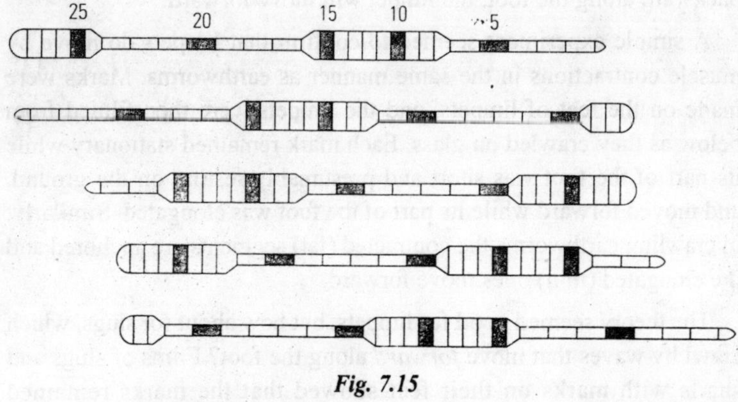

Fig. 7.15

The crawling of slugs and snails looks even more mysterious than that of snakes and worms, for the animals glide forward without obvious movement of the body. However, if you allow them to crawl on a sheet of glass and you watch them from below, you will see something happening on the undersurface of their bodies, or the sole of the foot. A pattern of darker and lighter bands moves along the sole as the animal crawls. Other gastropod molluscs such as limpets, whelks, and periwinkles also crawl on their feet and show similar patterns of bands, with some variations. In limpets the bands are split into left and right halves: bands having a dark side on the left and a light side on the right alternate with bands having the sides reversed. As the animal crawls forward, these bands travel backward along the foot. In slugs each band is light or dark across the whole width of the foot and moves forward. The bands move over the sole of the foot at a rate considerably faster than the mollusc travels. For example, the bands on the foot of a slug crawling at a typical speed of 2 millimeters per second move forward over the foot at 7 millimeters per second. Other molluscs show slightly different patterns of bands, but whether the bands move backward or forward, the effect is always to move the animal forward.

The moving bands suggest the possibility that limpets may crawl essentially as earthworms do. The bands would be waves of muscular contraction formed by alternate bands shortening and elongating. The shortening bands might swell, lifting the elongated ones off the ground. The lengthening of a band would push on the elongated band ahead of it and, unattached to the ground, the elongated band would move forward easily. A shortening band would pull the elongated band behind it forward. The same reasoning tells us that if the contractions travel backward along the foot, the limpet will move forward.

A simple experiment seemed to confirm that limpets do move by muscle contractions in the same manner as earthworms. Marks were made on the feet of limpets, and the limpets were then filmed from below as they crawled on glass. Each mark remained stationary while its part of the foot was short and presumably resting on the ground, and moved forward while its part of the foot was elongated. Similarly, in crawling earthworms the contracted (fat) segments are anchored and the elongated (thin) ones move forward.

The theory seemed good for limpets, but how about for slugs, which crawl by waves that move *forward* along the foot? Films of slugs and snails with marks on their feet showed that the marks remained stationary while their part of the foot was elongated and moved forward while it was contracted—just the opposite of what happens in limpets.

Perhaps some of the puzzle could be resolved by taking a closer look at the sole of the foot through a microscope. To catch the sole in midmotion, the slug would have to be frozen in the act of crawling. A good specimen was obtained by picking up a crawling slug and dropping it into a container of liquid nitrogen, where it froze immediately. It was allowed to thaw in a solution of a chemical fixative that coagulated its proteins, preserving the shape of the soft tissues. Part of the slug was then sliced into thin sections, and the sections were examined under a microscope. They showed that the surface of the foot had ridges running across it, as suggested in the theory of limpet crawling. However, the tissue of the projecting rides that presumably rested on the ground was elongated and that of the furrows contracted; again, this pattern is the opposite of the supposed pattern in limpets.

A little thought shows that the crawling of slugs could be explained by a mechanism that is only a little different from that of limpets. Suppose that the extended bands of the foot are anchored by resting on the ground. Suppose also that the rear edge of each anchored band is shortening and that the rear edge of each raised part is lengthening. The raised parts will be moved forward, but the waves will also travel forward because it is at its rear edge that each band is turning into the other kind.

This explanation of molluse crawling seemed to be further confirmed by the arrangement of muscle fibers within the foot. There are no distinct layers of fibers running in different directions, as in earthworms. Instead, the muscular foot consists of interwoven fibers running in various directions. In limpets, some of the fibers run vertically through the thickness of the foot and others transversely across its width. If the vertical fibers contract in part of the foot, squeezing it thinner, that part must grow longer or wider or both in order to retain constant volume. If at the same time the transverse fibers prevent that part of the foot from widening, it has no other alternative but to become longer. Thus we can expect contraction of the vertical fibers to lengthen part of the foot, and because the fibers pull upward they will lift it off the ground. The muscle fibers of slugs have a different arrangement; instead of running vertically and horizontally, they slope, some pulling obliquely up and forward and others up and back. When these muscles contract, they *shorten* their part of the foot and lift it off the ground. Each molluse thus has the arrangement of muscles that seems to be needed for its style of crawling.

The observations of the animal's movements, of the waviness of the feet when they were quickly frozen, and of the arrangement of muscles seemed to add up to a clear, coherent story that told how molluses crawl. The explanation seemed complete until the work of an American graduate student forced some drastic rethinking.

While still an undergraduate at Duke University, Mark Denny had made his name as a research worker through a beautiful study of the properties of spider silk. He went on to work for his doctorate in Vancouver, and his work on slugs there, published in the early 1980s, gave us a new understanding of mollusc crawling.

Denny saw two problems with the theory. First, it failed to explain the trail of slime (muscus) on which slugs and snails crawl. That slime consists of water and dissolved salts, with a small but significant proportion (3 to 4 percent) of glycoprotein, a compound of protein with sugar molecules. (The mucus that runs from your nose when you have a bad cold has a similar composition.) The mucus presumably did something useful, since the animal would save water and protein by not producing it, but the theory did not explain its function.

The second problem that Denny saw with the theory was that huge forces would be needed to lift the foot from the surface when ever a muscular wave passed. If you lay damp sheets of glass on top of each other, they are very hard to separate (except by sliding) because a thin layer of water forms between the sheets that glues them together very effectively. If the sheets were smeared with mucus instead of water, separating them would be even more difficult, because mucus is viscous like molasses. The flexible feet of molluses fit very closely onto rocks and other hard surfaces, and the short distance between the two surfaces ensures that the molluses are firmly glued down. Some limpets of 3 centimeters diameter can hold on to rocks against forces of more than 200 newtons (45 pound's force).

Denny performed a very simple experiment to check whether the foot really is lifted as the muscular waves move along it. He persuaded a slug to crawl on metal foil and then froze it suddenly with out detaching the foil. Only after the specimen had been chemically fixed and thawed was the foil peeled off. When the sole of the foot was sectioned and examined microscopically after this treatment, it was found to be flat. There were lengthened and shortened bands running across it, but they were not raised into ridges. Denny concluded that there are no ridges on the feet of crawling molluscs and that the ridges seen previously had formed only after the animal was pulled off the surface on which it had been crawling.

If the whole foot lies flat on the ground, why do some parts move forward while others remain anchored? There is no obvious ratchet like the bristles of earthworms. Denny looked for an answer by investigating the mechanical properties of mucus.

A layer of rubber is sandwiched between two steel plates and glued firmly to both. After fixing the lower plate rigidly to a table top, you push the upper one horizontally, distorting the rubber layer. The farther you push it, the bigger the force you feel on your finger. If you hold the plate in position the force remains, and if you release it the rubber springs back to its original shape, returning the steel plate to its original position. This is elastic behaviour.

Now imagine the same experiment with a layer of molasses (treacle, to British readers) between the plates instead of rubber. A small force will move the upper plate slowly and a large force will move it faster, but if it is kept moving at constant speed the force remains constant. If you stop moving the plate, in any position, the force disappears, and the plate does not spring back to its original position when released. This is viscour behaviour. The force does not depend on how far the plate has moved, but on how fast the plate is moving.

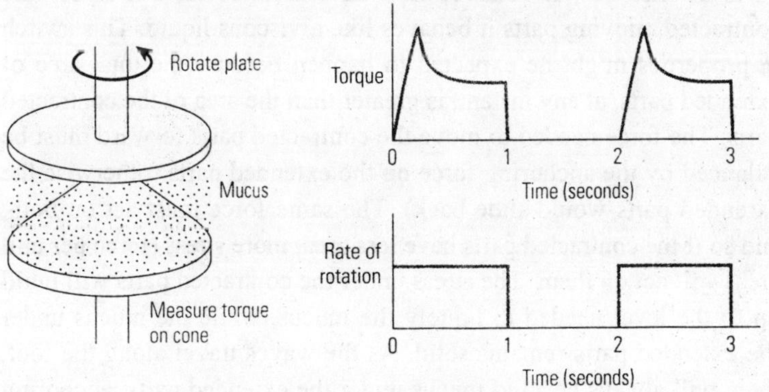

Fig. 7.16 : Denny studied the properties of mucus by testing it in a viscometer (left)

Denny collected a supply of slug mucus and performed essentially the same experiment on it. He put the mucus between two metal plates, one flat and one slightly conical, that formal part of an instrument known as a cone and plate viscometer. The flat plate was rotated, and the torque needed to prevent the cone from rotating was measured. In rotating the plate, Denny wanted to simulate the conditions under the foot of a crawling slug, where each point on the foot moves at one stage of each muscular wave and remains stationary at another. Therefore, he rotated the plate at constant speed for one second, then held it stationary for a second, then rotated it for an other second, and so on.

If the experiment had been performed with an elastic solid such as rubber instead of mucus, the torque would have increased when ever the cone was rotating and would have remained constant while it was stationary. If the experiment had been performed on a viscous liquid such as molasses, there would have been a constant torque whenever the cone was turning, and that torque would have fallen to zero whenever the cone stopped. The actual result resembled neither of these possibilities. In the early stages of each rotation, the torque rose progressively as if the mucus were an elastic solid. After a while, however, it fell a bit and then remained constant for as long as rotation continued, as if the mucus were a viscous liquid. The period of rest between rotations was apparently enough for the mucus to revert to its original elastic state.

Denny suggested that the state of the mucus under a slug's foot changes with the passage of each muscular wave. Under the extended parts of the foot, it behaves like an elastic solid, but under the contracted, moving parts it behaves like a viscous liquid. This switch in properties might be expected to happen because the total area of extended parts, at any instant, is greater than the area of the contracted parts. The force needed to move the contracted parts forward must be balanced by the anchoring force on the extended parts (otherwise the extended parts would slide back). The same force must act on both, and so if the contracted parts have less area, more stress (force per unit area) will act on them. The stress under the contracted parts will build up to the level needed to liquefy the mucus, while the mucus under the extended parts remains solid. As the waves travel along the foot, there will always be solid mucus under the extended parts, anchoring them, and liquid mucus under the contracted parts, allowing them to slide forward. The changing properties of the mucus have the same effect as would a ratchet under the foot.

The same mechanism can also work for the backward-moving waves of limpets, if the total area of the parts of the foot that are extended at any instant is less than the area of the parts that are contracted. In that case, the contracted parts will be anchored and the animal will move in the direction opposite to the waves.

The mollusc's manner of crawling seems to have severe disadvantages. A snail's pace is proverbially slow. 2.5 millimeters per second (one tenth of an inch per second) or less, a hundred times slower than some bettles. Snails are inevitably slow, because if they were moving fast very large forces would be needed to overcome the viscosity of the mucus under the moving parts of the foot. A second disadvantage is that the method of crawling is expensive of energy and materials. Measurements of oxygen consumption show that the metabolic energy cost per unit distance is about 12 times higher for a slug crawling than for a mouse of the same mass running. For every 10 meters that it travels, a 15-gram slug loses about 1 gram of water in its mucus and 30 milligrams (0.03 gram) of glycoprotein.

Costly though it may be, slug and snail crawling has one vary clear advantage. The animal is quite firmly struck to the surface it is crawling on, at all times. Limpets are not easily dislodged by waves, and a snail can climb up a plant stem without much danger of falling off.

CHAPTER 8
FLYING ADAPTATIONS

The flight adaptations are known as volant adaptation. Due to volant adaptations, animals become beautifully modified, due to which, they are offered the least possible resistance by air to the attainment of speed. Volant adaptations among the invertebrates occur in insects. Among the vertebrates, the volant adaptations are found in fishes, amphibians, reptiles, birds and mammals.

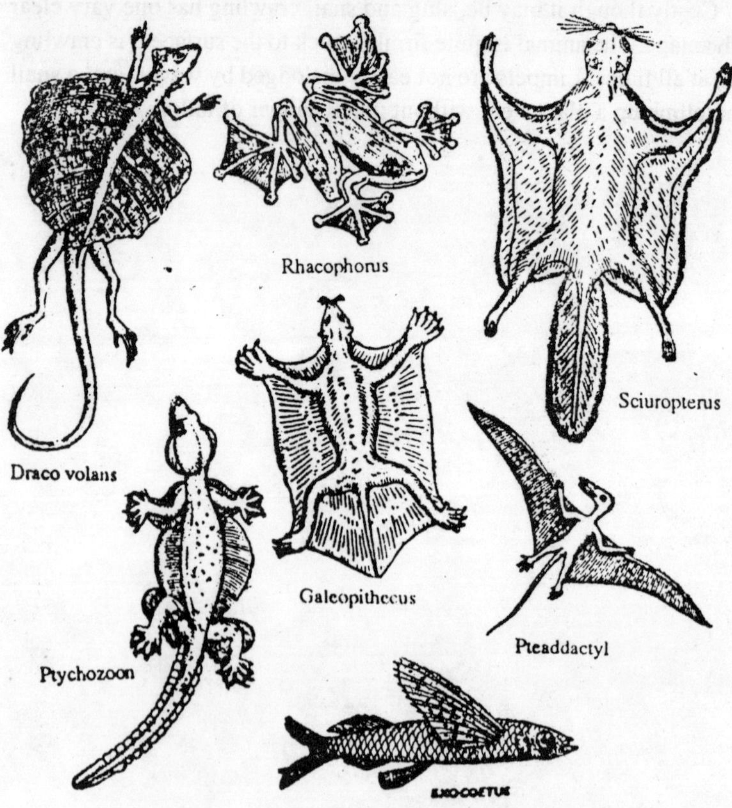

Fig: 8.1: Volant forms

Types of flight

Flight is of two sorts:

I. Passive or gliding flight

In this kind of flight, the animals only take an initial leap from a high point and are held up by certain sustaining–organs and impelled by gravity, they glide to a lower level in a horizontal plane. Sometimes, the distance covered by gliding may be many yards. This kind of flight can be compared to a gliding aeroplane without engine–power.

II. Active or true flight

In active flight, there is a sustained movement through the air. It covers long distances and involves locomotive force. True flight has evolved thrice among vertebrates: in the reptilian Pterodactyles, the birds and the bats. Whether flying fishes should be included is a much controversial question. However, the animals having true flight, move their wings with varying degrees of rapidity or after having gained their high altitude by flapping their wings, they may also soar or sail on motionless wings for several hours continuously.

Modifications due to the aerial mode of life

1 Body contour

The body becomes boat–shaped or spindle–shaped (stream–lined). This gives the least resistance to movement in the air.

2 Sustaining surface:

Except in flying fishes, in all the other flying vertebrates, the sustaining surface is a fold or a series of folds of the skin known as patagium, which is supported in various ways. In the flying dragon, Draco, it is supported by five or six elongated ribs. The patagium of most of the soaring mammals is supported between the skin–fold and extends to the neck and behind between the hind limbs and tail, as for example in Galeopithecus, the flying lemur. Here, the patagium extends even between the digits of the limbs and it appears as though, they are webbed, as in the case of aquatic animals. In Pterosaure and bats, the patagium extends from the arm to the lateral sides of the body and also to the front of the hind–limb. The inter–femoral membrace present in the bats is absent in pterosaurs.

The patagia supported by ribs and forelimbs can be folded like a fan against the sides of the body, when it is not used.

The traces of patagia, in–front of and behind the arms might have performed an important supporting functio before their function was taken over by the feathers. In birds, the buoyancy is mainly given by the broad vanes of the supporting feathers of the wings (remiges) and retrices of the tail.

3 Wing

The wing has been evolved thrice, twice with patagia and once with feathers.The bat's wing is the least modified. In this case, the humerus is well developed, the radius is long and curved while the ulna due to disuse has become vestigial. The pollex is free and bears claws for climbing and crawling. In the wing of Pterodactyle, the radius and ulna are more or less of equal size, the former is slightly smallere than the latter. The fourth metacarpal is well developed and bears the large wing finger. The three metacarpels are very thin and bear the small, free, clawed, first, second and third digits. The pteroid bone lying in front and directed inwards towards the shoulder presumably supported the anterior border of the small patagium lying in–front of the arm from the wrist to the neck. The fourth single huge wing finger supported the patagium entirely in the anterior region beyond the wrist. The wing of Pterodactyle was capable of being flexed between metacarpal and proximal phalanx, rather than at the wrist, as in the case of birds and bats.

The wings of the bird is the most specialized of all. It is a modified fore–limb of other vertebrates. The fore–limb bones have become very much modified here. The digits are reduced to three and they are fused together and their main function is flight. There are three unequal metacarpals which are firmly co–ossified. The wing expanse is provided with feathers as the patagium is small. Feathers can be replaced or renewed when injured or lost; while the injured patagium of Pterodactyles and bats cannot do so, thus impairing the flight of these flying creatures. The powerful flight muscles help in the vigorous up and down strokes of the wings, thus helping in the flight. The flight muscles are pectoralis major, pectoralis minor and coracobrachiales. Pectoralis major and coracobrachiales are responsible for the downward stroke of the wings while pectoralis minor is responsible for the upward stroke. The wing feathers help in the filght and they form a continuous sheath impervious to air. The tail feathers (retrices) help the bird to change the direction during flight.

The feathers are of three types–quill feathers, contour feathers and the filoplumes. The barbs bear barbules, which in their turn bear the hooks. This forms a good interlocking system giving stability to the vane. The contour feathers and filopumes act as insulators and maintain the high body temperature, so necessary for the maintenance of high metabolic rate, which is essential for flight.

4 Pnlumatic bones

In birds and pterodactyles, the bones are hollow, containing no marrow and are air–filled. Due to this pneumatism in their bones, lightness of the body is achieved. The pneumaticity of the bones is brought about by the communications of the air–sacs into the hollow cavity of the bones through the pneumatic foramina. In birds, together with pneumaticity of bones, air–sacs have also developed in the abdominal as well as in the other portions of the body. These store air and help in double respiration and reduce the specific gravity of the bird. The birds are very active animals and rapid fliers, so in them, respiration is double and is very perfect, to supply a great amount of oxygen constantly for rapid oxidation of the food liberating a great amount of energy. The large pneumatic bones of the bird offer a relatively greater surface for muscle attachment without increasing in weight.

5 Sternum and shoulder girdle

The sternum is well developed in true flying vertebrates, e.g. birds and bats. The sternum is provided with a broad plate like median keel of carina which provides surface for the attachment of powerfully built flight muscle. This structure (carina) is found in rudimentary condition in flightless birds. The shoulder girdle becomes very much rigid due to the development of the clavicles and coracoid in birds. This is to resist the contractile force of the pectoral muscles. In bats, the coracoids are wating and in the flying birds are movably–jointed. The two clavicles are fused together and form a 'V' shaped bone known as furcula, which acts as a spring and helps in keeping the wings apart.

6 Brain and sense organs

The eyes of birds have sclerotic plates for protecting it against pressure and for rapid focussing, respectively. Pterodactyles have only sclerotic plates. The brain of the true flying vertebrates are highly developed. The bats can see in twilight but are blind in bright light. They are renewed for detecting supersonic sound–waves and thus, avoid colliding against obstacles in a dark room.

Flying vertebrates:

Fishes

Several genera of flying fishes are found and each of them exhibit a separate volant adaptation. For example, in Exocoetus, which is found in all tropical and subtropical seas, the pectoral fins are greatly enlarged to form a flying organ, with the help of which, the fish can skin over the water surface–whose length varies from 200 to 300 yards. The lower, longer lobe of the tail helps in giving the final impetus, as the fish leaves the water. Whether the flight of Exocoetus is a true flight or simply a soar, is a controversial question. Other flying fishes. are various species of Dactylopterus, also called as the flying gurnets. Mosely compares the flight of the gurnets with that of grasshoppers. Pantodon, African flying fish, found in Congo and Niger rivers is 3 to 4 inches long. It leaps out of water and flutters through the air for some distance. Gastropelecus, found in the rivers of British Guyana is a small compressed fish, having long curved pectoral fins which are not large, It swims 40 or more feet along the water surface, beating the water with the help of its pectoral fins and then, it leaves the water, for a distance of 5 to 10 feet and when exhausted, it falls on its lateral side into the water again. *Pegasus volitans*, a small fish found along the coasts of India, Japan China and Australia, swims with the help of its broad pectoral fins for a short distance above the water surface. Several extinct forms were also "flying fishes", for example - Dollopterus (middle Trias), Thorapterus (upper Trias), etc.

Amphibia

In class Amphibia, the only volant adaptation is found in the frog, Rhacophorus. In this case, feet are webbed, which help in sustaining the prolonged leaps. It is found in oriental region, particularly in Borneo. It has feeble gliding powers because its wing expanse is nearly 3 square inches only - rudiment of patagia in-front of and behind the arms. The digits bear terminal adhesive-pad and are webbed.

Reptilia

The most remarkable example of flying lizards is the flying dragon, Draco which is commonly found in the Indo–Malayan region. In this animal, the patagium is supported by a number of ribs. The flying gecko of Malaya, Ptchozoon, has lateral expansions of skin along the sides of the neck, body, tail limb and between the toes. Chrysopelea, the flying snake of Borneo has a concave ventral side, which helps in sustaining the animal when it descends obliquely through the air from the top of the tree to the ground. The pterodactyles of Mesozoic are a very famous group of extinct reptiles. They possessed true flight (upper Triassic to upper Cretaceous) and because their's and bird's ancestors were common, so they showed much similarity with birds.

Birds

The perfection of aerial adaptation among carinatae (birds of flight) is very excellent and is comparable to man–made examples,of which are as follows:

White storks	48 miles per hour
Rooks	45 miles per hour
Geese	55 miles per hour
Gloden Plover	60 miles per hour

As regards the height of their flight, vultures rise from 7,000 to 15,000 feet and condor hovering has been recorded at a height of 20,498 feet.

Mammals

Bats are the only true flying mammals. Gliding flight has evolved among mammals of different orders, of which marsupials will be the first to be dealt with. The flying phalangers include several unrelated species such as Petaurus (Australia and New Guinea), Petauroides, volangs and Acrobates pygmaeus. All of them have a Well–developed fold of skin along the lateral sides of the body between the fore–limbs and hind–limbs and a little skin–fold, infront of the foreleg and a bushy tail and long fur on the body, giving a great buoyancy of the pataguium. The tail may or may not be furless at its tip and in prehensile.

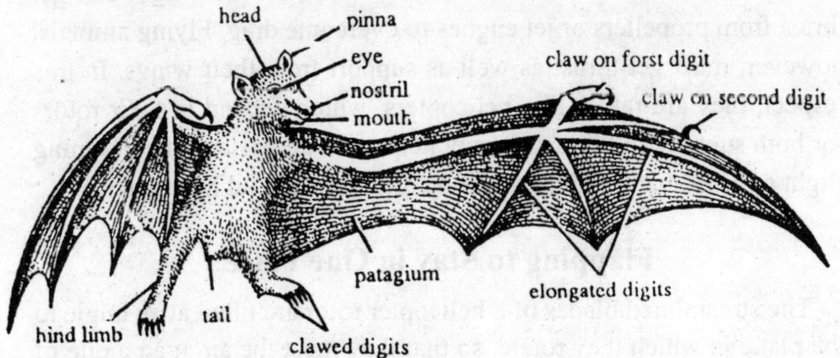

Fig.8.2: Bat

There are two families of rodents, e.g. Family (i) Anomaluridae— It includes genus Anomalurus, having six flying species, all found in

Africa, and Family (ii) Sciuridae—It includes three genera, such as Pteromys (in forests of tropical south–eastem Asia, Japan and some of the Malaysian islands–soars about 80 yards), sciuropterus (Northern America, Eurasia, India) and Eupetaurus (rock climbing; found in north–western Kashmir). All these have well–developed patagium, extending from the fore–limb to the hind–limbs and the hairy tail is also present. An inter–femoral membrane is present in some cases but absent in others.

Among the Insectivora, Galeopithecus is the only representative of the suborder Dermaptera. Its patagium shows the highest development. It extends from the rear of the head in–front of the arm (prepatagium), between the fingers to the base of the claws, between the fore and hind–limbs, between the toes and fingers, and between the hind–limb and entire tail (interfemoral membrane), as in the case of bats. Galeopithecus and bats were both derived from a common stock. Galeopithecus is noctural as are the other volant mammals. It rests suspended head–down from a branch. It possesses a great soaring power (about 70 yards). It has two species:

(i) *Galeopithecus volans,* found in Malaya, Sumatra and Borneo.

(ii) *Galeopithecus philippinesis,* found in Philippine islands.

Among the primates, in the family Lemutridae, only one genus Propithecus, including three species–all from Madagascar, shows volant adaptations. It has quite limited powers of flight (10 yards leap). It has also a skin fold extending between arms and body. It is diurnal.

Airplanes get lift from their wings to support their weight and get thrust from propellers or jet engnes to overcome drag. Flying animals, however, must get thrust as well as support from their wings. In this respect, they are rather like helicopters, which depend in their rotors for both support and thrust. It may help us to understand the flapping flight of birds, bats, and insects if we first think about helicopters.

Flapping to Stay in One Place

The streamlined blades of a helicopter rotor are tilted at an angle to the plane in which they rotate, so that they meet the air at an angle of attack. The blades drive air downward, and the air exerts lift on them in return. To hover steadily in one place on a windless day, the rotor is kept horizontal; the air is driven vertically downward and the lift acts vertically upward. To travel horizontally, the rotor is tilted so that the air is driven at an angle, downward and backward. Left then acts upward and forward, providing thrust as well as support.

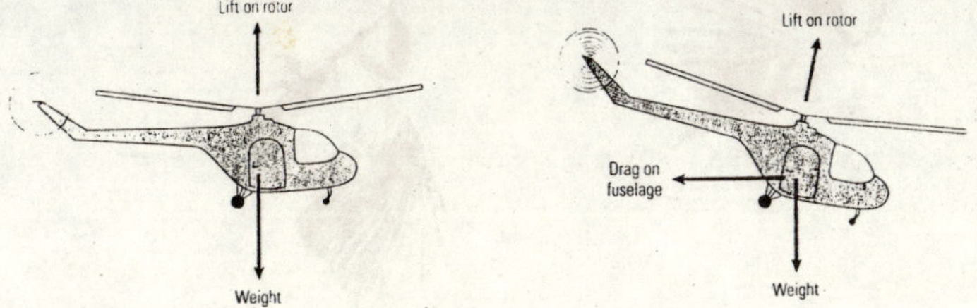

Fig: 8.3 : Hovering (left) and forward flight (right) of a helicopter.

If flying animals are indeed like helicopters, and their wings play the role of rotors, we might expect them to hold the wings horizontally while hovering and to tilt the wings at an angle while moving forward– and in fact many of them do. Hummingbirds hover like tiny helicopters as they suck nectar from flowers. The body is tilted up at a steep angle that positions the wings to beat horizontally. Instead of rotating like helicopter blades, however, the wings beat backward and forward, changing their angle at the end of each stroke to provide the angle of attack needed to give upward lift. In each forward stroke the wing faces what seems to be the right way up: the surface that is uppermost in the flight of other birds faces upward, and the anterior edge (where the bones are) is in front. However, high–speed films show that the wing turns upside down for the back stroke, so that the anterior edge faces backward. In both the forward and the backward stroke, the relatively thick, bony anterior edge of the wing is always in the lead with the feathers trailing behind. High–speed film is needed to see this because the wings beat at very high frequincies, between 15 cycles per second in large, 20–gram hummingbirds and 60 cycles per second in small 2–gram ones.

Moths, bees, and many other insects hover in the same way as hummingbirds. In insects, too, the stiffer edge of the wing always leads, and insects also turn their wings upside down for the backward stroke. Insect wings are not stiffened by bone, however, but by veins and by pleating. Like the bones in hummingbirds, the largest veins and deepest plest are found near the anterior edge. Hummingbirds are alone among birds in using this techinque to hover. They are, after all, the smallest of all birds, similar in size to the largest beetles and moths.

Most birds that can hover (and only small birds can) move their wings in a very different way, more like the tit in the illustration. Notice

Fig. 8.4 : A great tit hovering.

how the primary wing feathers (the large feathers in the outer part of the wing) are bent upward during the forward stroke, showing that large lift forces are then acting on them. In the backward stroke, however, the wings are partly folded and the feathers are not bent, which shows that any lift forces are small. When tits and most other birds hover, they seem to produce lift only in the forward stroke of the wings. The wings are tilted in the forward stroke at an angle of attack that produces lift, but are angled to give little or no lift in the that backward stoke, and they do not turn upside down.

This explanation of hovering flight may seem adequate, but problems emerge when zoologists try to calculate the forces on the wings. If you know the area of a wing and its speed of movement, you can estimate the greatest lift it can produce, by using the conventional aerodynamics that engineers apply to airplane wings and helicopter rotors. This is the lift that the wing would give if tilted almost to the angle of attack at which it would stall. The maximum possible lift, estimated in this way, increases and decreases in the course of each wing beat as the wing speeds up and slows down, but it can be averaged over a wing beat cycle. The calculations are quite complicated but lead to a clear conclusion: many hovering birds and insects connot (in theory!) hover. The calculations seem to tell us that the wing movements of a hovering flycatcher (a small bird) can produce no more than one third of the force needed to support the bird's weight, and similarly that the wings of a hovering hoverfly can support only one third of the weight of that small insect.

Since flycatchers and hoverflies do in fact hover, the theory must be wrong. The explanation is that the theory was developed for airplane wings and helicopter rotors that move steadily through the air. It does not work well for unsteady movements like those of the flapping wings of birds and insects, which keep stopping and starting.

To understand how stopping and starting can affect the lift on wings, we need to know more about how air flows around them. We saw in Chapre 4 how air flowing over a stationary wing is deflected downward if the wing has a positive angle of attack. The air traveling over the top of the wing speeds up and the air going under slows down: similarly, if two people run side by side around a bend, the one on the outside of the bend has to go faster. The flow of air *relative to the wing* is the same as before, but as seen from the ground the bulk of the air is now stationary, the air immediately over the wing is moving backward, and the air under the wing is moving forward: there is a circulation of air around the wing.

This circulation takes a little time to develop after the wing starts moving, and it does not cease immediately when the wing stops. For that reason, there is less lift on an accelerating wing than you might expect from the wing's speed, and more on a decelerating one. Birds are able to exploit unsteady effects such as these to gain more lift than conventional aerodynamics seems to allow.

One way to achieve additional lift involves clapping the wings together at the top of the upstroke, Pigeons often do this when they take off, producing quite a loud clapping noise, but the "clap and flying" mechanism was first demonstrated in research on a tiny insect. Torkel Weis–Fogh of Cambridge University had taken a high–speed film of *Encarsia,* a miniature wasp with wings only 0.6 millimeter long. It hovers rather like a bee or a hummingbird, holding its body vertical and beating its wings in a horizontal plane. The wings are clapped together back to back, then flung apart, separating first along their upper (anterior) edges. Air rushes into the expanding space between the separating wings, with the result that by the time separation is complete, there is a circulation of air around them. This circulation is much stronger than could be achieved if the wing were moving steadily at the same speed instead of stopping and starting, and the lift is correspondingly greater. This is only one of several tricks that flying animals use to exploit unsteady effects and improve their lift. Dragonflies and hoverflies seem to depend for their hovering flight on different unsteady effects, moving their wings quite differently from how other hovering insects move theirs.

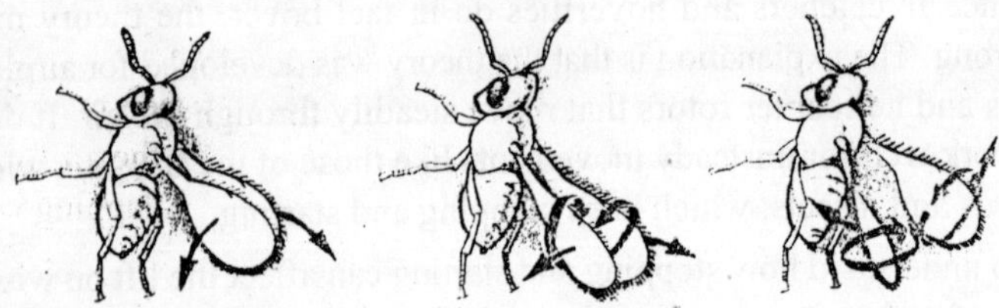

Fig. 8.5 : A "clap and fling" used by a hovering insect Encarsia.

Vortices and Airflow

When helicopters hover, stationary in the air, they produce the upward force that is needed to support their weight by driving air downward. To fly forward they must drive air backward as wel', to supply forward thrust. Similarly, animals that are flying forward rather than merely hovering must drive air both downward (for weight support) and backward (for thrust). The air movements in the wake of a flying animal cam tell us about the forces it exerts, much as force plates tell us about the forces on the feet of runners.

Blowing smoke rings was a favorite trick of smokers, in times when smoking was socially more acceptable than now. They would fill their mouths with smoke and then blow it out in a single, gentle puff. A ring of smoke formed and could be watched drifting slowly away. The creation of the ring depends on a property of moving air: as a puff of air moves through still air, it sets a surrounding belt of air roating as a "vortex ring". (Similarly, a crate being pushed along on rollers makes the rollers rotate.) Most of a mouthful of smoke is soon dispersed, but some is trapped in the core of the vortex ring by the air roating around it, and the ring becomes visible.

Vortices from whenever a body of air moves through still air (or through air with a velocity different from its own). Huge vortex rings form around rising thermals, for example. A whole complex of vortices forms around the air that is driven downward behind the wings of an airplane, outlining the shape of a rectangle. The two long sides of the rectangle consists of vortices that have formed around the wing tips and now trail behind; they sometimes become visible as vapor trails. A "starting vortex" running across the wake is left behind where the airplane took off. Finally, the rectangle of vortices around the downward–moving air is completed by the circulation around the wing itself.

Fig. 8.6 : A vortex ring (left), and the vortices behind an airplane (right). Arrows represent air movements.

When a hummingbird hovers, each wing stroke drives downward a puff of air, and that small air mass is presumably surrounded by a vortex ring. Successive wing beats should build up a stack of vortex rings, one on top of another, and the wake should be little different from the wake below a hovering helicopter, which produces a continuous stream of air rather than a series of puffs. No one has actually studied the airflow under a hovering hummingbird, but it seems pretty clear what it must be like.

It is less obvious what the pattern of airflow should be like in forward, flapping flight, but Nikolai Kokshaysky of Moscow obtained photographs of vortex rings by flying finches through clouds of brightly lit sawdust. His photographs seemed to show that vortex rings are produced only in the downstroke, but they could not give very accurate information about airflow because the sawdust particles did not simply move with the air, but sank through it. Many photographs of birds flying slowly or hovering show the main wing feathers bent only in the downstroke. Like Kokshaysky's photographs of vortex rings, these suggest that lift is obtained only in the downstroke.

Colin Pennycuick devised a much more sophisticated version of Kokshaysky's experiment and assembled a team in Bristol to perform it. HIs colleagues were Jeremy Rayner, who had devised a vortex theory of bird flight, and Geoff Spedding, then a graduate student, who did most of the practical work. They set out to make the vortices visible by having a bird fly through a cloud of tiny soap bubbles.

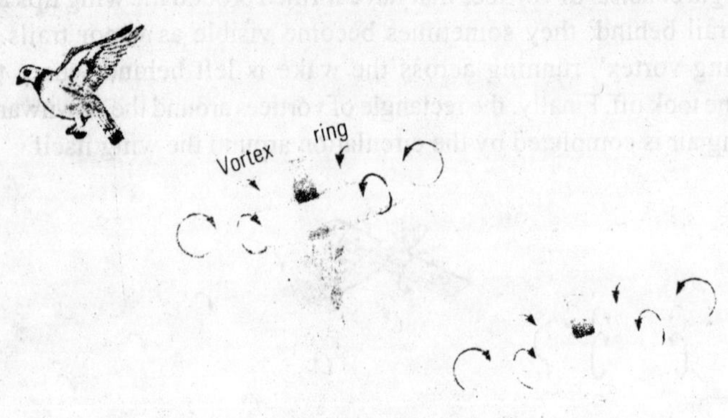

Fig. 8.7 : The pattern of vortices behind a pigeon flying at about 3 meters per second

To obtain soap bubbles that would not sink or rise. Pennycuick's team had to create bubbles that were exactly the same density as air. Soap solution is denser than air, so air–filled bubbles sink. Helium, however, is less dense than air, and helium–filled bubbles rise. The experimenters filled their bubbles with a mixture of air and helium, in just the right proportions for them neither to sink nor to rise.

Spedding set up a machine that would blow soap bubbles of this composition, and he filled a laboratory with clouds of bubbles. The room was darkened, and a bird was let loose. As the bird flew through the bubbles, two cameras took flash photographs of the wake behind it. The cameras were angled to give a stereoscopic effect so that the positions of bubbles could be recorded in three dimensions. During a single opening of the shutter, four flashes fired in succession; thus each photograph showed four images of each bubble, indicating its path and speed of movement. Conveniently, the flashes died away rather slowly, so that each image had a tail to it; otherwise it would not have been clear which was the first, and which the last, of the four images of each bubble.

The resulting photographs show huge numbers of bubbles and are difficult to interpret, but the team managed to make sense of them. Photographs of a pigeon and a jackdaw (a European crow) verified the predictions of Rayner's mathematical theory. Each downstroke of the wings produced a vortex ring (showing that lift was being produced), but the upstroke did not.

The wings of pigeon beat forward and down at an angle of 45 degrees to the horizontal. The vortex rings form behind the moving wings tips, but are tilted at only about 10 degress to the horizontal. The angle of the vortex rings is so much shallower because the rings are already moving downward while they are being formed. The back of the ring, formed behind the wings tips while the wings are at the top of their stroke, has moved down some distance by the time the front of the ring is completed.

The downward flow of air follows the tilt of the rings and is thus inclined at 10 degress to the vertical. This air is being driven downward and a little backward, so the aerodynamic forces on the bird act upward and a little forward, supporting the bird's weight and providing thrust.

The vertical component of this force can be calculated from the size of the vortex rings and the velocity of the air moving through them. Spedding measured and calculated as necessary, but reached a disturbing conclusion. He found that this supporting force, averaged over a wing beat cycle, was less than the bird's weight. An error in the calculations was suspected, but none could be found. The most likely explanation seems to be that at the stage of their flight when they were photographed, the birds were not fully supporting their weight, but were to some extent falling.

The lift acting perpendicular to the direction of movement of the wings and the drag acting backward along their path. Lift acts in the downstroke only, in an upward and for ward direction, and its forward component is sufficient to overcome the drag on the body.

The observations of pigeons and crows confirmed expectations in showing a series of vortex rings, one for each downstroke, but experiments with a kestrel brought a surprise. Its wake consisted not of vortex rings but of a pair of continuous vortices trailing from the wing tips. These were not straight like the wing tip vortices of airplanes but wavy, rising and falling with the beat of the wings. Moreover, the vortices were farther apart for the downstroke (when the bird's wings were fully spread) than for the upstroke (when the elbow and wrist joints in the wings were flixed a little, reducing the wing span).

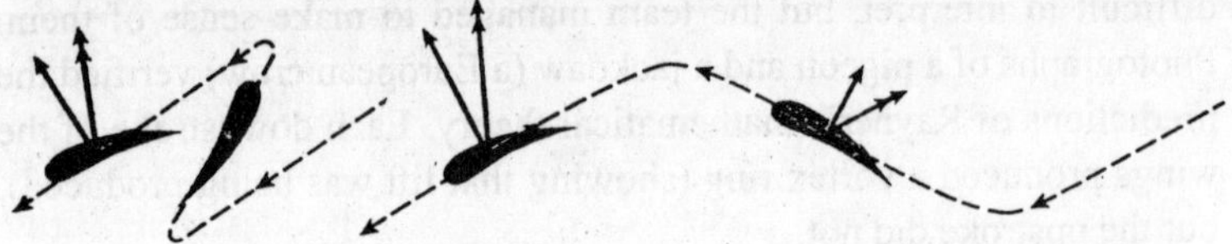

Fig. 8.8 : Aerodynamic forces on the wings of a pigeon flying slowly (left) and a kestrel flying fast (right).

The downward airflow between the wing tip vortices is every–where at right angles to them, downward and a little backward during the downstroke but downward and a little forward during the upstroke. The wings are pushing down on the air, getting the upward force needed for support, in both strokes. However, they are providing thrust only during the downstroke: during the upstroke the horizontal component of forces acts in the opposite direction. If the bird is to maintain its speed, the thrust priduced in the downstroke must not be cancelled out by this retarding force in the upstroke. This unfortunate effect is avoided by making the aerodynamic forces smaller during the upstrike than during the downstroke. The bird pushes strongly on the air during the thrust–producing downstroke and more gently in the retarding upstroke, so the net effect over a wing beat cycle is support plus adequate thrust.

The forces during the upstroke could be reduced by one of two means: either by accelerating the same amount of air as in the downstroke, but to a lower speed, or by driving less air at the same speed. The latter is what happens. Air flows downward between the wing tip vortices at the same speed in both strokes, but less air is moved in the upstroke because the wings are less widely spread, bringing the vortices closer together.

During the downstroke, the path of the wing is tilted downward, so lift acts upward and forward, but during the upstroke its path is tilted upward and the lift acts upward and backward. The forces on the wing must be reduced in the upstroke if the resultant of lift and drag, averaged over a complete wing beat cycle, is to have a forward thrust component as well as an upward weight--supporting component.

The experimental results suggested a connection between the speed of flight and the pattern of vortices. The pigeon and jackdaw flew slowly in the experiments, at about 2.5 meters per second (6 miles per hour), and produced vortex rings in the downstroke only. The kestrel flew much faster, at about 7 meters per second (16 miles per hour), and produced continuous wavy wing tip vortices. It seemed possible that birds in general might use one style of flight to go slower and the other to go faster: the two styles might be gaits used at different speeds, like walking and running. The question might have been settled if the pigeon and jackdaw had sometimes flown fast or if the kestrel had sometimes flown slowly, but (infuriatingly) the birds would not change speeds in the experimental setting. In the wild, both pigeons and jackdaws usually fly quite fast.

A noctule bat was more obliging. It usually flew slowly through the bubbles at 1 to 3 meters per second, but occasionally it went much faster, increasing its speed to 7 to 9 meters per second. When flying slowly it used the vortex ring gait like the pigeon and jackdaw, but when flying fast it used the continuous vortex gait, like the kestrel.

At low speeds, walking uses less energy than running, whereas at high speeds the reverse is true. It seems reasonable to guess that the vortex ring gait is more economical than the continuous vortex gait at low speeds and that the continuous vortex gait is more economical at higher speeds. A simple argument suggests why this might be true, although mathematical analysis is still needed to verify the argument.

In the vortex ring gait, air is driven downward only during the downstroke,so the bird pushes on less air than if lift were being produced continuously, and consequently that air has to be accelerated to a higher speed. We have already seen that it is more economical to obtain lift by accelerating a lit of air to a low speed than by accelerating less air to a higher speed. The characteristic is that the downstroke and the upstroke drive the air in different directions (downward and back, or downward and forward). As a consequence, more kinetic energy has to be given to the air than if it were all driven in the direction of the required force. At low speeds, the animal travels less far in each wing beat cycle, so the paths of the wings in the downstrokes and upstrokes slope more steeply down and up they do at higher speeds. If the continuous vortex gait were used at low speeds, the air would be driven in very different directions in the downstrokes and upstrokes, and so the extra kinetic energy needed because the air was not all driven in the required direction would be relatively larger. It seems likely that the vortex ring gait needs less power from the muscles at low speeds and the continuous vortex gait needs less power at high speeds.

Bounding Flight

Many birds rise and fall as they fly, taking a wavy path through the air. Careful observation shows that there are two distinct wavy styles of flight. Crows and gulls, and also some bats, flap their wings for a few cycles, make a brief glide with their wings spread, flap for another few cycles, and so on: this is called undulating flight. Sparrows and many other small perching birds use a different gait called bounding flight, as do woodpeckers, kingfishers, and small parrots. The wings beat for a few cycles, then are folded against the sides of the body for a while, then beat again. They beat while the bird is in each trough of its wavy flight path, and they are folded at the crests. In the case of sparrows, for example, bursts of about six wing beat cycles, lasting 0.3 second, alternate with 0.3–second periods of wing folding.

A few years ago, we had a theory of bounding flight that seemed convincing. While the wings are folded, the bird avoids the profile drag that would act on the wings if they were spread. However, if they are folded for part of the time, they must drive air down faster during the time that they are beating in order to increase the work done against induced drag. At low speeds, induced drag is large and profile drag is small, so the profile power that can be saved by folding the wings is less than the increase in induced power, but at high speeds the reverse is true. Thus birds should beat their wings continuously when flying slowly, but use bounding flight when flying fast. Mathematical analysis suggested that the change should be made at a speed just a little above the maximum range speed.

An alternative theory of bounding flight considers the metanolic energy used by the muscles rather than the work that they do. It has already been obsevaerd that the metabolic energy used while doing a given amount of work is least if the muscles shorten at about one third of their maximum rates. This is also the rate of shortening at which they can give the highest power outputs : muscles are most efficient and deliver most power at about the same shortening rate. Evolution matches the properties of muscles to the jobs they have to do, and it seems likely that bird muscles have evolved so that their most efficient rate of shortening, and the rate at which they can produce most power, is the rate at which they work in fast flight, when maximum power output is required.

The alternative theory proposes that bounding flight allows birds to use their muscle at the most efficient rate of shortening. Suppose that a bird is flying at a speed that does not require its muscles to produce maximum power. In that case it has a choice between beating its wings continuously and slowly, or intermittently and rapidly. If the muscles work more efficiently at the higher speed, the bird may save metabolic energy by using the intermittent gait, even it the work required is a little increased. In that case the bird should beat its wings continuouslsy only when a very high power output is needed; for example, when climbing fast or accelerating.

If the theory is correct, a sparrow in bounding flight would use less metabolic energy than if it flew the same distance in the same time, beating its wings continuously. However, if the wings beat for only half the time in bounding flight, they would have to do the same (or slightly more) work in half the time: during the bursts of flapping, their power output must be double what be needed in continuous flapping flight. Bounding flight is an option only for birds that have plenty of power in reserve. Small birds have more power in reserve than large ones, which seems to explain why only small birds bound. The power needed in the flapping phase of undulating flight is less than that needed for bounding flight, because the wings are kept spread and provide lift in the intervals between bouts of flapping. Because undulating flight needs less reserve power than bounding flight, undulating flight is possible for larger birds.

Another way to reduce the energy cost of flight is to reduce the number of vortices, a strategy that may explain the elegant V formation

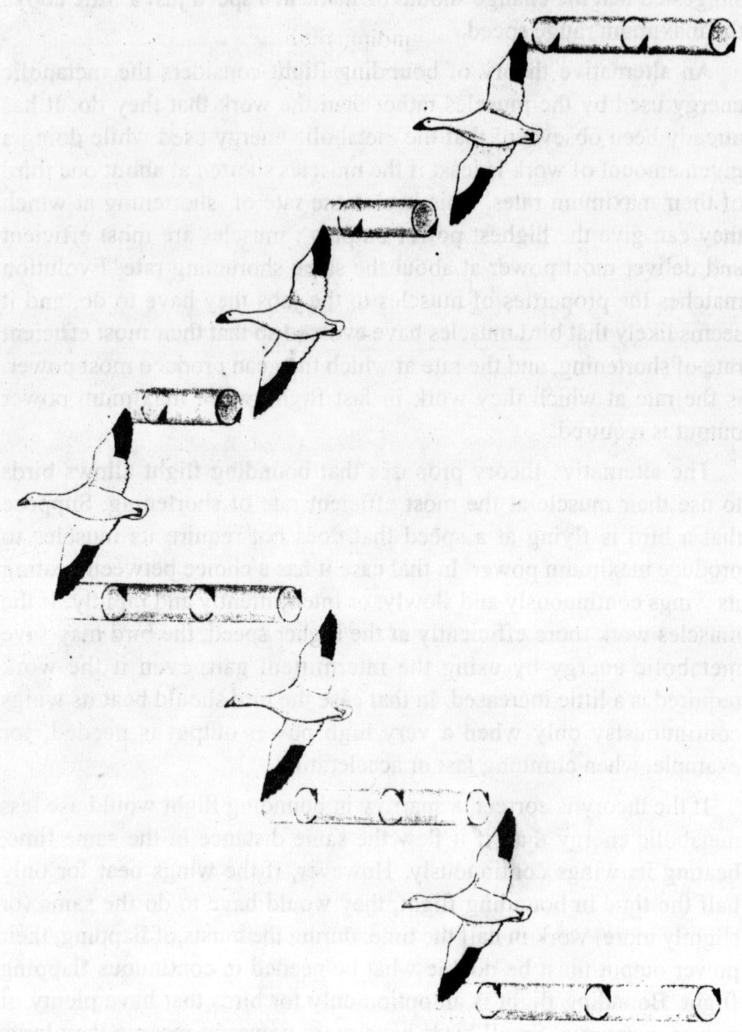

Fig. 8.9 : The wing tip vortices behind a formation of geese.

of flying geese. Fewer vortices mean less work, because the bird does not have to impart kinetic energy to the roating air. Birds might be able to eliminate vortices by flying in group, if they positioned themselves correctly. for example, they might fly side by side with their wing tips touching, so that their wakes of downward–moving air merged and there were wing tip vortices only at the edges of the group. The left wing of the leftmost bird and the right wing of the rightmost bird would leave wing tip vortices, but the others would not. Birds do not fly with wing tips touching (they might have some nasty accidents if they did), but groups of geese often fly in V formation with the left wing tip of one bird, for example, close behind the right wing of another. If the relative positions of the birds were right, the wing tip vortex of the bird in front would be cancelled by the effect of the wing tip of the bird behind. There would be vortices in the wake only behind the outer wing tips of the birds on the edge of the formation, and energy would be saved. This seems to happen.

Fig. 8.10 : Showing Flapping Flight.

Can Bumblebees Fly?

The forward flight of insects involves patterns of force on the wings that are intriguingly different from those found in the forward flight of birds. Scientists used to be fond of saying that aerodynamics had proved that bumblebees cannot fly. The behavior of the bees themselves shows us that this was nonsense, and in any case bee flight better than we did.

The bumblebee is the best example to take in a discussion of insect forward flight because the experiments that have been done on it are far more satisfactory than those done on other species. In these other experiments, the insect's body was glued temporarily to a wire, which was then fixed so that the animal was held supported in an air current, facing into the wind. In these circumstances many species beat their wings as if they were flying. This convenient behavior has been exploited in many investigations of insect flight mechanics: insects have been filmed, or the forces on their bodies have been measured, while they "flew" fixed in wind tunnels. The results have doubtful value because the movements may not be the same as if the insects were really flying; for exemple, locusts that beat their wings at 23 cycles per second in free flight made only 20 cycles per second in tethered flight.

The most detailed study of flight by free–flying insects has been made at Cambridge University by Charles Ellington and his student Robert Dudley. They filmed bumblebees flying in a wind tunnel, in a space enclosed by plastic netting. If the bees were put into the empty enclosure they flew irregularly, apparently because they missed the movement of the landscape through their field of view. Dudley and Ellington persuaded them to fly steadily by placing rotating cylinders with barber's–pole stripes on either side of the enclosure. When these were rotated so that the pattern of stripes seemed to move slowly downwind, the bees flew steadily into the wind, sometimes for as long as 1½ minutes.

Bumblebees beat their wings at 150 cycles per second, so filming with an ordinary movie camera would have been useless. Instead, Ellington and Dudley used a special high–speed camera running at 5000 frames per second, enough to give them about 30 pictures of each wing beat cycle. With their single camera they could not achieve a stereoscopic effect, but they were able to work out the movements of the wing in three dimensions by a method that depended on the assumption that the movements of the left and right wings were symmetrical. They filmed from an oblique angle and measured the positions of corresponding points on the left and right wings.

Bumblebees flying in the wind tunnel behaved in essentially the same way. When hovering, they held their bodies tilted at about 50 degrees to the horizontal, and their wings beat in a horizontal plane. At a low forward speed of 2 meters per second (bees fly between closely spaced flowers at about this speed), both the body and the plane of the wings were at 25 to 30 degrees to the horizontal, and in fast flight at 4.5 meters per second, the body was held at 10 degrees and the wings beat at 40 degrees. Whatever the speed, the wings beat at about the same frequency and through the same range of angles.

Dudley and Ellington made no attempt to observe the vortices in the wake; instead, they estimated the forces acting on the wings by measuring the angles of attack. In hovering, the wings moved like hummingbird wings, backward and forward in the same horizontal plane, turning upside down for the backward stroke and presumably producing lift in both stroke. In forward flight, the insect moved ahead through the air while the wings beat up and down, so the wings followed a sawtooth path, moving through the air at different angles in the down–and upstroke. This implies that the forces on the wings act at different angles in the two strokes. In fast flight the downstroke seems to supply most of the forward thrust. The wing moves faster relative to the air in the downstroke, because the forward velocity of the wings relative to the body is then added to the forward movement of the body through the air. The fast flight of the bee seems to be quite different from any gait of birds, which do not get thrust from their upstrokes.

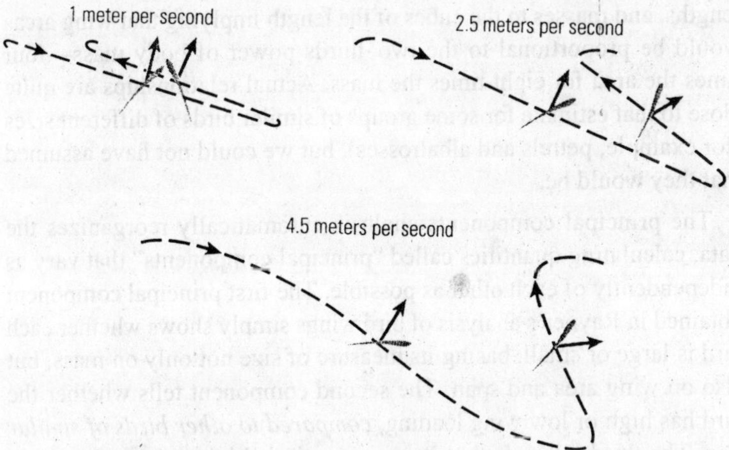

Fig. 8.11 : The path of the wings of a bumblebee flying at different speeds, and the aerodynamic forces on them. Each diagram shows two sections through the wing, one during the downstroke and one during the upstroke.

Wings to Suit Flying Habits

Birds, bats, and insects each have their own distinctive kind of wings: bird wings consist of feathers attached to the bones of a modified arm and hand; bat wings are made of skin stretched between immensely elongated fingers; and insect wings are thin sheets of cuticle stiffened by thickenings known as veins and by light pleating. Although each group conforms to its basic wing plan (for example, no insect or bat has evolved feathers), variations on the basic plan have evolved within each group to suit different flying habits. We have already seen that albatrosses have long but very narrow wings and that vultures have much broader ones: albatrosses have much higher wing loadings and aspect ratios than vultures of the same body mass. Similarly, the broad wings of a butterfly are very different from the narrow wings of a dragonfly. The relationship of body structure to flying habit is brought out sharply in a study by Jeremy Rayner.

Rayner studied the sizes and shapes of bird wings, using the statistical technique known as principal components analysis. He started with measurements of body mass, wing area, and wing span for hundreds of different species. Plainly, these measurements are related to each other (heavier birds generally have larger wings), but the relationships are not simple. If birds were geometrically similar to each other, wing areas would be proportional to the squares of the bird's lengths, and masses to the cubes of the length implying that wing areas would be proportional to the two–thirds power of body mass—four times the area for eight times the mass. Actual relationships are quite close to that estimate for some groups of similar birds of different sizes (for example, petrels and albatrosses), but we could not have assumed that they would be.

The principal components analusis automatically reorganizes the data, calculating quantities called "principal components" that vary as independently of each other as possible. The first principal component obtained in Rayner's analysis of bird wings simply shows whether each bird is large or small, basing its measure of size not only on mass, but also on wing area and span. The second component tells whether the bird has high or low wing loading, *compared to other birds of similar size.* The third tells whether its aspect ratio is high or low, compared to other birds of the same size and wing loading.

A graph shows the second and third components, the ones that are related to wing loading and aspect ratio. Different groups of birds occupy different parts of the graph, according to their different flying habits. Most seabirds have high aspect ratios and are placed in the upper half of the graph: gannets albatrosses, gulls, terns, and others are all found there. However, among seabirds, frigate birds (which soar on thermals) are well over to the right, with low wing loadings, and murres, or guillemots (which use their wings to swim underwater, as well as for flight), are well over to the right, with high wing loadings. Birds

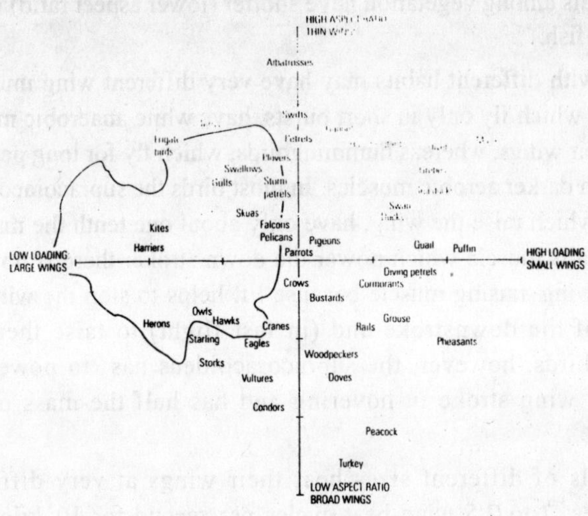

Fig. 8.12 : Principal components analysis of wing dimensions separates birds with different flying habits. The patch shows the area occupied by bats.

that soar in thermals over land (vultures, pelicans, stroks and others) lie in or near the lower left quarter of the graph: they have low wing loadings and low aspect ratios, as we have already seen. Birds of prey (owls, hawks, kites and eagles) have low wing loadings, so they can still fly well when their wings have to support a load of heavy prey. Game birds (grouse, pheasants, peacocks, and turkeys) have small, broad wings and do not fly well; they are found in the lower right quarter.

Though birds with similar flying habits are generally grouped near each other on the graph, groups with different habits sometimes mingle. The swifts ans swallows, which catch insects in the air, are mixed in with the seabirds because they too have long, narrow wings. The hummingbirds are quite near the diving birds because both have rather small, narrow wings, though for quite different reasons.

A patch of color shows how bats fit into the analysis: some have narrow wings and some rather broad once, but all have rather large wings for their size. Differences among bats (not shown on the graph) reflect their different habits. For example, the few bats that hunt for frogs, small mammals, or fish have lower wing loadings than most other bats, enabling them to carry heavy prey; but those that hunt for frogs or mammals among vegetation have shorter (lower aspect ratio) wings those that fish.

Birds with different habits may have very different wing muscles. Chickens, which fly only in short bursts, have white anaerobic muscle to beat their wings, whereas hummingbirds, which fly for long periods, have much darker aerobic muscles. In most birds the supracoracoideus muscles, which raise the wing, have only about one tenth the mass of the pectoralis muscle which power the down stroke: there is no need for large wing–raising muscle because lift helps to stop the wings at the end of the downstroke and (in fast flight) to taise them. In hummingbirds, however, the supracoracoideus has to power the backward wing stroke in hovering and has half the mass of the pectoralis.

Animals of different sizes beat their wings at very different frequencies: 2 to 2.5 wing beat cycles per second for 10–kilogram condors, about 20 cycles per second for 10–gram hummingbirds, 260 cycles per second for very large, 10–milligram mosquitoes, and even more for small midges. You can count the slow wing beats of a condor (if you are lucky enough to see one); you cannot see the individual wing beats of a hummingbirds, but their frequency is high enough that you can hear a low hum; and you can hear a high–pitched buzz from a mosquito.

It may seem obvious that condors cannot flap their wings fast enough to buzz, but let us think about the reason why. Imagine two geometrically similar flying animals, one 10 times longer than the other and 1000 (10 × 10 × 10) times heavier. The larger one has 1000 times more wing muscle than the smaller one, able to do 1000 times more work to accelerate the wings at the beginning of each stroke. Work is needed to give the wings kinetic energy, ½ × mass × (speed)^2. But even though 1000 times the work is available, the wings have 1000 times the mass. The two factors cancel each other, and the larger animal can accelerate its wings to only the same speed as the smaller one.

To work out the effect of size on wing beat frequency, we next consider the span of the wings. The large animal has wings 10 times longer than the small one, if they beat through the same angle. Thus, if the wing speed is the same as for the small animal, the wings will take 10 time longer to go through a cycle, and the wing beat frequency is divided by 10. This argument (which is admittedly oversimple) says that if a 10–milligram mosquito beats its wings at 260 cycles per second, a 10–gram hummingbird (1000 times heavier) should beat at 26 cycles per second and a 10–kilogram condor (a further 1000 times heavier) at 2.6 cycles per second. These frequencies are not far from the truth. A similar argument explains why elephants take fewer strides per second than mice.

Wings That Oscillate

The 260–cycle–per–second wing frequency of a mosquito is high, but other small insects beat their wings even faster, at up to 1000 cycles per second in a midge. This amazingly high rate would not be possible if the insects did not have special wing muscles. Those in sects whose wing muscles have properties like those of human muscles, though capable of shortening faster, can beat their wings at frequencies up to only 100 cycles per second. Flies, bees and wasps, beetles, and the insects that are properly known as bugs have a very special kind of muscles called fibrillat flight muscle, and some of them have very much higher wing beat frequencies.

It become clear that was something odd ablout fibrillar flight muscle when electrodes were put into the wing muscles of tethered flies to record the electrical signals (called action potentials) that travel through muscles when the are stimulated by their nerves and make them contract. Ordinary mucles need an action potential to strat every contraction, but these muscles were passing only three action potentials every second, even though they were making 120 contractions each second to beat the wings—there was only one action potential for every 40 contractions. It seemed that the muscles were somehow going into a state of oscillation that needed only occasional stimuli to sustain it.

The nature of the oscillations was made clearer by an ingenious experiment performed by Ken Machin and Johan Pringle at Cambridge University. They wanted to avoid the special problems of experimenting with tiny muscles, whose high frequencies of oscillation would be difficult to observe, and for that reason they used the largest insects they could get, rhinoceros beetles and giant water bugs up to ll centimeters long. Their experiment was equivalent to connecting the wing muscle to a tuning fork has a resonant frequency of vibration—when struck it will vibrate at that frequency, emitting its particular musical note. The vibrations of a struck tuning fork gradually die away, but a muscle capable of very rapid contractions could start a fork vibrating and *keep it indefinitely* by making a contraction in each cycle of the fork's resonant vibrations. In Machin and Pringle's experiment, however, the muscle did not sustain the vibrations of a simple fork, but of a complex electronic device that had a resonant frequency that could be altered as the experimenters chose. They found that a muscle from a beetle, which would in life have made 40 wing beat cycles per second, could be made to work at frequencies at least from 30 to 70 cycles per second by adjusting the resonant frequency of the apparatus. There was no need to change the frequency of the electrical stimuli: the muscle automatically matched its freequency of contraction to the resonant frequency of the apparatus. The muscle would be capable of driving tunning forks of a wide range of frequencies, each at its own specific frequency.

To appreciate the signigicance of Machin and Pringle's experiment we need to know more about resonant systems. Any system that has mass and elastic compliance (meaning that it can be distorted and will spring back elastically) has a resonant frequency and can be made to vibrate at that frequency much more easily than at any other. To demonstrate this, take a rolled magazine and attach to it a chain of long, thin rubber bands. The demonstration will work well if the weight of the magazine is enough to stretch the rubber bands by about 25 centimeters (10 inches). Take the free end of the chain of rubber bands in your hand and hold the magazine suspended. Now move your hand repeatedly up and down through a distance of about 5 centimeters (2 inches). If you do this very slowly, the magazine will mimic the movements of your hand, rising as it rises and falling as it falls, through about 5 centimeters. If you do it very fast, the magazine will remain almost stationary as your hand moves up and down. However, there is an intermediate frequency (about 1 cycle per second if you perform the experiment exactly as suggested) at which small up and down movements of the hand will cause larger up and down movement of the magazine than at any other frequency. This is the resonant frequuency, and its value depends on the mass and the elastic compliance. If you use two copies of the magazine, doubling the mass, the frequency is reduced by about 30 percent. If you use twice as long a chain of rubber bands, doubling the compliance, the frequency is also reduced by 30 percent.

Machin and Pringle's experiment showed that the body of an insect with oscillating wings contains a resonating system similar in principle to the magazine and rubber bands. Insect bodies have threee parts: the head, the thorax to which the wings and legs are attached, and the abdomen. The wings and thorax together have a resonant frequency at which they can be made to vibrate most easily. The most important mass in this system is the mass of the wings; they are only a tiny fraction of the total mass, but it is their mass that matters because they are the moving parts. The elastic compliance is partly in the cuticle of the thorax wall and partly in the muscle themselves. The importance of resonance in setting the frequency of oscillation is very easily demonstrated by a simple experiment. If the wings of a tethered fly are cut short, reducing their mass, they beat at a higher frequency. If, on the other hand, they are loaded with small weights, their frequency of beating is reduced. In the intact fly, just as in Machin and Pringle's experiment, the muscles adjust their frequency of contraction to match the resonant frequency of the system that they are driving.

The thorax could be constructed as a nonresonant system. In that case the muscles would have to do work to accelerate the wings at the beginning of each stroke and negative work (acting as brakes) to stop them at the end. The quantities of work would be large: Charles Ellington's calculations for various inscts show that the positive work needed to accelertarate the wings is typically about twice the work that has to be done against aerodynamic drag. If the system is resonant, this extra work is not required because as the wings are stopped at the end of a stroke, their kinetic energy is converted to elastic strain energy in the muscles and thorax wall. Later, this energy is returned in an elastic recoil for the next stroke. Thus energy is saved in much the same way as it is saved by tendon elasticity in running. The resonance of the thorax has two important functions: it saves energy, and it enables fibrillar flight muscles to operate at extremely high freqencies in small insects.

No such resonant system has been found in birds and bats. Their wings beat at frequencies far below the range at which fibrillar flight muscles are needed, but there does seem to be scope for energy saving. Torkel Weis–Fogh estimated that if hummingbirds had resonant wings, the work that their muscles have to do when they hover would be reduced by 43 percent.

The Hard Work of Flying

Gliding is flight powered by gravity. It has been described that gliding birds and bats lose height at rates between about 1.0 and 2.5 meters per second. As they descend, they lose an amount of potential energy equal to (mass × gravity × height change). A 1–kilogram mass descending 1 meter loses $(1 \times 10 \times 1) = 10$ joules potential energy, so gliding birds lose potential energy at rates between about 10 to 25 joules per second (that is, between 10 and 25 watts) for every kilogram of their mass. The lost potential energy is the power that propels the birds through the air, doing work against drag. In level flight the power has to be supplied by muscles that are unlikely to work with better than 25 percent efficiency—4 joules of metabolic energy must be used to perform 1 joule of work, metabolic energy cost of flight to fall between 40 and 100 watts per kilogram.

The actual power has been measured by experiments with animals flying in wind tunnels. Vance Tucker of Duke University was the first to make such a measurement. He trained budgerigars (small Australian parrots) to fly in a wind tunnel wearing masks connected to oxygen analysis equipment: this was the aerial equivalent of Richard Taylor's measurements of the oxygen consumption of mammals running on moving belts. When they were resting, the budgerigars used 20 watts of metabolic energy per kilogram of body mass, but when they flew at 10 meters per second they used oxygen six times faster, at a rate corresponding to 120 watts per kilogram. The difference of 100 watts per kilogram (shown in the graph) is the energy cost of flight and is at the top of the range suggested by our initial calculation. The birds used more power when they flew more slowly and also when they flew faster. This had been expected from studies of aircraft flight, because the power needed to propel an aircraft is high at low speeds (when induced power is high) and high at high speeds (when profile power is high), and is lowest at an intermediate speed, the minimum power speed.

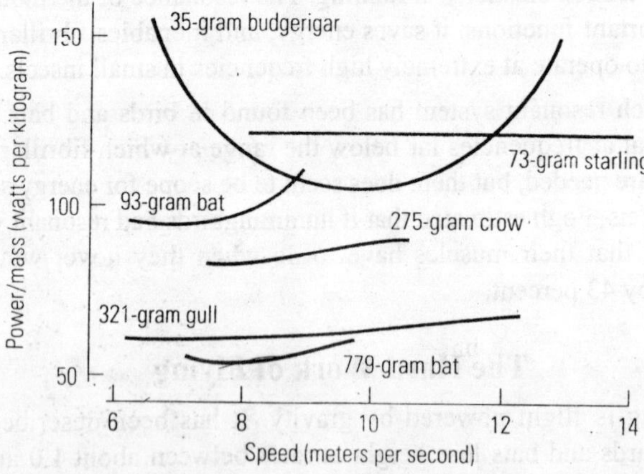

Fig. 8.13 : The metabolic energy cost of flight at different speeds, for some birds and bats. The resting metabolic rate has been subtracted from the rate during flight to obtain the power (rate of energy use) required for flying.

Since Tucker performed his pioneering experiment, he and others have made similar measurements on various birds and bats. They have found metabolic energy costs ranging from about 50 watts per kilogram (for a 800–gram bat) to 120 watts per kilogram (for 70–gram starlings), in good agreement with our rough calculation. Budgerigars and pigeons used less power at a moderate speed than when flying faster or slower, as expected, but starlings, crows, and gulls gave flat graphs showing little change of power consumption with speed. So far, this anomoly has not been explained satisfactorily.

Charles Ellington and colleagues at Cambridge have recently succeeded in measuring the oxygen consumption of bumblebees. Many measurements had previously been made of the oxygen consumption of tethered insects beating their wings, but these measurements were the first for insects in free flight.Insects breathe through spiracles (holes in their sides), not through the mouth or anything corresponding to a nose, and the Cambridge group decided that it would be too difficult to attach a mask to collect the air that the bees breathe out. Instead, they repeatedly circulated the same small volume of air through their wind tunnel. The bee removed a little of the oxygen during each circuit of the air, causing the oxygen concentration to decrease gradually (but only by a very little). Indeed, it decreased so little that they had to use a specially modified oxygen analyzer to record the changes. The measurements showed that the bee used oxygen at the same rate at all speeds between hovering and fast flight at 4 meters per second. The rate of consumption was much higher (in proportion to its mass) than for the birds and bats: 300 watts per kilogram. The rough calculation of likely energy cost that we made for birds and bats is inappropriate for bees because their beating wings move through the air many times faster than the body.

The Kori bustard (*Ardeotis kori*) seems to be the largest flying animals: Geoffrey Maloiy and I weighed one in Kenya that had a mass of 16 kilogram (35 pounds). Larger modern birds such as ostriches and Emperor penguins cannot fly, but a few fossil bones have been found of *Argentavis,* a gigantic bird of prey whose mass has been estimated as 80 kilogram (nearly 180 pounds), and which had well developed wings.

It is difficult for very large animals to produce enough power for flight.We used the rates of loss of height for gliding birds to estimate that a gliding bird requires mechanical power equal to 10 to 25 watts per kilogram. Model gliders lose height at similar rates to full–sized ones, which suggests that the power requirement is the same for aircraft of all sizes. In that case, it would presumably also be the same for small birds as for large ones. However, another argument seems to tell us that larger aircraft need more power per unit mass for flight than smaller ones. Thus flying birds of all sizes must produce *at least* 10 watts per kilogram. To do so, a hummingbird's muscles beating 26 times per second would have to do at least 0.4 joule of work per kilogram of body mass in each beat; a condor's muscles beating 2.6 times per second would have to do even more. There must be a size limit above which the muscles cannot do enough work.

The study of falpping flight has presented formidable challenges to research workers. The aerodynamic forces on wings that keep stopping and starting cannot be calculated by the conventional aerodynamics of fixed–wing aircraft, so novel thorìès have been needed. These relate the forces on wings to the swirling movements of air in the vortices created by the flapping wings, but these vortices could not be photographed sartisfactorily and measured until the helium bubble technique had been devised. Similarly, the oxygen consumption of insects in free flight could not be measured until Ellington had perfected his astonishingly sensitive equipment.

Great progress has been made in the past 15 years, but our understanting of flapping flight remains very imperfect. One of the insights that seems most likely to be helpful in future studies of flight is that several gaits are involved. There seems to be a continuous gradation in the flight styles of bees, from hovering to the fastest forward flight, but the slow vortex ring gait of birds and bats seems distinct from their fast continuous vortex gait, and bounding flight is another distinct gait that we will not understand fully until we know more about the physiology of the wing muscles.

Wings and Flight in Birds

The part of skeletan of a bird wing which corresponds to the forefeet of other tetrapods is grossly modified. Its largest component is the carpometacarpus, which develops from three metacarpals and some carpals that fuse together as a single as a bone. The three metacarpals are those of the second, third and fourth digits, and a few phalanges of these digits articulate with them. The first digit (thumb) and fifth digit (little finger) can be found as vestiges in some embryos, but not in adults.

The radius and ulna cannot cross over each other but remain always parallel. If they could cross, muscles would have to be used to prevent twisting of the wing during flight. The elbow and wrist joints are so constructed that when the elbow is bent the wrist automatically bends as well, by a parallel–rule machanism. These movements occur when the bird folds its wings.

The main flight feathers are the primaries and secondaries, which are attached along the posterior edges of the hand bones and of the ulna, respectively. The group of smaller feathers known as the alula or bastard wing is attached to the second digit (i.e. to the most anterior of the three digits). Other small feathers are arranged to give the wing a streamlined surface around the bones and muscles.

The pectoral girdle and sternum are almost as peculiar as the wing itself. To understand them, one must know how the wing muscles are arranged. The largest muscle is the pectoralis, which is responsible for the downstroke of the wings. In reptiles, which stand with the humerus more or less horizontal, pointing laterally, the pectoralis has an important weight–supporting function. It runs from the humerus to the sternum and interclavicle, and tension in it keeps the humerus horizontal and so keeps the animal's chest off the ground. The coracoid, runnig between the sternum and the shoulder joint, prevents the pectoralis from pulling the shoulders in towards the mid–line. In flying birds, the pectoralis is enormous and the sternum is greatly enlarged, with a deep keel, to provide much of its origin. The coracoids, which prevent the pectoralis from pulling sternum and shoulders towards each other, have become stout pillars of bone. The scapulae are long, but attached by muscle to the ribs as in reptiles. The clavicles have apparently joined to from the furcula (wishbone) which, though generally a single bone, has separate halves in some parrots and owls.

The pectoralis muscle which is responsible for the downstroke of the wing runs directly from the sternum and furcula to the humerus. The supracoracoideus muscles , which is responsible for the upstroke, work less directly. It also originates on the sternum but its tendon runs over a notch in the pectoral girdle which serves as a pulley, so that though the muscle lies ventral to the wings it serves to serves to raise them. The surface of the notch is covered by cartilage, and the sliding tendon is presumably lubricated in the same way as joints between bones. In many birds that fly strongly including pigeons, budgerigars and hummingbirds (these examples are chosen because experiments on

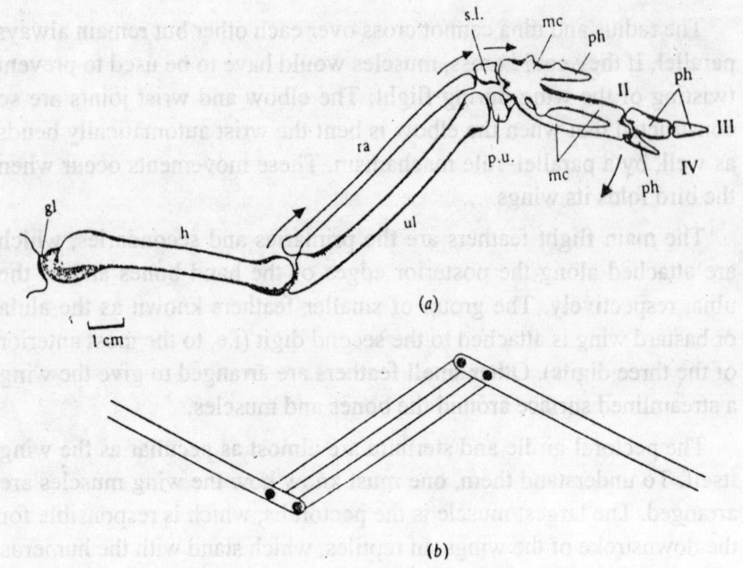

Fig: 8.14: (a) Right wing skeleton of domestic fowl. The arrows indicate movements which occur as the wing is folded. gl, glenoid ; h, humerus; mc, carpometacarpus; ph, phalanx; p.u., s.l., carpals; ra, radius; ul, ulna; II,III, IV, digits.

them are described later in this section) the pectoralis and supracoracoideus muscles together make up 25% or more of the weight of the body. In running birds such as the domestic fowl and waterfowl such as the coot (*Fulica atra)*, they tend to be smaller, making up, in the coot, only 11% if the weight of the body. It will be shown later that the downstroke is the main power stroke in normal flight but that the upstroke produces important forces in hovering. In most birds the supracoracoideus muscle which produces the upstroke is only around one tenth the weight of the pectoralis, bit in hummingbirds (which hover a great deal) it is often half the weight of the pectoralis.

Anyone who has eaten chicken and pigeon will have noticed that chicken breast meat is white, while pigeon breast meat is much darker. The difference is due to the different proportions of red and white muscle fibres in them. The pectoralis muscle of the domestic fowl contains 67% white fibres, 22% intermediate fibres and 11% red fibres. That of the pigeon contains only 14%% white fibres, but 86% red fibres. The supracoracoideus muscles differ in the same way, though the percentages are not exactly the same. The red and white fibres differ in properties in the same way as in fishes. The domestic fowl and its wild relatives do not generally fly far, so white muscle is appropriate for beating their wings. Pigeons are capable of long flights and depend on red fibres for sustained power output. The pectoralis and supracoracoideus muscles of the humming bird *Archilochus* consist entirely of red fibres, which provide the power for sustained hovering.

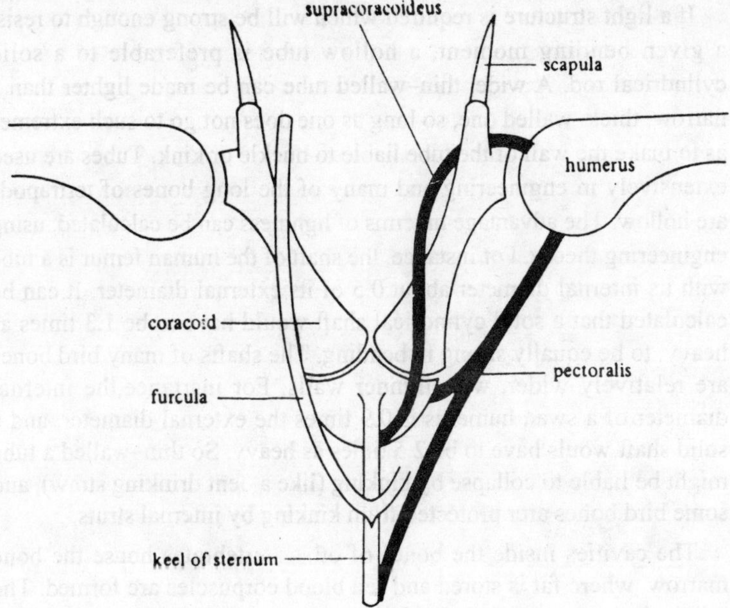

Fig. 8.15: A diagrammatic anterior view of the pectoral girdle and sternum of a bird, showing the arrangement of the main flight muscles.

Though the breast muscle of domestic fowl is white, other muscles including leg muscles are darker. They presumably include a higher proportion of red fibres. Red leg muscles seem apppropriate, since fowls run about a good deal. It is less easy to explain why the small muscles in the wing itself, which are reponsible for adjustments of wing position rather than for actually beating the wings, also contain substantial proportions of red fibres.

Bone is the densest tissue in birds, and lightness is important because the power needed for flight depends on the weight of the body. Lightness is particularly important in the wings because the work which has to be done accelerating them in every wing beat is proportional to their moments of inertia about the shoulder joints. The long bones both of the legs and of the wings, while light, must be strong enough to withstand the forces which can be expected to act on them. Partcularly dangerous are forces acting at right angles to bones, tending to bend them: readers who doubt this are reminded that it is easier to break a stick buy bending it than by pulling on its ends. An example of a bone subject to large bending moments is the humerus, which has to withstand large aerodynamic forces acting on the wing.

If a light structure is required which will be strong enough to resist a given bending moment, a hollow tube is preferable to a solid cylindrical rod. A wide, thin–walled tube can be made lighter than a narrow, thick–walled one, so long as one does not go to such extremes as to make the wall of the tube liable to buckle or kink. Tubes are used extensively in engineering and many of the long bones of tertrapods are hollow. The advantage in terms of lightness can be calculated, using engineering theory. For instance, the shaft of the human femur is a tube with its internal diameter about 0.5 of its external diameter. It can be calculated that a solid cylindrical shaft would have to be 1.3 times as heavy, to be equally strong in bending. The shafts of many bird bones are relatively wider, with thinner walls. For insrtance,the internal diameter of a swan humerus is 0.9 times the external diameter, and a solid shaft wouls have to be 2.5 times as heavy. So thin–walled a tube might be liable to collapse by kinking (like a ɔent drinking straw), and some bird bones arer protected from kinking by internal struts.

The cavities inside the bones of otł ertebrates house the bone marrow, where fat is stored and red blood corpuscles are formed. The relatively larger spaces in bird bones are in general only partly filled by marrow, and partly by air. There are air–filled cavities in the skull which are extensions of the nasal and middle–ear cavities, but the cavities in other bones are continuous with the air sacs.

Fig. 8.17 shows the movements involved in ordinary horizontal flight. The wings beat up and down. During the downstroke (*c*) the primary feathers at the wing tip are bent upwards and forwards, indicating that an upward, forward force is acting on them. During upstroke, they are not obviously bent, indicating that forces on them are small. Fig. shows how this might happen. Since the wings rise and fall while the bird moves forward, their path through the air is sinuous. The figure shows longitudinal sections through the wing, during the downstroke and during the upstroke. In the downstroke the outer part of the wing is set at a substantial angle of attack, so that lift and drag act as indicated. The resultant force on the wing has an upward

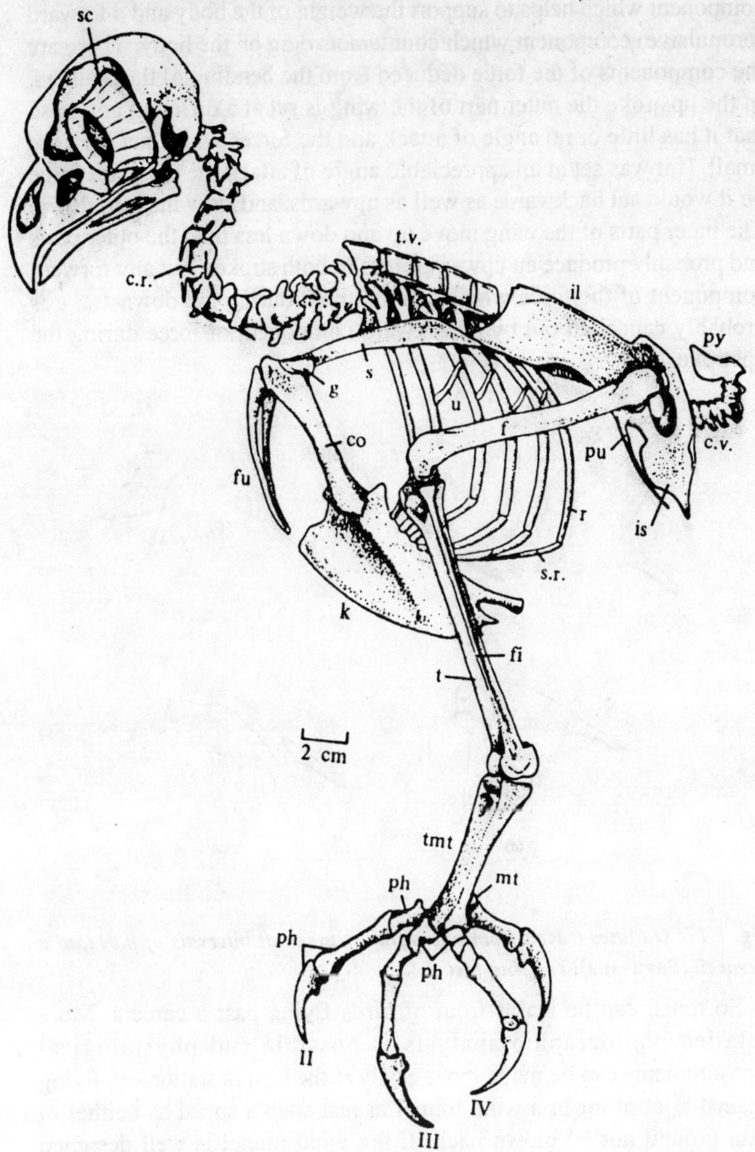

Fig. 8.16: Skeleton with wings removed of an eagle owl (*Bubo bubo*). co, coracoid; c.r., cervical rib; c.v., caudal vertebrae; f, femur; fi, fibula; fu, furcula; g, glenoid; il, ilium; is, ischium; k, keel on sternum; mt, first metatarsal; pu, pubis; py, pygostyle; r, rib; s, scapula; sc, scleral ossicles; s.r., sternal rib; t, tibiotarsus; tmt, tarsometatarsus; t.v., vertebrae; u, uncinate process; I–IV, digits.

component which helps to support the weight of the body and a forward (propulsive) component which counteracts drag on the body. These are the components of the force deduced from the bending of the feathers. In the upstroke the outer part of the wing is set at a different angle, so that it has little or no angle of attack and the forces which act on it are small. If it was set at an appreciable angle of attack the resultant force on it would act backwards as well as upwards, and slow the bird down. The inner parts of the wing move up and down less than the outer parts and probably produce an upward forces in both strokes, but any forward component of the force which acts on them during the downstroke is probably cancelled out by a backward component of force during the upstroke.

Fig. 8.17: Outlines traced from photographs taken at intervals of 0.01s, of a Great tit (*Parus major*) flying fast.

So much can be learnt from of birds flying past a camera. More detailed photographic analysis is possible and physiological measurements can be made more easily if the bird is stationary, flying against a jet of air in a wind tunnel at just such a speed as neither to gain ground nor be blown back. If the wind tunnel is well designed airflow (in the absence of the bird) will be smooth and uniform throughout the jet. The movements of air relative to a bird flying and keeping station in such a wind tunnel against a wind velocity *u* will be the same as if the bird were flying at velocity *u* through still air. Several recent investigations have involved training birds to fly in wind tunnels. Air is sucked in at the left and blown out at the right by the large fan *c*. The vanes *e* and the 'honeycomb' *f* stop the jet from swirling around. Any unevenness in air velocity in this part of the tunnel is swamped by the sudden uniform acceleration of the air as it passes through the contraction. The bird flies in the position shown. The whole tunnel is suspended, pivoted at *b*, so that ascending and descending flight can be simulated.

Birds have been trained to fly in wind tunnels by various means. An end view of the working end of the tunnel is shown as an inset. A tube fixed diagonally across the opening has a teaspoon bowl, soldered to its end. Maple peas were rolled down the tube into the bowl, and the bird could only get them by flying in the required position. In other experiments with other species the working sections of wind tunnels have been enclosed by grids which could be electrified, and the birds learned that they would suffer a mild electric shock if they landed while an experiments was in progress.

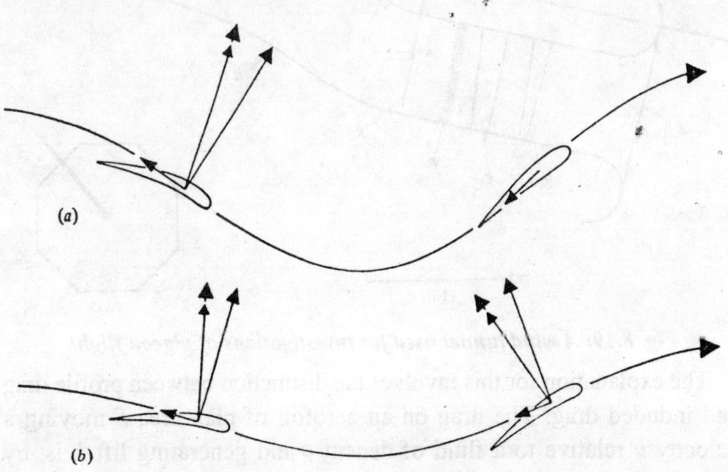

Fig. 8.18: A diagram showing the paths through the air of (a) the outer and (b) the inner part of a bird's wing in normal fast flight, and the forces which probably act on these parts of the wing.

Dr. V.A. Tucker has measured the oxygen consumption of budgerigars (*Melopsittacus*) flying in a wind tunnel. Transparent plastic masks were used, as in the experiment with a lizard. The masks were rounded and fitted the head much more closely than the ones used on lizards, and probably did not interfere at all severely with streamlining. They show as one would expect that oxygen is used more rapidly in ascending than in descending flight, and also that in horizontal flight oxygen is used less rapidly at 35 km h^{-1} than at higher or lower flying speeds. In this, the budgerigar resembles man–made aircraft. There is a particular speed at which an aircraft needs least power for horizontal flight. If is flies faster or slower than this minimum power speed, its engines must supply more power.

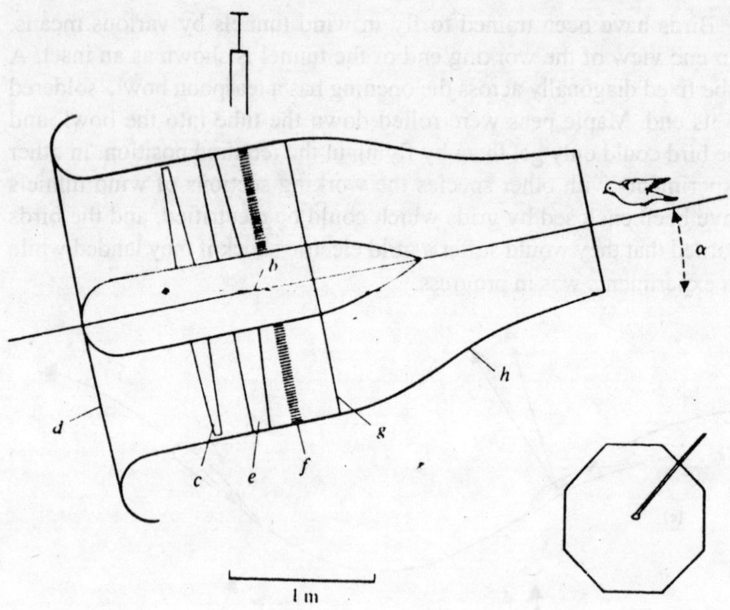

Fig. 8.19: A wind tunnel used for investigations of pigeon flight.

The explanation for this involves the distinction between profile drag and induded drag. The drag on an aerofoil of plan area *S* moving a velocity *u* relative to a flùid of density ρ and generating lift *L* is, by equation 4.3.

$$\frac{1}{2}\rho u^2 SC_{D0} + \left(2KL^2 / \pi R\rho u^2 S\right).$$

C_{Do} is the zero–lift drag coefficient, *R* is the aspect ratio (wing span/ mean chord) and *k* is a factor which depends on how the wings taper towards their tips, but is normally about 1. The first of the two terms in the above expression represents the profile drag which would act even if the angle of attack were set so that no lift was produced. The second term is the induced drag which is associated with lift production. If a complete aircraft is being considered rather than just and aerofoil, drag on the fuselage has also to be considered. It is known as parasite drag. If the frontal area of the fuselage is A and its drag coefficient C_{DF}, the parasite drag is $\frac{1}{2}\rho u^2 AC_{DF}$.. Thus

Total drag = parasite drag + profile drag + induced drag

$$= \frac{1}{2}\rho u^2 (AC_{DF} + SC_{D0}) + (2kL^2 / \pi R\rho u^2). \qquad (11.3)$$

The power required to propel the aircraft is this total drag multiplied by the velocity

$$\text{Powere} = \frac{1}{2}\rho u^3 (AC_{DF} + SC_{D0}) + (2kL^2 / R\rho uS). \qquad (11.4)$$

The first of the two terms on the right–hand side of equation 11.4 (representing power needed to overcome parasite and profile drag) increases as u increases. The second term (power to overcome induced drag) decreases as u increases. There is therefore a particular value of the velocity u, at which the power required for flight is least.

Equation applies to fixed–wing aircraft, with wings and fuselage moving at the same velocity. A bird's wings, beating up and down, move rather faster than the rest of the body, but the explanation of the minimum power speed is otherwise the same as for fixed–wing aircraft.

The budgerigar, whose oxygen consumption varies with speed so nicely in accordance with theory, may be unusual among birds. Subsequent experiments with other species have not revealed a minimum–power speed. In one of these experiments starlings (*Sturnus*) flew freely in a wind tunnel, without masks. This tunnel was designed so that air flowed repeatedly around a closed circuit. Samples of the air were taken at intervals, and analysed. The starlings (of mean mass 73 g) used 23 cm^3 oxygen $g^{-1}h^{-1}$ at all speeds between 8 and 18 ms^{-1}.

Though an aircraft uses least power at the minimum–power speed, a rather higher speed is the most economical in terms of miles per gallon of fuel. It is known as the maximum–range speed since it allows the longest possible flight without refuelling. Consider the oxygen consumption of budgerigars per kilometre travelled in level flight, At the minimum–power speed, 35 km h^{-1}, 22 cm^3 oxygen g^{-1} h^{-1} is used, or 22/35 = 0.63 cm^3 g^{-1} km^{-1}. At 39 km h^{-1} the consumption would apparently be 23.5 cm^3 $g^{-1}h^{-1}$ or 0.60 cm^3 g^{-1} km^{-1}, which would allow a slightly longer range.

Fat is the main fuel for long flights by birds, which often accumulate fat stores of more than 25% of the weight of the body. Metabolism of

1 g fat uses 21 oxygen, so that the oxygen consumption of 0.6 cm^3 g^{-1} km^{-1} which has just been calculated implies use of fat at a rate of only 3% of the body mass per 100 km. This helps to explain why birds can make long migrations. Migrating birds fly 1400 km across the Sahara Desert apparently without an opportunity to feed, and others make even longer flights over the sea.

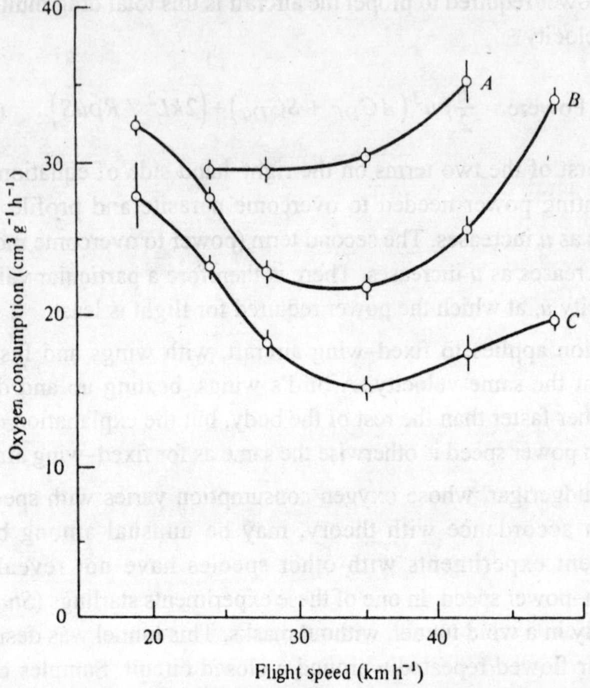

Fig. 8.20: Graphs of oxygen consumption against speed for budgerigars (**Melopsittacus*****) flying in a wind tunnel. (*****A*****) refers to ascending flight at 5° to the horizontal, (*****B*****) to level flight and (*****C*****) to descending flight at 5°.***

Hovering is stationary flapping flight. Kestrela (*Falco tinnunculus*) keep more or less stationary 10 m or so above the ground, watching for prey on the ground. They do this by flying into the wind at just such a speed as neither to make headway nor to be blown downwind. The next few paragraphs are not about this sort of hovering but about hovering in still air, which is a much more strenuous activity. The wings have to be moved fast enough to produce enough lift, althouth the body is not moving through the air. Hummingbirds weighing 2–20 g can hover indefinitely but pigeons weighing 0.4 kg can hover only for a very few seconds and large birds cannot hover at all.

Birds hovering in still air keep their bodies more or less vertical and beat their wings horizontally. Two techniques are used, one by hummingbirds and one by other birds. In the latter technique the wing surface is never very far from vertical, and if the wing were acting as a single aerofoil its angle of attack would be so high that it would stall. However, the primary feathers are spread so that they no longer overlap and are free to be twisted by the aerodynamic forces which act on them. They are twisted to lower angles of attack because each has its shaft much nearer the anterior than the posterior edge of the vane. Separation of the feathers so that they act as separate aerofoils increases the total aerofoil area by eliminating overlap, and so reduces the wing speed needed to produce sufficient lift.

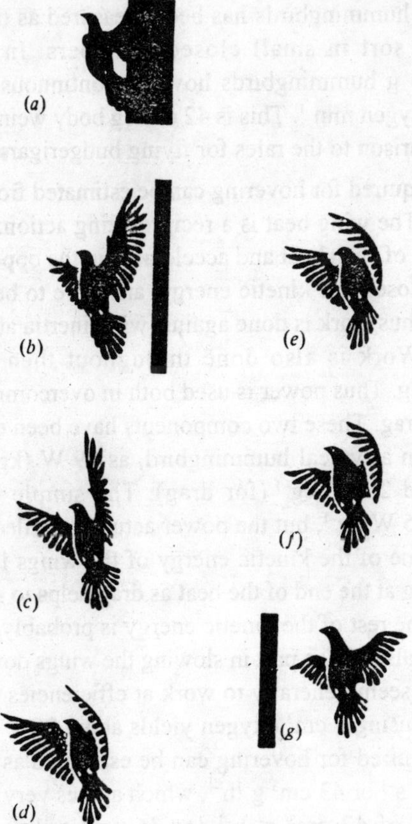

Fig: 8.21: Silhouettes traced from a cine film of a domestic pigeon hovering briefly before landing. The interval between successive silhouettes was about 0.017s.

Feather separation apparently prevents stalling at the wing tip. Further in on the wing, the tendency to stall may be reduced by the alula, which can be seen projecting. This supposition is based on the principle of the slotted wing as used in aircraft, and on experiments with bird wings in wind tunnels. It has been found that at high angles of attack bird wings may produce up to 25% more lift with the alula in the extended position than with it fixed in the normal position flat against the main wing surface.

Hummingbirds hover in front of flowers, using their long slender tongues to take the nectar on which they feed. They play the same role as pollinators of some flowers, as nectar–feeding insects do of many others. Theu will also hover to take sugar solution from suitably designed bottles, and are fed in this way in aviaries. Oxygen consumption of hummingbirds has been measured as they hovered at bottles of this sort in small closed chambers. In one of these experiments a 3 g hummingbirds hovered continuously for 35 min, using 2.1 cm^3 oxygen min^{-1}. This is 42 cm^3 (g body weingt)$^{-1}$ h^{-1}, which is high in comparison to the rates for flying budgerigars.

The power required for hovering can be estimated from engineering considerations. The wing beat is a reciprocating action: the wings are halted at the end of each beat and accelerated in the opposite direction. Each time they lose their kinetic energy, ans have to be given kinetic energy afresh. Thus work is done against wing inertia at the beginning of each beat. Work is also done throughout then beat, against aerodynamic drag. Thus power is used both in overcoming inertia, and in overcoming drag. These two components have been estimated from measurements on a typical hummingbird, as 29 W (kg body mass)$^{-1}$ (for inertia) and 26 W kg^{-1} (for drag). The simple total of these components is 55 W kg^{-1}, but the power actually needed is only 46 W kg^{-1} because some of the kinetic energy of the wings is converted to work against drag at the end of the beat as drag helps to slow the wings down. Most of the rest of the kinetic energy is probably lost as heat in the muscles that also play a part in slowing the wings down. Vertebrate striated muscles seem generally to work at efficiencies of about 20%, and metabolism using 1 cm^3 oxygen yields about 20 J, so the oxygen consumption required for hovering can be estimated as about $46 \times 5/20 = 12$ cm^3 kg^{-1} s^{-1} or 43 cm^3 $g^{-1}h^{-1}$, which agrees very welll with the measured valus of 42 cm^3 $g^{-1}h^{-1}$ (or 36 cm^3 $g^{-1}h^{-1}$ if the resting metabolic rate is subtracted, as representing energy being used for puoposes other than hovering).

It can be shown by further analysis that more power is needed for hovering than for forward flight at moderate speeds. Also, (power for hovering)/ can be shown to be larger for large birds than for small ones. As size increases, power required for hovering must increase faster than power available, so that birds above a certain size cannot hover. The largest hummingbirds weight only about 20 g, and this may be about the limit of size for birds which can hover without incurring an oxygen debt. Larger birds hover more briefly: 400 g pigeons are apparently unable to hover for more than about a second, and substantially larger birds cannot hover at all.

Small and medium–sized birds often take off by hovering and then gradually gathering speed, changing gradually from the hovering action to the normal flying action. Brids too large to hover cannot do this, and can only take off in still air by diving from a high perch or by taking a taxiing run. They must get up enough speed for their wings to lift them. The relationship between body size and take–off speed has already been discussed inconnection with a very large pterosaur. The largest birds that can fly, such as the Kori bustard (*Ardeotis kori*) weigh up to about 16 kg. Some flightless birds such as the ostrich (*Struthio*) and the Emperor penguin (*Aptenodytes*) are of cours much heavier.

The gliding ability of birds has been investigated in wind tunnels inculding the one illustrated in C.J.Pennucuick trained pigeons to glide and fly in this tunnel. With the wind speed fixed he varied the angle at which the tunnel was tilted. He found the angle at which the bird could only just glide. It could glide more steeply at the same speed if it used its feet to brake but it could not glide at shallower angles and had to resort to flapping flight. Observations were made at different wind speeds, finding the shallowest gliding angle for each. The results of these experiments and of similar ones with other species. It is not the angle itself which is plotted but the sinking speed, which is the rate at which the bird would lose height in still air. The falcon and vulture were much better gliders than the pigeons.

Fig. show that for each species (and for a man–made glider) there is a particular gliding speed at which the sinking speed is least. This is easily explained. Fig *a* shows the forces acting on a bird of weight mg which is gliding at an airspeed u and an angle θ, so that its sinking speed is $u \sin\theta$. The lift is L and drag D. At equilibrium

$$D = mg \sin\theta$$

$$L = mg \cos\theta$$

mg if θ is small.

Using these equations and equation 11.3

Sinking speed = $u \sin \theta$

$= Du / L$

$= \left(\rho u^3 / 2L\right)\left(AC_{DF} + SC_{D0}\right) + \left(2kL / \pi R \rho u S\right)$. (11.5)

The first of the two terms on the right–hand side of the equation increases as u increases, and the second decreases. There is therefore a particular airspeed u at which the sinking speed is least.

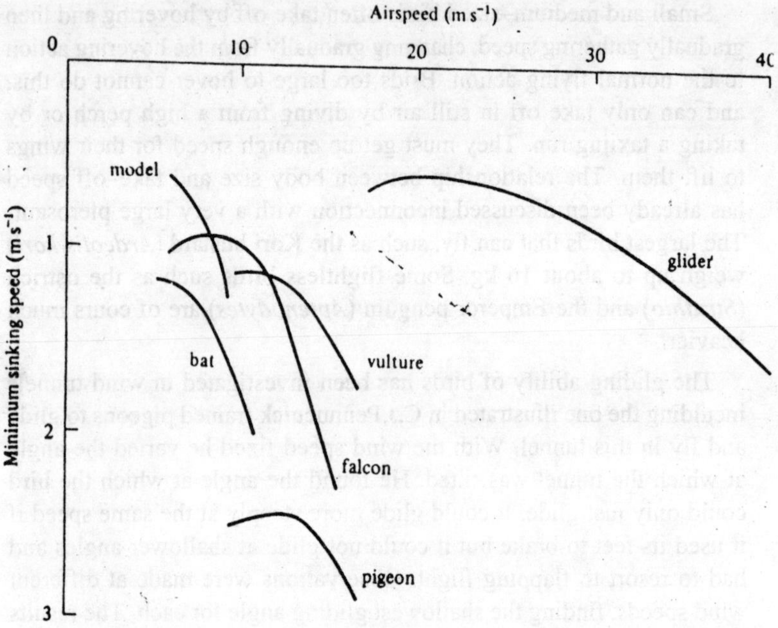

Fig. 8.22: Graphs of minimum sinking speed for gliding at various airspeeds for a glider (wing span 17 m), a model glider (wing span 1.1 m), a vulture (*Coragyps atratus*), a falcon (*Falco jugger*), the pigeon (*Columbia livia*), and a tomb bat (*Rousettus.aeghuptiacus*).

Just as there is an ideal area of selachian pectoral fins which enanbles them to give the required lift at a particular speed with least drag, so for a bird gliding at a particular speeds with least drag (p. 102), so for a bird gliding at a particular speed there is an ideal wing area which gives minimum drag and leadt rapid loss of height. The higher the speed the smaller the ideal area, and birds partially fold their wings when they are gliding fast. Just as fines of high aspect ratio give lift with least drag, so do long narrow (high aspect ratio) wings. However, too large a wing span could be a disadvantage. Long wings might be awkward at take–off or when fluing among branches. Also, the longer the wings the greater the bending moments at their bases and the stronger (and heavier) the wing bones to be. Aspect ratios of about seven are usual among. birds, but the albatross *Diomedea* has an exceptionally high aspect ratio of about eighteen.

Flapping flight uses a lot energy but birds can remain airborne with relatively little expenditure of energy by soaring. Many brids, like human pilots in gliders, make use of thermals. These are masses of hot air which rise periodically from griuyns heated by the sun. They form where the ground becones hottest: for instance, from dry dark areas and fron slopes which face the sun. A glider or gliding bird in a thermal will rise, if its sinking speed is less than the upward velocity of the air. It can gain considerable height by circling in the thermal and then glide away, losing height but possibly covering a substantial distance before it needs to gain height again. Vultures, some crows and gulls, and various other birds keep airborne for long periods in this way. For instance, a vulture returning to its nest after feeding was followed by a motor glider in East Artica. It travelled 75 km by soaring in 96 min, pausing to circle in five thermals.

Albatrosses also remain airborne for hours, hardly ever flapping their wings, but their technique is quite different, It depends on the wind speed being lower near the surface of the sea than it is higher up. They glide down gaining speed, and then facc into the wind and rise. A glider can always gain height at the expenes of speed, until it is moving too slowly for its wings to produce enough lift. As the albatross rises it loses speed, relative to the ground. However, its speed relative to the air (on which lift depends) may actually increase, because the wind it is facing is faster at higher levels.

It can be shown by applying aerodynamic theory that a vulture can rise in small weak thermals only if it glides slowly, but that an albatross can only rise in a weak wind gradient without losing air speed if it glides fast. It has been shown that the ideal wing area is greater at low speeds than at high ones. Vultures appropriately have wing areas about double those of albatrosses of the same weight.

Flying brids use oxygen estremely rapidly. Fig. shows that budgerigars use oxygen at up to 36cm^3 g^{-1}h^{-1} in ascending flight, and rates up to 55 cm^3 g^{-1}h^{-1} have been measured in turbulent air. The oxygen consumption of the Black duck (*Anas rubripes*) has been measured in much shorter flights, not in a wind tunnel, and found to be 14 cm^3 $g^{-1}h^{-1}$. These are very much faster than the maximum metabolic rates of raptiles. At high body temperatures lizards similar in mass to the black duck could use no more than 0.5 cm^3 $g^{-1}h^{-1}$. Even mammals (except perhaps bats, see p. 440) cannot use oxygen as fast as birds similar mass. Kangaroo rate (*Dipodomys merriami*) similar in mass to budgerigars use only 12–14 cm^3 $g^{-1}h^{-1}$ when running at top speed. Birds achieve their very high rates of oxygen uptake by means of peculiar lungs, which are described in the next section of this chapter.

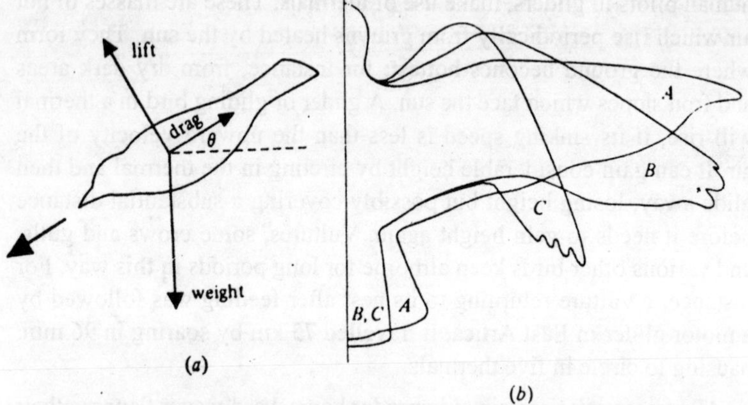

Fig. 8.23: (a) A diagram showing the forces which act on a gliding bird. (b) Outlines of a falcon gliding at (A) 6.6 m s^{-1} and (C) 14.3 m s^{-1}.

Even in horizontal flight at the minimum power speed, budgerigars were found to use oxygen 10–15 times as fast as they would probably have done at rest. Hovering hummingbirds used oxugen seven times as fast as when resting at the same temperature. Since the resting rate is enough to maintain the body temperature, flying and hovering birds might be expected to overheat. However, the movements of air past the body which are involved must help heat loss by convection, and spreading the wings exposes the flanks which are rather sparsely covered with feathers. Heat loss by evaporation is important in mammals but relatively unimportant in birds, which do not sweat. Evaporation is mainly from the respiratory system. It accounted for only 15% of heat loss from budgerigar flying in Tucker's wind tunnel in air at 20°C and 50% relative humidity. If birds were more dependent on heat loss by evaporation they would have to drinks more often on long flights, and migrations across the Sahara of broad seas might not be practicable. The budgerigars in the experiment just mentioned were losing water by evaporation at a rate of 6% of body mass per 100 km, but some of this must have been replaced by water produced by metabolism. Metabolism of fat produces an approximately equal mass of water so if fat were being used at the rate of 3% of body mass per 100 km the net rate of loss of water would beonly 3% of body mass per 100 km.

CHAPTER 9
GLIDING

Flying squirrels, which lack the refined as design of birds, use gliding as a fast and saving means of travel through the forest *Glaucomys volans*, photographed in.

Due to flying adaptations, animals become beautifully modified, due to which, they are offered the least possible resistance by air to the attainment of speed. Volant adaptations among the invertebrates occur in insects. Among the vertebrates, the volant adaptations are found in fishes, amphibians, reptiles, birds and mammals.

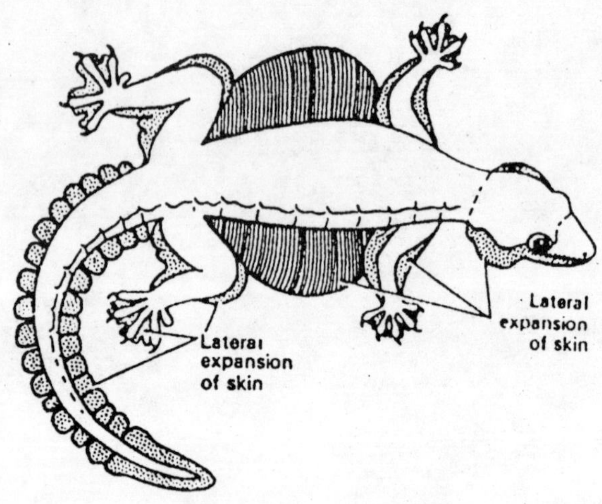

***Fig.9.1: Flying gecko* (Ptychozoon).**

Volant adaptations are concerned with the flight. This flight may be : (1) *passive* or *gliding type* characterized by leaping (jumping) from a high point and held up by certain sustaining organs, then glides to lower level. There is no locomotive force other than the gravity (ii) *True flight* is the aerial flight caused by the action of wings. It is found in insects, pterodactyles, birds and bats. In them the nature of development and structure of wings are quite different which suggest that the flight has evolved independently in different groups. In such flights the power is implied and the movement in air is sustained.

Types of Flight

Flight is of two sorts :

Passive or Gliding Flight

In this kind of flight, the animals only take an initial leap from a high point and are held up by certain sustaining-organs and impelled by gravity, they glide to a lower level in a horizontal plane. Sometimes, the distance covered by gliding may be many yards. This kind of flight can be compared to a gliding aeroplane without engine-power.

The gliding flights are performed by various lizards (flying dragon-*Draco volans*); fishes (*e.g. Exocoetus* etc.); birds (ostriches etc.); mammals (lemurs) and amphibians (*Rhacophorus*). The following are noticable adaptation met with in such forms:

(a) Development of Patagia: The sustaining surface for the gliding is a fold or series of folds of the skin known as *patagium*. This may be supported by ribs as in *Draco* (flying dragon). It lies between forelimbs and hindlimbs. This patagium can be folded like a fan against the sides of the body when not in use. Another example among reptiles is *Ptychozoon*. "The flying or fringed gecko", in which lateral expansion of skin (patagium) extends along the side of neck, body, tail and limbs and between toes.

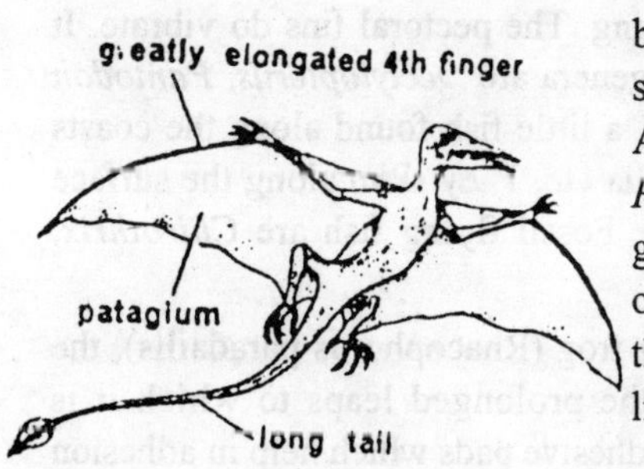

Fig.9.2: Pterodactyl.

Flying snakes (*Chrysopelea*) also leap by the concave ventral side.

The *pterodactyls* of Mesozoie era were volant creatures akin to birds. They possessed true flight patagia. Patagia were extensions between limbs supported by ribs. In flying lemur (*Galeopithecus volans*), which is a soaring mammal, patagium extends from side of the neck to the tip of tail even including digits, which are webbed as for aquatic life. In the bats, the patagium is supported mainly by the elongated forelimbs and the second, third, fourth and fifth digits. The first digit is free.

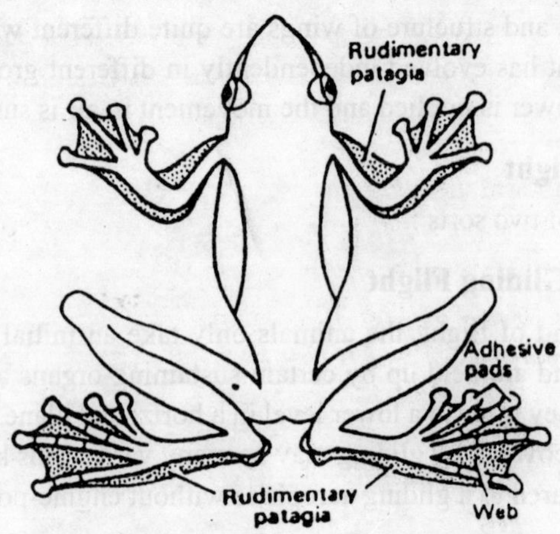

Fig.9.3: **Rhacophorus *(Flying frog).***

Traces of patagia are also found in front of and behind arm is birds which have adequate supporting function before the feathers usurped their place.

(b) Enlargement and High Insertion of Pectoral Fins : Among fishes (*Exocoetus* etc.) pectoral fins become enlarged in the form of parachutes and are highly inserted on the body. The lower lobe of tail is also invariably longer, helping in leaping. The pectoral fins do vibrate. It flies upto 200-300 meters. Other genera are *Dectylopterus, Pantodon* (African flying fish) and *Pegasus*, a little fish found along the coasts of Japan, China, India and Australia etc. They skim along the surface of the water for 40 feet or more. Fossil flying fish are *Chirothrix, Gigantopterus*.

(C) Webbing of Feet: In flying frog (Rhacophorus paradailis), the feet are webbed which sustain the prolonged leaps to which it is addicted. The digits terminate in adhesive pads which help in adhesion to trees. There are also rudiments of patagia in front and behind the arms.

Active or True Flight

In active flight, there is a sustained movement through the air. It covers long distances and involves locomotive force. True flight has evolved thrice among vertebrates: in the reptilian Pterodactyles, the birds and the bats. Whether flying fishes should be included is a much controversial question. However, the animals having true flight, move their wings with varying degrees of rapiduty or after having gained their high altitude by flapping their wings, they may also soar or sail on motionless wings for several hours continuously.

Modifications

Body contour in volant animals has been emphasized and is second only to that of the purely aquatic forms in its degree of perfection for the lessening of resistance.

Sustaining Surface: The sustaining surface is primitively, except in the fishes, a fold or series of folds of the skin known as the *patagium* (Lat. *patagium*, an edge or border). This may be supported in various ways, but with one exception, in the little lizards (*Draco spp.*) which inhabit the Indo-Malayan region, the limbs form the chief supporting agents. In the "flying dragons," *Draco*, just mentioned, the body is depressed and the sides extend outward into a pair of large, winglike membrances, supported by five or six elongated ribs. The entire device can be folded like a fan against the sides of the body when not in usc. The soaring powers are not very great but when resting among these luxuriant foliage of their habitat the animals are said to resembles butterflies in their habit of opening and closing the wings.

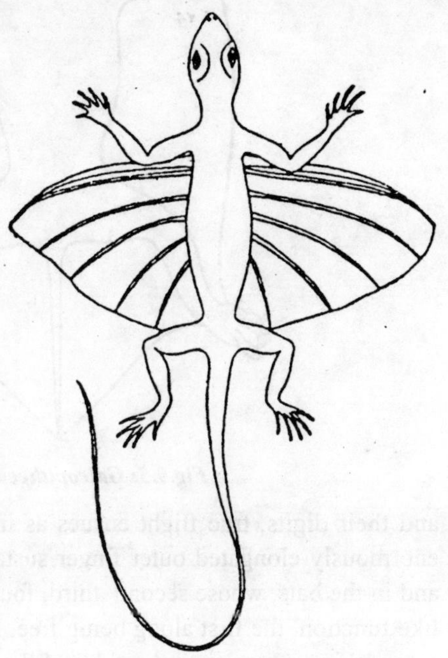

Fig. 9.4: "Flying dragon", **Draco volans.**

Most soaring mammals have the patagium supported between the fore and hind limbs and sometimes the skin-fold extends in front of the fore limb to the neck and again between the hind limbs and the tail. Perhaps the extreme of development may be seen in *Galeopithecus*, the so-called "flying lemur," for here the patagium extends from the sides of the neck to the tip of the tail, even including the digits, which are webbed as though for aquatic life.

Where the patagium is supported mainly by the elongated fore limbs

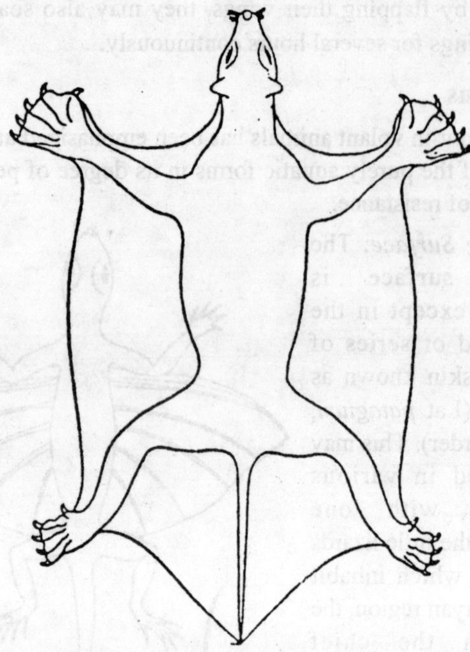

Fig.9.5: Galeopithecus volans.

and their digits, true flight ensues as in the pterosaurs, wherein the enormously elongated outer finger sustains over half the membrane, and in the bats, whose second, third, fourth, and fifth digits perform a like function, the first along being free. In both groups the membrane extends from the arm to the sides of the body and also to the front of the hind limb. An interfemoral membrane, which, however, may have existed, has not been demonstrated in the pterosaurs but is variable present in bats.

Gliding Vertebrates

Fishes: There are enumerated several genera of flying fishes, each of which represents a separate volant adaptation. Of these the first to be mentioned are the several species of the genus. *Exocaetus*, allied to the skippers and garfish, which live in all tropical and subtropical seas where the fly in shoals in their efforts to escape the relentless tunny and albicore. These flying fishes are trim-built creatures with large pectoral fins, which are the main organs of flight, and variably developed but much smaller pelvics. The lower lobe of the tail is invariably the longer and aids in giving the final impetus to the fish as

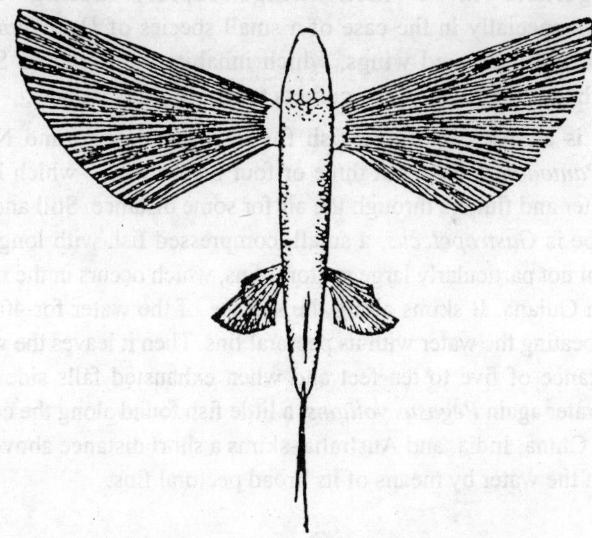

Fig.9.6: Flying fish, Exocaetus spilopterus.

it leaves the water and also in accelerating its speed if in the course of its flight it comes near enough to the surface of the sea. The length of flight is said to vary up to 200 or 300 yards and it is sometimes sufficiently high to strand the creature on the deck of an ocean-going craft. Whether the flight of *Exocaetus* is true flight or merely a soar is a much disputed question and the evidence, briefly summarized, is as follows:

The wing expanse is hardly sufficient for such extended soaring. The wings (pectoral fins) do vibrate, but whether due to muscular effect or to friction, as a flag is flapped in the wind, is not clear. The muscular development seems insufficient for true flight, but on the other hand it is more highly developed than in allied non-flying fishes. It may well be that while true flight as such does not exist among fishes, rapid wing vibration insufficient in itself to support or drive the animal may aid in maintaining or prolonging a soaring flight of which the main propulsive effort is acquired by the tail before having the water. At all events, their flight is remarkable and the creatures are one of the most interesting features of the storied tropical seas.

The various species of *Dactylopterus* are known as the flying gurnets and while the flying is by no means as sustained as in *Exocaetus,* of the former fishes Moseley writes: "I have distinctly seen species of flying gurnets move their wings rapidly during their flight.........especially in the case of a small species of *Dactyolpterus* with beautifully colored wings, which inhabits the Sargasso Sea." Moseley likens the flight of the gurnets to that of grasshoppers.

There is an African flying fish found in the Congo and Niger rivers—*Pantodon,* a form but three or four inches long—which leaps out of water and flutters through the air for some distance. Still another flying type is *Gastropelecus,* a small, compressed fish with long and curved but not particularly large pectoral fins, which occurs in the rivers of British Guiana. It skims along the surface of the water for 40 feet or more, beating the water with its pectoral fins. Then it leaves the water for a distance of five to tea feet and when exhausted falls sideways into the water again *Pegasus volitans,* a little fish found along the coasts of Japan, China, India, and Australia, skims a short distance above the surface of the water by means of its broad pectoral fins.

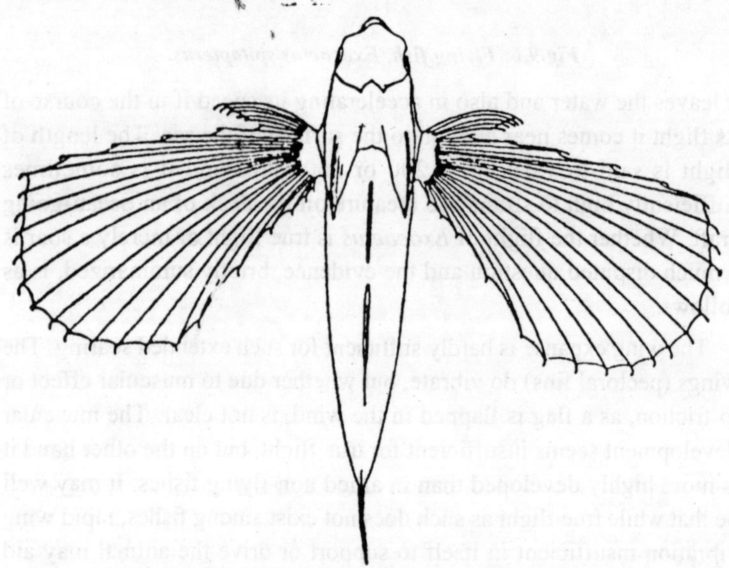

Fig. 9.7: Flying fish, gurnet, Dactylopterus volitans.

Several extinct forms have been described as "flying fishes." These are: *Dollopterus* from the Middle Trias of Jena, *Thoracopterus* and *Gigantopterus* from the Upper Trias of Austria, *Exocaetoides* and *Chirothrix* of the upper Createaceous of Mt. Lebanon, Syria. The last mentioned is of particular interest on account of the huge size of the pectoral fins, which seem to imply powers of flight fully equal to those

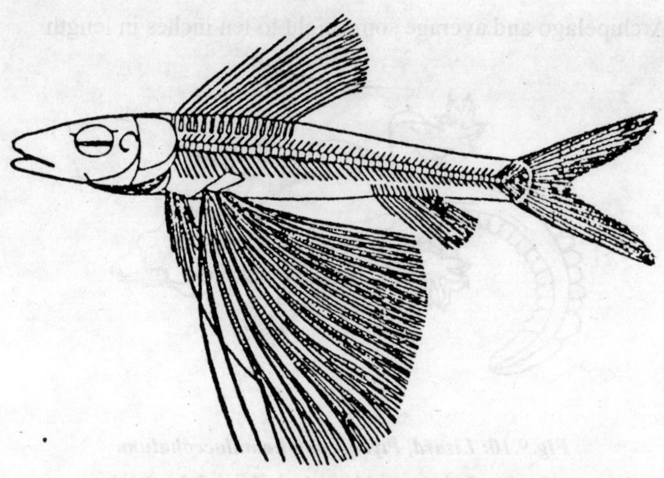

Fig.9.8: Fossil flying fish, Chirothrix libanicus.

Fig. 9.9: Flying frog, Rhacophorus reinhardtii.

of the living *Exocaetus* and *Dactylopterus*. We have therefore among fishes no fewer than ten separate adaptations to aerial conditions, one or two, possibly three of which approach very near to, if they have not attained, true flight.

Amphibia: The only volant adaptation among amphibia is that of the tree-frog, *Rhacophorus* whose webbed feet sustain it in the prolonged leaps to which it is addicted. This genus includes a large number of species in the Oriental realm, especially in Borneo. The digits terminate in adhesive pads, in common with those of other tree-frogs, and are connected by web-like expansions of the skin. There are also rudiments of patagia in front of and behind the arms. In *Rhacophorus pardalis* the total alar expanse is about three square inches, which would imply rather feeble gliding powers.

Reptilia: Lizards includes at least two genera and several species of gliding forms, of which the most remarkable is the flying dragon, *Draco,* already referred to in which the patagium is supported by a number of extended ribs. They occur principally in the Malay Peninsula and Archipelago and average some eight to ten inches in length.

Fig.9.10: Lizard, Ptychozoon homalocephalum.

Ptychozoon is the flying or fringed gecko of the Malay countries, which is bedecked with lateral expansions of skin along the sides of the neck, body, tail, and limbs, and between the toes. While these may aid in breaking the creature's fall, they may also, coupled with the colour, serve a cryptic function and render the animal less conspicuous against the back of the tree upon which it rests.

Several so-called flying snakes are recorded, such as *Chrysopelea,* the flying snake of Borneo, which descends obliquely through the air, its body right, and the ventral side concave to sustain the creature in its fall.

The pterodactyls or flying dragons of the Mesozoic were a very remarkable group of reptiles whose first recorded appearance is in rocks of the Rhaetic or uppermost Triassic period. They range through the Jurassic and on into the Upper Cretaceous, when they become extinct through racial death. They were undoubtedly skin to the birds, but that simply means, in all probability, derivation from a common, possibly Permian ancestry; nevertheless the two groups show a number of highly comparable, homoplastic characters, some of which have already been referred to. The remarkable thing is that, like the turtles, they first appear fully developed and characteristic of their order, with no record thus far discovered of their antecedent evolution, and the subsequent changes are mainly increases in size, perfection of the shoulder girdle articulation and loss of tail and of teeth. In size they range from that of a sparrow to the mightiest of nature's airplanes, for the replica of the late Cretaceous *Pteranodon* mounted at Yale measures 13 feet 6 inches in alar expanse and Eaton is authority for the statement that at least one individual, judging from the relative proportions of the bones which have been preserved, had an estimated breadth of 26 feet 9 inches from tip to tip. The pterodactyls possessed true flight, which in those from the Kansas chalk must have been sustained, as their remains are found in association with marine reptiles, fishes, and invertebrates, apparently far from the ancient shore. Three of the principal horizons whence these pterodactyls come, the Lias of Lyme Regis of England, the lithographic limestone of Bavaria, and the Kansas chalk, are all marine in origin, which is also true of the source of the Mesozoic birds. It is highly probable therefore that in each instance we have not as yet knowledge of the great bulk of the group, but only of a few of aberrant habits and adaptation.

Fig.9.11: Pterodactyl, Rhamphorhynchus phyllurus.

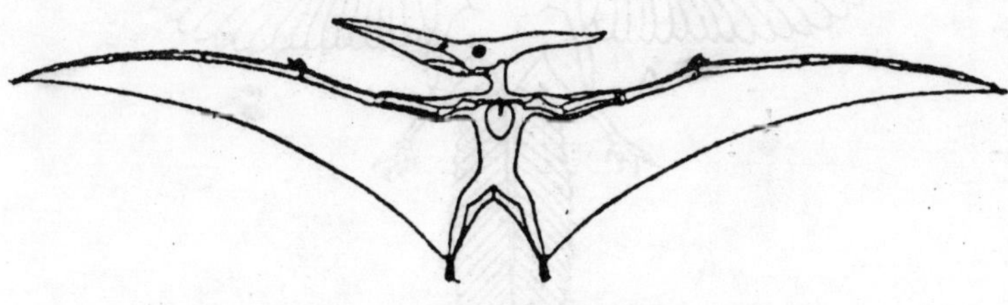

Fig. 9.12: Pterodactyl, Pteranodon longiceps.

Birds : The birds in all probability include but a single evolution to aerial life, although certain excellent authorities "believe more or less firmly that possibly birds had not one, but two points or origin, and feet that if we could follow back their lines of descent we should find that the ostriches came from one and the birds of flight from another".

If this be true, the ratite or ostrich group as a whole, with a single exception, have degenerated and lost the power of flight, although an examination of the skull and skeleton shows them to have been descended from flying normal birds; whereas in the carinate or flying birds loss of flight, while it has occurred (flightless rail, penguin, dodo, etc.), is relatively extremely rare. Flightless pterosaurs and bats, on the other hand, are inconceivable, as their flight mechanism involves the hind limbs which have, as a consequence, largely lost their terrestrial locomotion function; whereas birds, being double adapted, can lose their flying powers and still progress easily on the ground or in the water, as their legs are not thus involved.

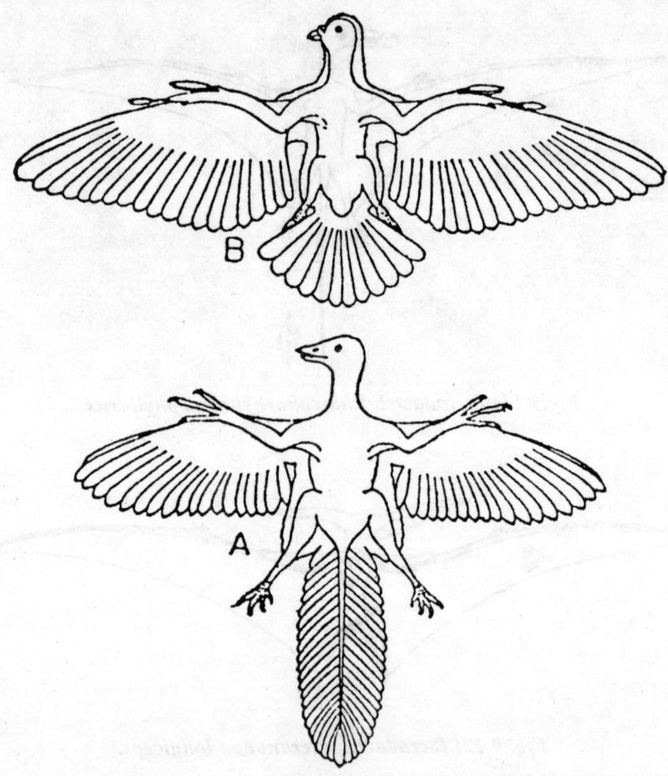

Fig.9.13 Reptilian bird, Archaeopteryx (A), compared with pigeon, Columbalivia (B).

Birds first appear in time in the Upper Jurassic (Solenhofen limestone) long after the initial record of the pterodactyls. These first birds, of which but two or three specimens have been recovered, are known as *Archaeopteryx* and *Archaeornis* and are so reptile-like that were it not for the preserved feathers it is doubtful whether they could be surely proved to have been birds. The reptilian traits are teeth, free clawed fingers in the hand, feeble breast-bone, abdominal ribs, etc.

Mammals : Among the mammals there are upward of thirteen separate volant adaptations, one of which, that of the bats, attained the power of true flight. Among the flying forms the first to be mentioned are the *marsupials*, of which the flying phalangers include several unrelated species: *Petaurus spp., Petauroides volans,* and *Acrobates pygmaeus*. These are characterized alike by having a well developed skin fold along the sides of the body between fore and hind limbs, and a feebly developed one in front of the fore leg. In each genus the flying form is especially related to a separate type of non-flying phalanger.

In *Petauroides* the flying membrane extends from wrist to ankle, but is very narrow along the distal segment of each limb. The tail is very bushy except for its prehensile tip, which is naked on the under side. The tail, together with the long fur of the body, must supplement to a considerable extent the buoyancy of the patagium. This genus, with its single species, includes the so called Taguan flying phalanger found in Australia from Queensland to Victoria.

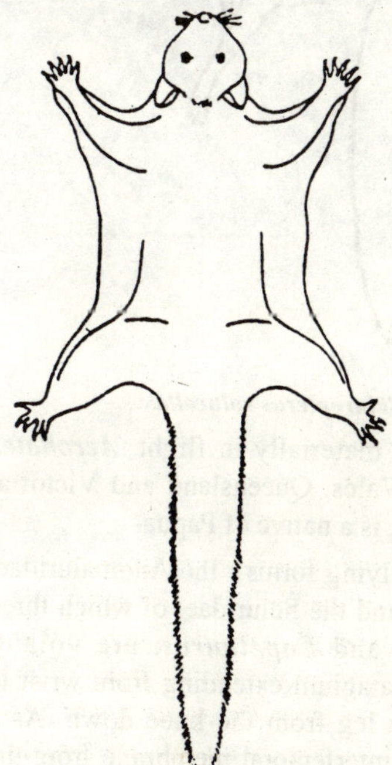

Fig.9.14: Flying phalanger (marsupial). Petaurus sciureus.

Petaurus has a much broader patagium and there is a naked prepatagium as well. The tail is very large and bushy, but lacks the naked prehensile tip of the preceding form. There are three species of this genus, ranging over New Guinea and part of Australia.

The genus *Acrobates* includes two small species of flying phalangers which have narrow patagia extending from the elbow to the knee along the flank. The long fringing hairs borne by the patagium, together with

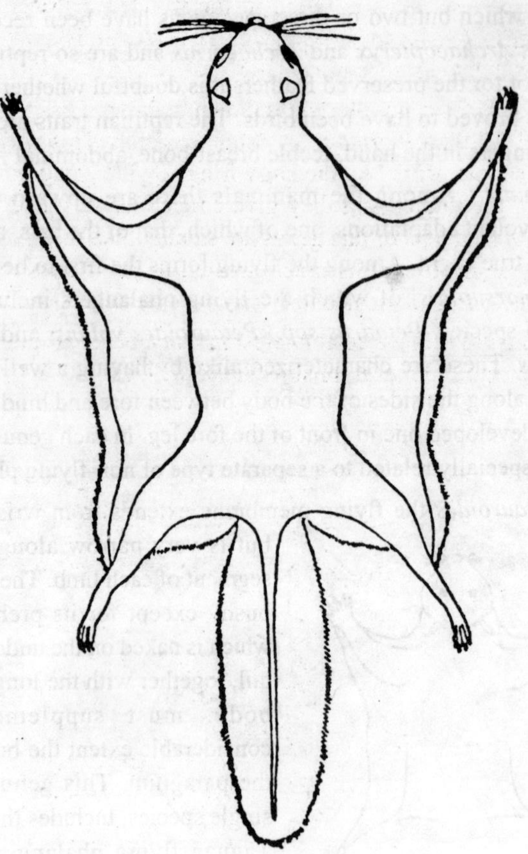

Fig.9.15: Flying squirrel, Sciuropterus volucella.

those on either side of the tail, aid materially in flight. *Acrobates pygmaeus* is found in New South Wales, Queensland and Victoria, while a second species, *A. pulchellus*, is a native of Papua.

Two families of *rodents* contain flying forms : the Anomaluridae, including the genus *Anomaluridae*, and the Sciuridae, of which three genera, *Pteromys, Sciuropterus,* and *Eupetaurus,* are volant. *Anomalurus* has a well developed patagium extending from wrist to ankle but narrowing in front of the leg from the knee down. As a compensation, however, there is an interfemoral membrane from the heel to slightly beyond the base of the tail. The genus includes six flying species, all found in Africa.

Ptermys has a highly developed patagium extending as far a the digits. There are also prepatgia and an interfemoral membrane, and the tail is large. The genus is found in the wooded districts of tropical southeastern Asia, Japan, and some of the Malasian Islands, and is said to soar through a distance of nearly 80 yards.

Sciuropterus has no interfeormal membrane, but has a much better developed tail than *Ptermoys* as a compensation. This wide hairy tail is further supplemented by the hairy fringe on the patagium and along the rear of the thighs, and they give collectively a broad supporting area. While members of this genus are smaller than those of *Pteromys*, their geographical range is much greater, as it includes the northern part of the North American and Eurasian contents, and India.

*Eupetaurus*n is of especial interest in that it is not arboreal but rock-and precipice-climbing. It inhabits the high elevations of northwestern Kashmir.

Fig.9.16: Bats.A, insectivorous, Vespertilio noctula; B, Frugivorous, Pteropus sp.

Among the *Insectivora, Galeopithecus,* the sole representative of the suborder Dermoptera, stands alone in its adaptation, for it exhibits the highest degree of aviation of any of the Mammalia except the bats, and while not of course ancestral to the latter, it is evidently derived from a common stock and gives a very clear idea of the manner in which the evolution of the bats was accomplished. In this genus the patagium reaches its highest development, as it extends from the rear of the head along the front of the arm (prepatagium), between the fingers to the base of the claws, between the fore and hind limb, webbing the toes as well as the fingers, and between the hind limbs and the tail (interfemoral membrane), including the entire length of the latter organs as in the insectivorous bats (Microcheiroptera). The musculature and innervation of the patagium resemble those of the bats and differ decidedly from those of all other volant mammals. The hand is much larger than the foot, but the finger show no trace of elongation. If they did the entire creature would be still more batlike. As it is, the brain is midway in its development between that of a typical insectivore and that of a bat, and the alimentary canal is also, batlike except for an elongated colon or large intestine, which in the bats and birds is very short.

Galeopithecus is nocturnal, as are most volant mammals, resting suspended, head down, from a branch. Its soaring powers are very great, for Wallace tells us of a record of 70 yards with a descent of not more than 35 or 40 feet, or less than one in five. The genus includes two species : *Galeopithecus volans,* from the Peninsula, Sumatra, and Borneo, and *G. philippinensis*, which inhabits the Philippine Islands.

The Gliding Skill of Birds

Much of our understanding of the gliding flight of birds comes from the work of Colin Pennycuick, an English zooligist who is now a professor in Miami. He set out to discover how well birds can glide by comparing their gliding abilities with those of human-made gliders and model aircraft. He started by observing seabirds from a cliff top but soon decided to attempt more accurate measurements in his laboratory, using domestic pigeons. There was not much flying space his laboratory, so instead of having his birds glide through still air he decided to attempt the flying equivalent of running on a moving belt : he would have a stationary bird gliding in moving air. Instead of the bird moving forward and downward through still air, the air would blow backward and upward past a stationary bird.

Pennycuick built a wind tunnel with a powerful electric fan that drove a jet of air through a nozzle of 1 meter diameter. The nozzle was tapered and grids arranged within to ensure that the air flowed out very smoothly, at almost exactly the same speed everywhere across the width of the nozzle. Now all that Pennycuick needed to perform his experiment was to train a bird to fly in the jet of air, at just the right speed relative to the air to keep it stationary relative to the laboratory. It may seem hard to imagine how a bird could be trained to do that, but Pennycuick achieved it with the help of a teaspoon bowl soldered to the end of a metal tube. He held the teaspoon in the jet of air where he wanted the bird's has to be and rolled chick peas down the tube so that they landed in the bowl. The only way the bird could get the food was by flying where Pennycuick wanted it to fly.

The wind tunnel was quite large and could blow air very fast, at speeds of up to 20 meters per second (45 miles per hour). Unable to find a suitable room for a tunnel of the size and power, Pennycuick hung it in the stairwell of a Victorian building at the University of Bristol, where he was working at the time. When he first switched it on it blew out a few windows, but the damage was soon repaired and the experiment worked well.

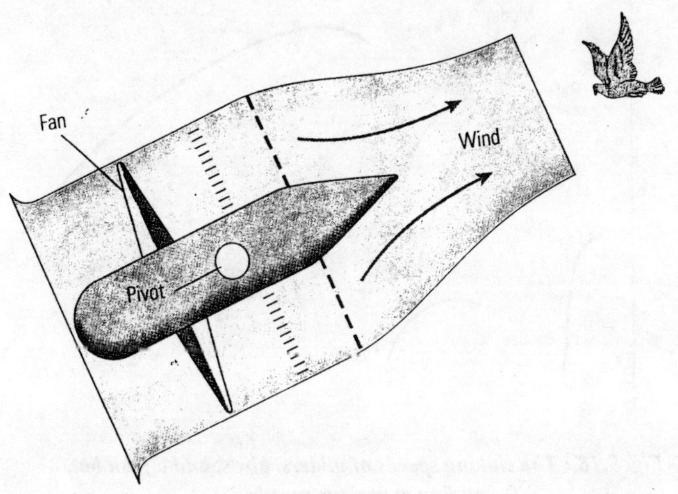

Fig. 9.17 : A wind tunnel used for testing the gliding ability of birds.

Pennycuick wanted to test birds ads they glided, not as they flapped their wings. If the wind tunnel was tilted up at a sufficiently large angle, which it had been constructed to do easily, and if it was blowing air fast enough, the bird did not need to flap its wings: it could simply glide into the wind (still remaining stationary relative to the laboratory). By varying the speed of the jet and the angle of tilt, Pennycuick was able to find the shallowest angle at which the bird could glide, at each speed. From that he calculated the rate at which it would have lost height if it had been gliding in still air.

The bird could not glide slower than its stalling speed of 8 meters per second (18 miles per hour). At high speeds it lost height rapidly. Like artificial gliders it had a minimum sink speed at which it lost height least quickly, but it performance was poor by engineering standards. Good gliders, both full-size ones and models, lose only about 0.5 meter of height per second at the minimum sink speed, but the pigeon apparently could lose no less than 2.5 meter per second. It is not clear whether pigeons really are very poor gliders or there was something about the experiment that prevented the pigeon from gliding well.

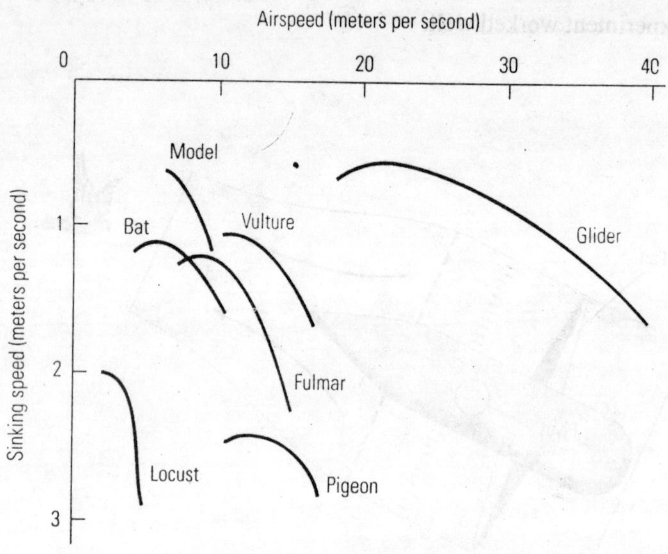

Fig. 9.18 : The sinking speeds of gliders, birds, and a fruit bat, gliding at various speeds.

Since that pioneering study, several similar experiments have been performed on other flying animals. Pennycuick himself moved his wind tunnel to Kenya and repeated the experiment on fruit bats (of about 0.55 meter, or 22 inches, wing span). Vance Tucker of Duke University repeated it with a small species of vulture and with a buzzard. All these, and fulmar petrels that Pennycuick had observed from a cliff top, did much better than the pigeon but slightly less well than artificial gliders : all lost height at minimum rate of around 1 meter per second. Pennycuick has pointed out that comparisons with gliders are really rather unfair: a gliding bird is not a proper glider, but is move like a powered aircraft with its engine switched off.

Pennycuick is a keen pilot, which is probably why he is so interested in bird flight. He was able to turn his piloting abilities to good advantage during several years he spent in Kenya, where he had a motor glider (an aircraft designed to glide, but with an engine that could be used to get it airbone or to get out of difficulties). Aloft in his aircraft, he glided with the vultures, storks, and pelicans over the East African plains, observing their behaviour in the air. By this means he was able to study the ability of these birds to glide for great distances in weather conditions that are especially favourable for that means of travel.

Thermal Soaring

Although vultures fly for most of the day, they flap their wings very little, keeping themselves airbone by soaring in thermals. These are columns or patches of rising air, formed over ground that has been heated by the sun: the hot ground heats the air, which expands, becomes less dense, and so rises. Thermals are often reasonably easy to find because they form in predictable places (for example, over bare rock). Their tops are often marked by the presence of cumulus clouds (the ones that look like cotton wool). Strong thermals do not appear until the sun is fairly high in the sky, and they disappear in the evening, so thermal soaring is possible only between those times. Vultures spend the night roosting on cliffs or trees.

Vultures and other birds circle upward in thermals to gain the height they need to start a straight glide. They circle by banking, tilting so that one wing tip is higher than the other. The tilt gives the lift a horizontal component that serves as centripetal force. If the birds tried circling in still air, they would lose height—indeed, they would lose height faster than if they were gliding in a straight path, because circling requires more lift (and therefore more induced drag) that does straight gliding. A bird circling in a thermal is able to rise if the air in the thermal is rising faster than the bird would sink in still air. For example, if a bird in a typical thermal is sinking relative to the air at 1 meter per second, and the air in the thermal is rising at 4 meters per second, the bird will gain height at a rate of 3 meters per second. It is common for soaring birds to gain height at such rates.

Vultures travel by gliding from thermal to thermal, sometimes over long distances. For example, Ruppel's griffon vulture nests on cliffs at the edge of the Serengeti and, in the breeding season, commutes daily between its nest and the herds of wildebeest and other mammals from which it takes its food. The vultures soar over the herds, watching for deaths so that they can feed on the careasses.

Pennycuick used his glider to follow vultures on their daily journeys. On one occasion he stayed with a Ruppel's griffon vulture for 96 minutes while it travelled back to its nest from the herds. In that time it covered 75 kilometers (47 miles) entirely by soaring, stopping to circle in only six thermals and rising to heights of up to 1500 meters (5000 feet) above the ground.

Even more impressive are the much longer journeys of white storks. These birds glide from the thermal to thermal for much of their annual migrations between Europe and southern Africa. A straight route would take them over the Mediterranean, where there are no thermals, but they manage to get help from thermals almost all the way by taking a detour. The western European populations travel over the Straits of Gibraltar and the eastern ones over suez.

Slope Soaring

Slope soaring is an alternative to thermal soaring, much used by seabirds. Gulls can often be seen soaring over a hillside or cliff, gliding backward and forward without beating their wings. They are held aloft by the upward air movements formed as the slope of the ground deflects wind upward. If the air is rising as fast as they would sink in still air, they can maintain their height.

The same technique works well over the open ocean, on the windy side of large waves. Albatrosses soar over the Antarctic Ocean, watching for fish or squid swimming near the surface. When they spot one, they often land on the sea to feed. Although they depend largely on slope soaring, they also use another soaring technique that involves swooping up and down between the strong wind high above the sea and the slower wind near its surface.

Pennycuick has made many observations of slope soaring using an instrument that he invented, called the ornithodolite. It is a combination of theodolite and rangefinder that will measure the compass bearing of a flying bird, its angle of elevation above the horizon, and its distance from the observer: with this information the bird's position can be fixed in three-dimensional space. Pennycuick has only to keep the instrument focused on the bird and press a button whenever he wants to record the bird's current position and the time in his portable computer. In that way he gets the information needed to calculate the speeds of soaring birds and their rates of change of height. He also measures the speed and angle of the wind, using an anemometer (wind gauge) attached to a pole as near as possible to where the birds are flying.

Pennycuick usually encountered the largest of albatrosses, the so-called wandering albatross, soaring between 2 and 12 above the waves. The wing span of this species measures more than 3 meters, and these impressive wings hold aloft an adult bird weighing from 8 to 10 kilograms. The bird would typically glide in still air at speeds of about 12 meters per second (27 miles per hour), but when it was slope soaring, gliding into the wind, it gravelled faster than 12 meters per second relative to the air and slower than that speed relative the sea. For example, an albatross gliding into a 10-meter-persecond headwind would typically travel at 16 meter per second relative to the air, gaining around at 16 – 10 = 6 meters per second. In very calm conditions, slope soaring could not keep these birds airborne and they had flap their wings occasionally, but unless it was very calm, they flapped their wings very seldom.

To stay in the upwardly deflected wind over a wave, the bird must soar along the wave—which may not be the direction in which it wants to travel. It can travel in other directions by taking a zigzag course, soaring for a while along a wave and gaining height or speed, then gliding for a while at an angle to the waves before joining another wave and soaring about it. Pennycuick observed albatrosses following a ship in this way and found that the total length of their zigzag paths averaged 1.5 times the straight-line distance that they travelled. On one occasion, for example, his ship was sailing directly into a 6-meter-persecond wind when a wandering albatross overtook it from astern. The bird changed direction every 10 to 15 seconds to follow a zigzag path and achieved a straight-line speed relative to the sea of 6 meters per second, entirely without flapping its wings.

The energy cost of gliding is much less than that of flapping flight, but even when the wings are not being flapped, metabolic energy is needed to maintain tension in their muscles. The main muscle involved is the pectoralis muscle, which pulls the wings down in the downstroke of flapping flight and holds them in position against the lift forces that act on them in gliding. Vultures, storks, and albatrosses have the pectoralis muscle divided into two distinct parts: a large superficial part that is dark red and a much smaller deep one that is paler. It is believed that the superficial part consists of fibers capable of shortening rapidly to beat the wing and the deep part of slowly contracting fibers that can maintain tension at little energy cost during gliding. Only in soaring birds is the muscle divided into these distinct parts. Albatrosses differ from the others in having what seems to be an even more effective device for saving energy while soaring. A fan-shaped ligament locks the shoulder joint when the wings are horizontal and fully spread. Consequently, the wings cannot be lifted above the horizontal until they are moved slightly back from the fully spread position. If the shoulder lock holds an albatross's wings horizontal the bird may be able to soar with very little tension in its muscles.

Whereas albatrosses travel long distances by slope soaring, small falcons known as kestrels have a different use for the technique. They hang stationary in the air, ready to pounce on any mice or voles that they see moving on the ground below. To hold themselves in place over level ground they fly into the wind, matching their speed exactly to that of the blowing air. Often, however, they slope soar over sloping ground, keeping themselves stationary without having to flap their wings.

Fig. 9.19 : Albatrosses

John Videler of Groningen University in the Netherlands has made many studies of Kestrel flight. He showed how precisely they stay in position by filming them with a camera on a tripod, locking the camera rigidly to the tripod as soon as it was focused in the bird. His films show that during 20 or 30 seconds the bird's head often moves less than a centimeter from its initial position relative to the ground. Those observations were of birds flapping their wings against the wind, but Videler has also made careful observations of slope soaring over a sea dike, against onshore winds avderaging 9 meters per second. The windward side of the dike sloped at 14 degrees to the horizontal, and at the height where the birds flew (on average, 6.5 meters above the ground), the wind was angled upward upward at about 7 degrees to the horizontal. When the bird remained stationary in this situation, its movements *relative to the air* and the forces on it were exactly the same as if it had been gliding in still air at 9 meters per second and at an angle of 7 degrees to the horizontal.

Some soaring birds seem to walk on water. These are the storm petrels, small seabirds of 30 to 40-grams. They hand suspended just above the sea surface with their wings spread but not flapping, dipping their feet into the water from time to time and looking out for the small fish and squid on which they feed. They are using their own peculiar soaring technique, seaanchor soaring, which works by the same principle as the flight of a kite. Their weight is supported by lift on their wings, while the drag exerted on them by the air is balanced by the forces that the water exerts on their feet. The bird is blown slowly backward over the water, so the drag on the feet is directed forward.

A bird that is good at slope soaring will not be particularly good at thermal soaring, and vice versa, because these two soaring techniques work best at different speeds. Many thermals are only a few tens of meters in diameter, and therefore thermal soaring requires the ability to turn in small circles. If thermal soarers circled at high speed, they would have to increase the lift on their wings greatly to obtain the necessary centripetal force. As a result, induced drag would increase and the birds would sink faster relative to the air. Accordingly, the best thermal soarers are birds that can circle very slowly. In contrast, good slope soarers can glide fast. To remain stationary as kestrels do, they must be able to glide as fast as the wind. To make headway against the wind as albatrosses do, they must be able to glide faster than the wind. Thermal soarers, then, should be good at slow gliding and slope soarers at fast gliding.

The two different soaring strategies are reflected in the areas of the wings. Both the minimum sink speed and the maximum range speed are proportional to the square root of wing loading, so thermal soarers should have low wing loadings (large wing areas for their weights) and slope soarers should have higher wing loadings. As expected, vultures have much bigger wing areas than albatrosses of the same weight; for example, Ruppel's griffon vultures averaging 7.6 kilograms mass (74 newtons weight) had an average wing area of 0.83 square meter, making the wing loading 90 newtons per square meter. Wandering albatrosses averaging 8.7 kilograms mass had 0.61 square-meter wings, giving a wing loading of 140 newtons per square meter.

Comparisons of this kind must be made between birds of similar size, because different-sized birds of similar habits have different wing loadings. To understand this, imagine two geometrically similar birds, one twice as long as the other. The longer bird will also be twice as wide and twice as high, and so it will be eight times as heavy. Its wings are twice as long and twice as broad, so have four times the area. Four times the wing area has to support eight times the weight; as a consequence, wing loading is twice as high for the longer bird as for the shorter. A smaller marine slope soarer like a 1-kilogram white-chinned petrel might be expected to have half the wing loading of an 8-kilogram albatross, which indeed it has. Similarly, small vultures have lower wing loadings than large ones. For any particular body mass, albatrosses and other slope soarers have smaller wing areas (and so large wing loadings) than vultures and other thermal soarers.

Wing span is the distance from one wing tip to the other, with the wings spread, and the chord is the distance from the front to the rear edge of the wing. Although vulture's wings have much larger areas than those of albatrosses of equal mass, their spans are a little less. The 7.6-kilogram vultures that we have been discussing had wing spans averaging 2.4-meters, and geometrically similar 8.7-kilogram vultures would have had 2.5-meter spans. In contrast, the 8.7-kilogram albatross had a span of 3.0 meters. Thus vultures have short broad wings and albatrosses have long narrow ones. The difference in shape can be described be calculating the aspect ratio, which is the span divided by the mean chord. The aspect ratio is 7 for the vulture that we have been discussing and 15 for the albatross.

As a general rule, the higher the aspect ratio the better the aerodynamic performance of a wing. The reason is that the bigger the span, the more air is driven downward each second. If two wings of different spans traveling at the same speed have to produce the same lift, the one with the longer span is able to do it by driving a larger mass of air downward, with a smaller downward velocity. It gives less kinetic energy to the air than the wing of smaller span, which gives a larger downward velocity to a smaller mass of air. For this reason, a bird with a longer wing span suffers less induced drag. A glider with a longer wing span will lose height less fast than a similar glider with a shorten span, especially at the low speeds at which induced drag is larger than profile drag.

Simple theory suggests that vultures would fly better if they had longer, possibly narrower wings. However, if, say an 8.7-kilogram vulture were to be given an albatross like aspect ratio without reducing its wing area, it would need wings with a 3.7-meter span, 22 percent more than the span of an albatross of the same mass. Wings that long might be difficult to manage when a vulture leaving a kill was taking off from the ground.

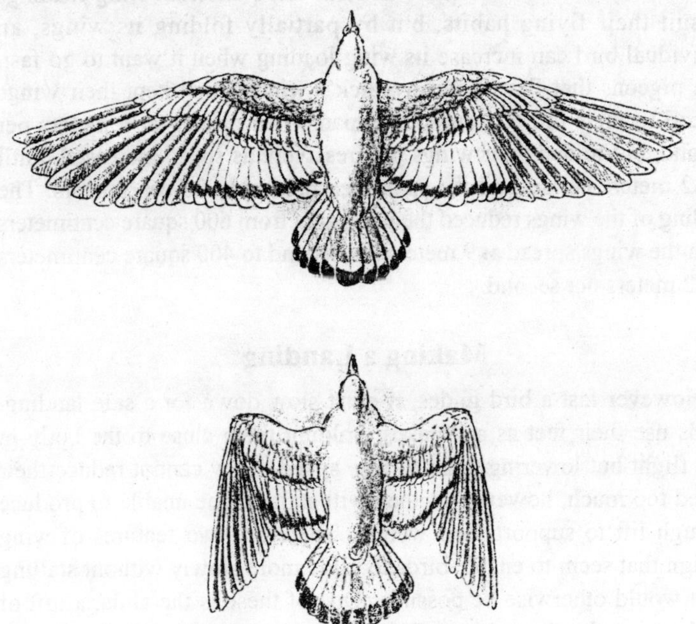

Fig. 9.20 : Birds spread their wings more when gliding slowly than when gliding fast. These drawings are based on photographs of pigeons gliding in Pennycuick's wind tunnel.

There is one group of birds that combines low, vulture-like wing loadings with high, albatross like aspect ratio, but the largest of them have a mass of only about 1.5 kilograms and a wing span of no more than about 2.3 meters, less than some vultures. These are the frigate birds, tropical seabirds with exceptional soaring habits. They live in the latitudes where the trade winds carry cool air from temperate regions over the warm tropical sea. Here, unlike at other latitudes, there are thermals over the sea, created as air is warmed by the sea surface and rises upward. Frigate birds soar in these thermals, apparently remaining airborne day and night, for they travel far out over the oceans but apparently never land on the sea surface. They feed on flying fishes and so squids that leap out of the water, and they also chase other birds and steal food from them. They need low wing loading for thermal soaring, and long wings are probably less of a problem than they would be for vultures, because frigate birds have no need to take off from the ground. They nest in treetops and can gain speed, when they take off, by diving out of the tree.

Not only do different kinds of birds have different wing loadings to suit their flying habits, but by partially folding its wings, an individual bird can increase its wing loading when it want to go fast. The pigeons that flew in Pennycuick's wind tunnel kept their wings spread to their full 65-centimeter span when gliding at 9 meters per second, but folded their wings progressively as speed increased, until at 22 meters per second the span measured only 25 centimeters. The folding of the wings reduced the wing area from 600 square centimeters with the wings spread at 9 meters per second to 400 square centimeters at 22 meters per second.

Making a Landing

However fast a bird glides, it must slow down for a safe landing. Birds use their feet as air brakes, holding them close to the body in fast flight but lowering them to lose speed. They cannot reduce their speed too much, however, or they will stall and be unable to produce enough lift to support their weight. There are two features of wing design that seem to enable birds to glide more slowly without stalling than would otherwise be possible. One of these is the alula, a tuft of feathers on the front edge of the wing supported by the bone of a rudimentary index finger. It lies flat against the rest of the wing in fast flight, but at low speeds is lifted and may help to keep the air flowing smoothly over the wing at angles of attack at which stalling would otherwise occur. The same principle is used in aircraft, in the device called a leading-edge slot. The stalling speed is probably also reduced by separating the large feathers at the wing tip, giving a multislotted effect. The feathers at the wing tips of crows, vultures, and many other birds separate in slow flight.

Even birds that have such devices seem to stall deliberately when landing: increasing the angle of attack until the wings stall is a very effective way of increasing drag. Stalling can be detected in photographs of landing birds because the irregular airflow over the upper surface of the wing ruffles the feathers.

There are other ways of slowing down, besides using the feet as air brakes or stalling the wings. Ducks land at high speed on ponds and slow themselves by holding their feet in the water. Guillemots landing on the cliffs where they nest approach at a lower level than the nest and veer upward, slowing themselves by converting their kinetic energy (energy of speed) into potential energy (energy of height) and stalling just before landing.

The birds that we have discussed so far all glide well: they can glide at shallow angles, losing height only slowly. There are other animals that glide much less well but are particularly interesting because they give us hints suggesting how birds and bats may have evolved the power of flight.

Flying Squirrels

These less good but still very effective gliders include flying squirrels which look much like ordinary squirrels but have flaps of skin between their fore and hind legs. By spreading their limbs they can open out this skin and stretch it tight, making a surprisingly effective airfoil. Much of what we know of the habits of flying squirrels come: from the observations of Keith Scholey, who while a Bristol University graduate student studied a species that live in Borneo. His base was a house in a forest clearing, from which he would watch the flying squirrels moving through the surrounding trees. They spend the day in their nests, in holes in dead trees, but came out at dusk to feed on leaves throughout the night. To find young leaves, which they prefer, they often had to travel some distance through the forest gliding from one tree to the next.

As dusk approached each evening, Scholey settled down on chair on the sun deck, waiting for the flying squirrels to emerge from their

to the bottom of the next tree quickly, at little energy cost.

Birds and bats must have evolved from ancestors that flew less well than they do, possibly gliding from tree to tree like flying squirrels and flying lizards. It is easy to imagine bats evolving from an ancestor that looked like a flying squirrel and that used its gliding ability either as flying squirrels do, to commute through the forest to trees bearing young leaves or fruit, or as *Draco* does, to get from tree to tree while searching for insects. There are two distinct groups of bats, and they are believed by some zoologists to have evolved separately. The largest number of species (including all the North American and European ones) are microbats. Most of these small bats feed on insects. Microbats usually catch their prey in the air, but they may have evolved from an ancestor with feeding habits like *Draco's*. The megabats of Africa and Asia are generally larger than microbats, as their name suggests: the biggest have a mass of about 1.4 kilograms and a wing span of 1.2 meters. They eat only plant food (largely fruit) and probably evolved from an ancestor that behaved like a flying squirrel, as well as looked like one.

Fig. 9.21 : Gliding Draco.

There are two theories about the evolution of birds. One is that ancestral birds lived in trees, moving from tree to tree by gliding. The other is that they were fast-running animals that lived on the ground. This theory suggests that these ground-dwelling birds leapt from time to time as they ran, spreading their wings and gliding to extend the leap. The problem with that suggestion is that wings were probably small in the early stages of evolution, so the wing loading would be high and the animals would have to travel very fast indeed to be able to glide. For example, it can be calculated from the wing loading of the giant red flying squirrel that it would probably stall at 13 meters per second. No flying bird or small mammals can run at such a speed, which is quite fast even for antelopes.

This chapter has been almost entirely about gliding vertebrates, because few insects make much use of gliding. The reason seems to be that small flying animals have low wing loading, and therefore their minimum sink and maximum range speeds are also low. They glide too slowly to be slope soarers, for they cannot make headway against any but the slowest winds. Although a slow speed is suitable for gliding up thermals, thermal soarers must be able to glide reasonably fast from one thermal to the next. It seems impossible to design a glider that will sink more slowly than 0.5 meter per second in still air, so a slow glider cannot travel very far before hitting the ground and might find it difficult to reach the next thermal. Vultures, storks, and albatrosses are large, and few soaring birds are small.

Despite these impediments, there are a few large butterflies that soar. Monarch butterflies (wing span 11 centimeters, 4.3 inches) soar while migrating between Canada and Maxico, although their maximum range speed is less than 3 meters per second and they lose a meter of height for every 4 meters gliding in still air. If the wind is against them they use flapping flight, keeping close to the ground where the wind speed is low, or else they do not fly at all: if they tried to soar against the wind, they would be blown in the opposite direction. On good days when the wind is behind them, however, they may spend as much as 80 percent of the time soaring, circling in thermals and slope soaring over building, letting the wind carry them along.

This chapter has shown how much animals can achieve, simply by gliding and soaring. Vultures soar in the thermals over the African plains, and storks migrate by thermal soaring between Europe and Africa. Slope soaring supports kestrels as they watch for prey and enables albatrosses to remain airborne day and night over the Antarctic Ocean. Even the lesser gliding skills of flying squirrels enable them to travel faster through the forest than if they had to run. Though so much can be done by gliding alone, even more is possible for animals that are capable of powered flight.

CHAPTER 10
SMALL-SCALE LOCOMOTION

The movements described in previous chapters are all driven by muscles, but very small organisms are powered by other motors. Many of these tiny creatures can be seen only through a microscope, and most have bodies that consist only of a single cell.

The bodies of animals are divided into cells, each controlled by a nucleus whose chromosomes carry a full set of genes, the inherited instructions for the processes of life. Typical cells have diameters between 10 and 100 micrometers (between one hundredth and one tenth of a millimeter).

The smallest living things consist of just one cell. We are not concerned in this chapter with the smallest living things of all, but mainly with the protistans, most of which are similar in size to the single cells of many-called animals. Some protistans (for example, amoebas) are traditionally thought of as animals, and others (such as diatoms) as plants, but the current fashion is to put all the protistans into a kingdom Protista, separate from both the animal and the plant kingdoms.

A few of the protistans are much larger than most cells. The giant amoeba *Pelomyxa* is about 2 millimeters long, much too large to be controlled easily by a single central nucleus: it has many nuclei, scattered through its single-celled body. Some ciliate protozoans such as *Spirostomum,* belonging to the group that we call the ciliates, are as much as 1 millimeter long, but in addition to an ordinary nucleus they have a long macronucleus extending through most of the length of the body and containing multiple copies of the genes.

This chapter describes the locomotion of these single-celled organisms, as well as that of some small many-celled animals that move in similar fashion. They include rotifers, which are common in ponds. Rotifers are similar in size to ciliate protozoans but have bodies made up of large numbers of tiny cells: about 1000 in the case of *Epiphanes*, which is about 0.5 millimeter long.

The Crawling of Amoebas

The amoebas are the most famous of the protistans. They look very simple under the microscope: the observer sees a transparent, irregularly shaped blob, and within that blob are visible several structures—a nucleus (or several), a few vacuoles containing food that is being digested, and a pulsating structure called the contractile vacuole, which pumps excess water out of the body. Although there is no permanent body part designed for locomotion, amoebas can crawl, but only slowly. The highest speed of an amoeba is only 5 micrometers per second, or 18 millimeters per hour.

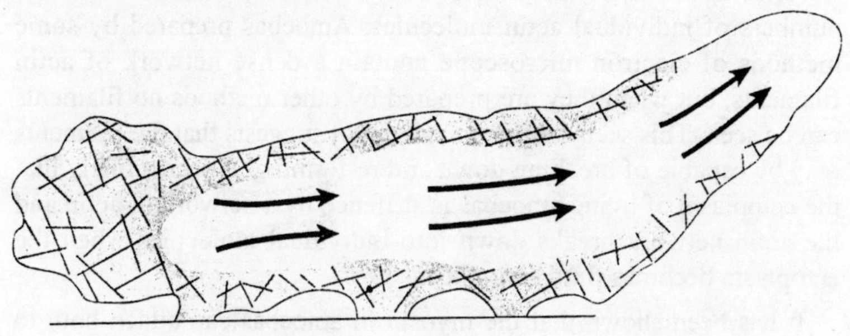

Fig. 10.1 : As an amoeba crawls (seen in profile), jellylike ectoplasm becomes fluid endoplasm at the rear end, and endoplasm becomes ectoplasm at the front.

This slow movement happens in a way that looks mysterious. A lobe (called a pseudopod) extends from the side of the body in the direction on motion, and the body contents flow into it. This empties the rear end of the body, which retracts so that the whole cell moves forward. We can see the flow of body contents by following the movements of the various kinds of particles that float within the amoeba. The core of the body (the endoplasm) flows like a liquid, with its particles swirling around freely. The outer layer (the ectoplasm), however, seems to be jellylike and anchored to the ground : particles in it remain stationary relative to each other and to the ground. New ectoplasm must form at the front of the amoeba as the pseudopod extends, and ectoplasm must be lost at the rear. Presumably the old ectoplasm at the rear is transformed into endoplasm, which flows forward and forms new ectoplasm at the front.

Although we cannot explain the movement of amoebas by pointing to structures that we would call muscles, these organisms do contain the major components of muscle. In Chapter 1 I explained how muscle contraction depends on the interaction between thick filaments of the protein myosin and thin filaments that consist largely of the protein actin. Amoebas contain similar proteins—so similar that myosin from vertebrate muscles attaches readily to amoeba actin.

The thick filaments of muscle are bundles of myosin molecules, each molecule shaped like a letter Y with a very long stem. The stems of the Y's form the filament itself and the arms project to form the crossbridges. The myosin of amoebas is not bundled into filaments, and the molecules are not Y-shaped. Each molecule consists of a single arm, similar to a muscle crossbridge, and a short stem.

Actin filaments, both of muscle and of amoebas, are built from large numbers of individual actin molecules. Amoebas prepared by some methods of electron microscope contain a dense network of actin filaments, but when they are prepared by other methods no filaments can be seen. This seems rather puzzling but suggests that the filaments may be capable of breaking down and re-forming. It seems likely that the ectoplasm of living amoebas is stiffened by a network of actin and the actin network breaks down into individual molecules when the ectoplasm becomes fluid endoplasm.

It has been shown that the myosin in amoebas can attach both to actin and to cell membranes. This ability suggests two possible mechanisms for extending the pseudopod. Myosin may form crossbridges between adjacent actin filaments and make them slide past each other, much as crossbridges in muscle slide thin filaments past thick ones. We saw in Chapter 1 how this process makes muscles shorten. Pseudopods may be made to extend by a similar process, working in reverse. Alternatively, the myosin may form bridges between actin filaments and the cell membrane and may make them move relative to each other. Either of these possibilities could be the basic mechanism of amoeboid crawling.

Yet another, very different mechanism has been suggested, and this mechanism does not involve the myosin at all. In a resting amoeba, all of the ectoplasm is presumably jellylike and filled with a network of actin filaments. Where a pseudopod is to form, the network must break down to allow the cell to bulge out. As the filaments break down into their constituent molecules, the concentration of molecule-sized particles in that part of the cell will be increased. Consequently, water will be drawn in by osmosis from other parts. The incoming water will make the cell membrane bulge out, forming a pseudopod which can be stiffened by allowing the network to reform. Amoebas are traditionally thought of as the simplest of animals, yet our knowledge of their movement is so uncertain that we cannot decide between these very different theories.

The Swimming of Flagellates

We have a much better understanding of the movement of another group of protistans, the flagellates. These tiny organisms are as important to life in the sea as green plants are to life on land. They are green like land plants, for they also contain chlorophyll and capture some of the sun's energy by photosynthesis, combining carbon dioxide with water to form sugars. Just as lice on land depends on the

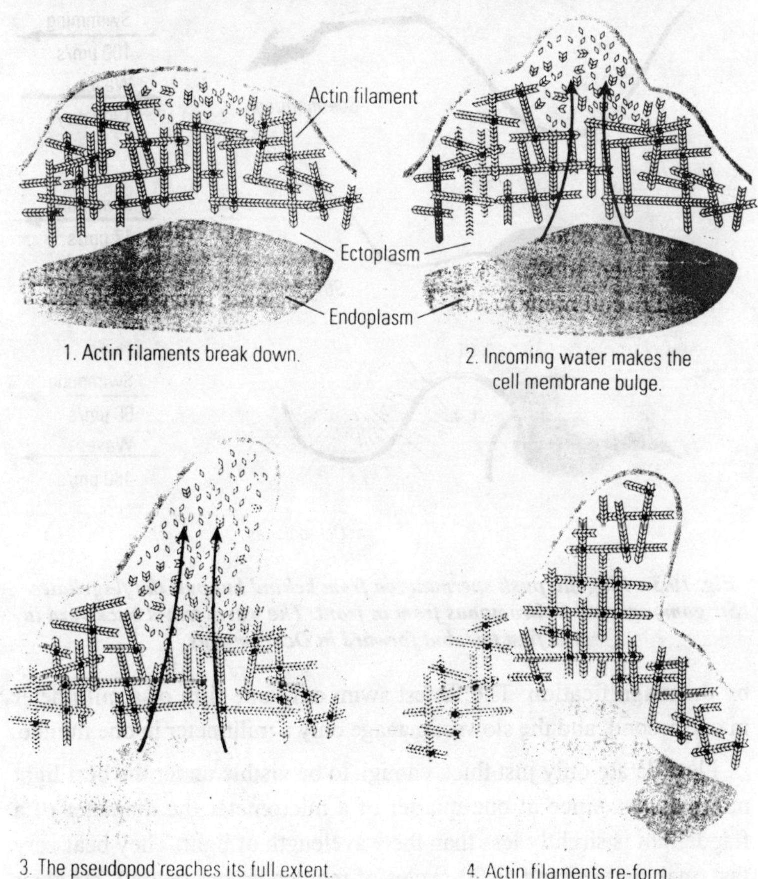

Fig. 10.2 : According to one possible mechanism for pseudopod formation, water would be drawn in osmotically when the actin filaments break down, extending the pseudopod.

photosynthesis of plants, life in the sea depends on photosynthesis by flagellates and other green protistans that float as plankton it its upper layers.

Flagellates swim by beating their flagella, which are one, two, or occasionally more long whiplike projections from their bodies, Spermatozoa swim in the same way: their tails are flagella. Both flagellates and spermatozoa may seem to swim quite fast when they are watched through a microscope, but their speed is an illusion created

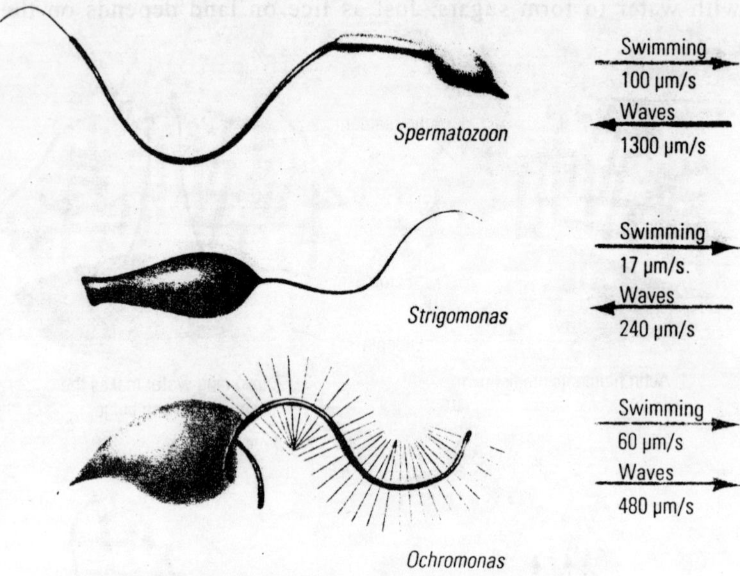

Fig. 10.3 : Flagella push spermatozoa from behind but pull the flagellates. Strigomonas and Ochromonas from in front. The waves travel backward in the first two, but forward in Ochromonas.

by the magnification. The fastest swim only one fifth of a millimeter in one second, and the slowest manage only a millimeter in one minute.

Flagella are only just thick enough to be visible under the best light microscopes, since at one quarter of a micrometer the diameter of a flagdellum is slightly less than the wavelength of light. They beat very fast, making as many as 70 cycles of movement per second, but their movements can be recorded either by high-speed moving photography or by multiple-flash still photography : in either case the camera must be attached to a high-quality microscope.

Flagellates swim in several different ways, but the flagellum usually undulates like the body of a swimming eel. Spermatozoa swim with the head in front, pushed along by the tail behind. Most flagellates swim with the flagellum in front, towing rather than pushing them.

Though the movements of spermatozoa look very like those of eels, there is an important mechanical difference. An eel or any other fish that suddenly stops making swimming movements will glide forward for some distance, slowing down only gradually. In contrast, a spermatozoon that stops undulating its tail comes to an immediate halt. The reason for the difference is that spermatozoa are so small and so slow. Because they are slow, they have little momentum to carry them forward : and because they are small, they have a large ration of surface area of volume, so the viscosity of the water slows them down very effectively. A shark that was shrunk to the size of a spermatozoan could still swim in water, but it would feel as if it were swimming in molasses. Sharks cannot stop suddenly, and many bony fishes spread their fins to slow themselves down, but spermatozoa have no need of brakes.

The dominant forces on swimming fishes and whales are inertial forces—that is, forces needed to accelerate the body or the water around it. In contrast, visosity dominates the forces on small, slow swimmers such as spermatozoa. This difference between fishes and spermatozoa can be highlighted by calculating the Reynolds number, the equivalent for swimming of the Froude number that was so important in our discussion of running. Remember that

$$\text{Froude number} = \frac{(\text{speed})^2}{\text{length} \times \text{gravitationalacceleration}}$$

It can be thought of as the ratio of inertial forces to gravitational forces. If gravity is important, different-sized bodies can move in dynamically similar fashion only if their Froude numbers are equal. Rather similarly, if viscosity is important, flow around different-sized bodies can be dynamically similar only if the bodies have equal Reynolds numbers:

$$\text{Reynold number} = \frac{\text{speed} \times \text{length} \times \text{fluid density}}{\text{fluidviscosity}}$$

The Reynolds number represents the ratio of inertial forces to viscous forces.

Spermatozoa and eels are both long, thin swimmers, but the patterns of flow of water around them cannot be the same (the flows cannot be dynamically similar) unless their Reynolds number are equal. For water, density/viscosity is 1 million seconds per square meter. Thus, a spermatozoon 100 micrometers (0.0001 meter) long swimming at 100 micrometers per second (0.0001 meter per second) would have a Reynolds number equal to 0.0001 × 0.0001 × 1,000,000 - 0.01. For a 1-meter eel swimming at 1 meter per second, the Reynolds number would be 1 × 1 × 1,000,000 = 1,000,000. Consequently, the patterns of flow of water around swimming spermatozoa and eels are very different, although the patterns of movement of their undulating tails are rather similar.

To swim forward, most flagellates send waves backward along their flagella. The flagellate *Ochromonas* is a peculiar exception : it sends waves forward along its flagellum, but nevertheless travels forward.

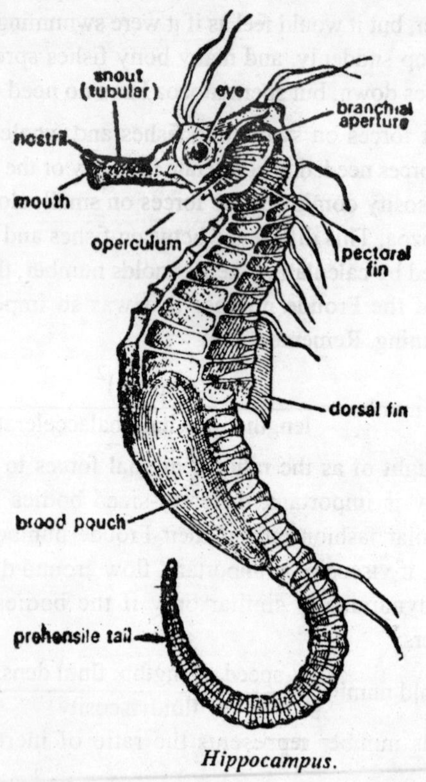

Hippocampus.

Fig. 10.4

The reason became apparent when its flagellum was examined by electron microscopy and found to be fringed by an enormous number of exceedingly fine "flimmer filaments." In aggregate, these filaments are many times longer than the flagellum, so the hydrodynamic forces that act on them have much more effect on the organism's motion than do the forces on the main axis of the flagellum.

Detailed analysis of the hydrodynamics of flagellate swimming shows that it works rather similarly. At low Reynolds numbers there is only half as much drag on a thin rod moving lengthwise through the water as there is on the same rod moving broadside-on at the same speed. The flimmer filaments lie in the plane of movement of the flagellum, at right angles to its main axis, so a movement that is lengthwise for the main axis is broadside-on for them, and vice-versa. For that reason, movements that would move an ordinary smooth flagellum backward drive one that has flimmer filaments forward.

Chlamydomonas and some other flagellates row with their flagella instead of undulating them. The flagella are held stiff during their backward power stroke so that they move broadside-on through the water and exert as much force as possible. Then, during the forward recovery stroke, they are allowed to bend so that they pull themselves lengthwise through the water and exert as little force as possible. The Reynolds number is too low for the organism's momentum to keep it moving forward during the recovery stroke, and it actually moves a little backward. The organism moves forward during the power stroke and a little backward during the recovery stroke, but the overall effect of each complete cycle of movement is to advance it through the water

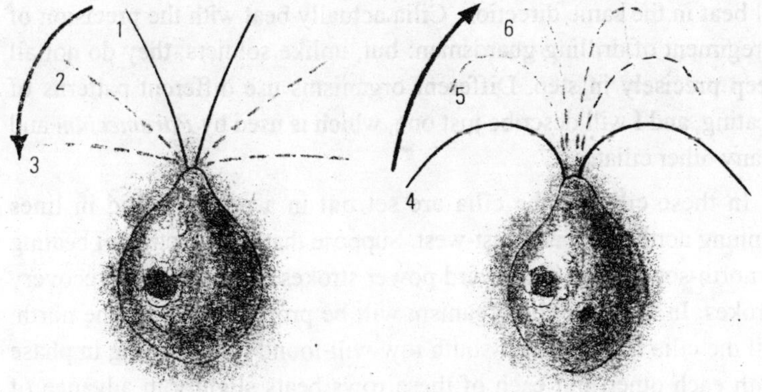

Fig. 10.5 : Chlamydomonas uses its two flagella to row itself along.

Rowing with Cilia

The ciliates are another important group of microorganisms. Many of them are planktonic but (unlike most flagellates) they have no chlorophyll. They do not photosynthesize like plants but feed like animals, eating flagellates and other microorganisms. The motors that power them are similar to the flagella of flagellates in some ways, but very different in others.

Flagella are generally longer than the cell that bears them, and each cell has only one flagellum or a few. Ciliate microorganisms bear cilia instead : these are just like flagella except that they are much shorter (considerably shorter than the cell) and very much more numerous, often covering the entire body. Indeed, they are so much more numerous that ciliates have much bigger motors that flagellates of similar size. For example, the flagellate *Euglena* and the ciliate *Tetrahymena* are each about 50 micrometers long. *Tetrahymena* has about 500 cilia, each 7 micrometers long, giving a total length of 3500 micrometers. The cilia of *Tetrahymena* have the same diameter as the flagellum of *Euglena* and the same internal structure, so *Tetrahymena* has 70 times more material to drive it than *Euglena* has. Not surprisingly, ciliates swim much a fater than flagellates. Whereas most flagellates swim at 0.02 to 0.2 millimeter per second, ciliates generally move at speeds between 0.4 and 2 millimeters per second.

Cilia make rowing movements rather like those of the flagella of *Chlamydomonas*, swinging stiffly back in their power stroke and bending much more in the recovery stroke. They are closely spaced and would collide if their beating were not coordinated, even if they all beat in the same direction. Cilia actually beat with the precision of a regiment of drilling guardsmen: but, unlike soldiers, they do not all keep precisely in step. Different organisms use different patterns of beating, and I will describe just one, which is used by *Paramecium* and many other ciliates.

In these ciliates, the cilia are set out in a regulate grid in lines running north-south and east-west. Suppose that the direction of beating is north-south, with southward power strokes and northward recovery strokes. In that case the organism will be propelled toward the north. All the cilia in each north-south row will found to be beating in phase with each other, but each of these rows beats slightly in advance of the next row toward the east. Waves of movement seem to travel from west to east like the ripples seen when wind blows over a field of wheat. The cilia seem to maintain coordination without any control being exerted by the organism: in stead, this coordination seems to depend on hydrodynamic interaction between the cilia—that is, on the tendency of one cilium to drag its neighbors along with it. The east-west waves are a consequence of the cilia beating asymmetrically, bending a little toward the east in the recovery stroke.

The viscosity of the surrounding water has some important consequences for swimming with cilia. If molasses is sandwiched between two flat plates, its vescosity resists sliding movements of one plate over the other. The faster the movement, the greater the resisting force. I might have added that the thinner the layer of molasses, the greater the resistance to movement at the same speed.

A field of cilia beating backward and forward sets up the same sort of movement in the water as the sliding plates set up in the molasses. We can think of the cell surface as one plate, and we can imagine the other plate at the level of the tips of the cilia. The major part of the hydrodynamic force on a beating cilium is the force created by the viscosity of the thin layer of water between the cell surface and the tips of the cilia. This is the force the cilia must overcome to move the organism. Just as less force is needed to slide one plate over the other (at the same speed) if the molasses layer is thick, long cilia need not exert as much force (at the same swimming speed) as short ones.

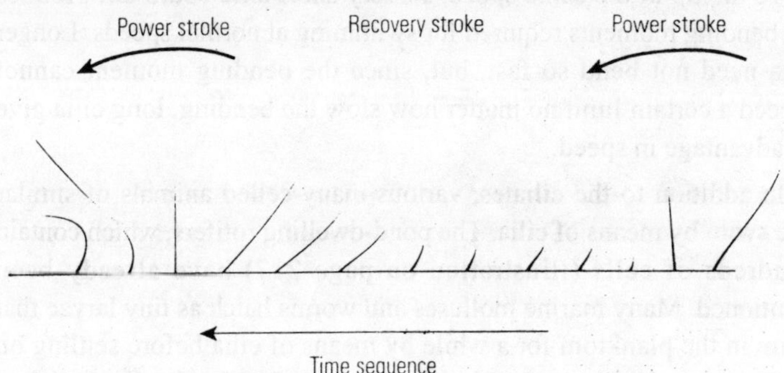

Fig. 10.6 : The rowing movements of a cilium.

You might conclude that the longer the cilia, the faster the organism should be able to swim, but this is true only up to a certain limit. The reason depends on the concept of bending moments. Suppose you are a rioter engaged in the dangerous sport of pulling down statuses of discredited politicians. You are more likely to succeed if you attach your rope to a statue's neck than if you tie it to the ankles. The higher the rope is attached, the more leverage you will have about the base, and the more likely you are to be able to topple the statue. The quantity that matters in this context is the "bending moment" about the base: the force in the rope multiplied by the distance from the rope to the base of the statue. In ciliates, doubling the length of the cilium halves the force required (for the same swimming speed), but it also doubles the distance from the cilium base at which the force must act. The bending moment at the base of the cilium is the force multiplied by the distance, so halving the force and doubling the distances leaves it unchanged. The bending moment that the culium can exert is limited, and consequently there is a limit to swimming speed that cannot be passed simply by making the cilia longer.

Although very long cilia cannot increase swimming speed, very short ones would reduce it. Remember from Chapter 1 that muscles cannot exert as much force when shortening fast as when shortening slowly. Similarly, cilia and flagella cannot exert bending moments as large when bending fast as when bending slowly. That faster bending movements are weaker was demonstrated by Kazuhiro Oiwa and Keiichi Takahashi of the University of Tokyo in an experiment of remarkable technical skill performed on the flagella of sea urchin spermatozoa. The shorter the cilium, the faster the base must bend to move the tip at the same speed, so very short cilia could not produce the bending moments required for swimming at normal speeds. Longer cilia need not bend so fast, but, since the bending moment cannot exceed a certain limit no matter how slow the bending, long cilia give no advantage in speed.

In addition to the ciliates, various many-celled animals of similar size swim by means of cilia. The pond-dwelling rotifers, which contain hundreds of cells (illustration on page 217) have already been mentioned. Many marine molluscs and worms hatch as tiny larvae that swim in the planktom for a while by means of cilia before settling on the bottom and developing their adult form. Unlike the single-celled ciliates that we have discussed so far, whose cilia cover the entire surface of their bodies, many-celled organisms may have only patches or bands of cilia. Rotifers, for example, have just a ring of cilia at the front end of the body. Many planktonic larvae have bands of cilia, and the rest of the body is bare.

Having bands of cilia instead of a complete covering can probably save energy. Most of the work required for swimming with cilia is needed to overcome the viscosity of the thin layer of water between the cell surface and the tips of the cilia. If the layer covers only part of the body, less work is needed.

Most of the many-celled animals that use cilia for swimming are no larger than the largest ciliates, 2 millimeters long or less. This is probably because ordinary cilia seem unable to move their possessors faster than about 2 millimeters per second, because the bending moments they can exert are limited. Larger animals can generally swim faster by using other means of propulsion. There is, however, one group of larger animals that swim faster by means of cilia, but these cilia are very remarkable. These animals are the ctenophores, gelatinous animals so transparent that they are difficult to see in water but extremely beautiful when you do see them. One example is the sea gooseberry *Pleurobrachis*, which has a diameter of about 15 millimeters and swims at about 8 millimeters per second. Instead of having separate cilia, it has comb plates, bundles of about 100,000 cilia that are packed very tightly together and beat as a unit. These cilia are up to 0.7 millimeter long, immensely longer than ordinary cilia, yet because there are so many in each comb plate, large bending moments can be exerted. The essential point about comb plates is that very large numbers of cilia are mounted on a very small area of the animal's surface.

Inside a Bending Cilium

It's clear how the bending of cilia and flagella propels organisms through liquid, but what makes these fibers bend in the first place? One possibility is that they are passive whips, moved from the base by some motor in the body of the cell. That suggestion was disposed of by Ken Machin of Cambridge University when he showed by mathematical analysis that waves of the shape seen in photographs of swimming spermatozoa could not be produced by waving a passive structure from one end. Also, flagella broken off their cells by centrifugation will swim in suitable solutions, suggesting that the mechanism moving the flagella lies inside these processes.

Cilia and flagella are barely thick enough to be visible by light microscopy, so electron microscopy is essential to reveal their internal structure. Sections examined in this way show exceedingly thin tubular structures, called microtubles, running lengthwise along the cilium or flagellum : two single microtubules in the center are surrounded by a

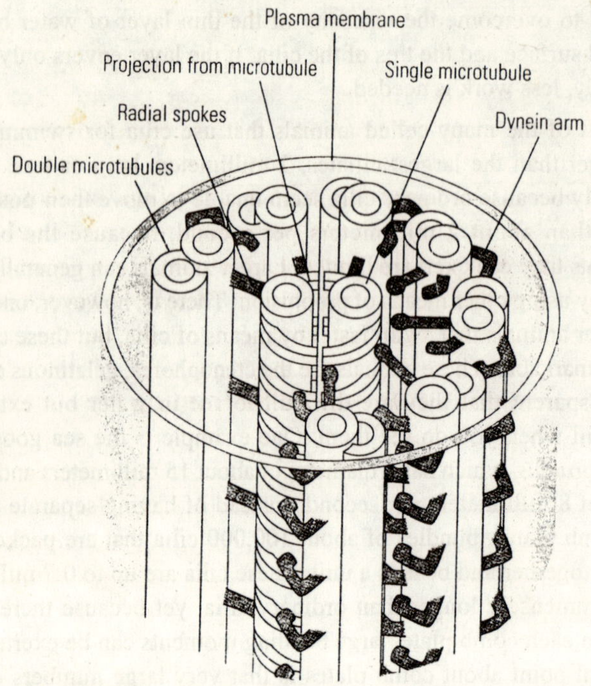

Fig. 10.7 : The internal structure of a cilium or flagellum.

ring of nine double microtubules. This nine-plus-two pattern is exactly the same in cilia as in flagella. Short projections called dynein arms extend from each double microtubule toward the next one: and longer projections called radial spokes extend in toward the center.

With this structure, cilia and flagella could work in various ways. One possibility is that the double microtubules lengthen and shorten like muscle fibers : a flagellum would bend to the left if the microtubules along its left side shortened. Another is that the microtubules may remain constant in length but slide past each other, driven by the dynein arms in much the same way as the crossbridges of muscle slide the thin filaments past the thick ones. This second possibility was confirmed by Fred Warner and Peter Satir of Syracuse University, New York, and the University of California, Berkeley, who studied cilia from the gills of freshwater mussels. From electron micrographs of sections showing cilia cut longitudinally, they showed that the radial spokes on each double microtubule are arranged in groups of three. When the cilium is straight, groups of spokes from microtubules on opposite sides of the cilium are in register. When it bends, the spacing of the spokes remains unchanged (showing that the microtubules are not lengthening or shortening), but groups of spokes from microtubules on the inside of the bend are not always directly opposite those from microtubules on the outside. For example, in the diagram on this page the groups of spokes numbered 1 are in register but groups 2, 3, and upward are progressively more staggered so that group 7 on the inside of the bend comes into register with group 8 on the outside. The sliding of neighbouring filaments past each other is presumably driven by the dynein arms.

Sinking Down and Swimming Up

The flagellates and other green protistans that live in the sea, getting their energy by photosynthesis, must stay reasonably near the surface to survive. If they sink too deep, the light will be too dim for photosynthesis and they will perish. They can survive only down to 100 meters or so in the clear water of the oceans, and they must remain even nearer the surface in murky coastal waters.

These organisms face a challenge in remaining near the surface: because they are denser than seawater, commonly about 5 percent denser, they sink in still water. You might think that they must swim

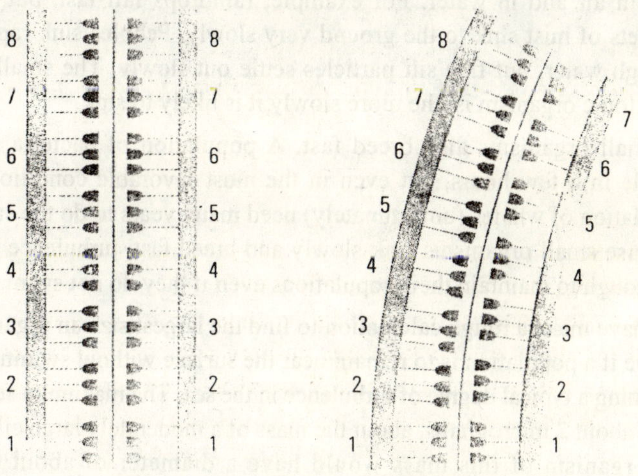

Fig. 10.8 : Groups of radial spokes on opposite sides of a cilium move out of register when it bends.

constantly upward to keep themselves in the brightly lit upper layer of the sea, where photosynthesis is possible. Unexpectedly, that is not necessarily the case.

The water in the sea is not still but turbulent, full of swirling eddies driven by wind, waves, and currents. In any particular interval of time a protistan is most likely to sink, but it could rise instead if it were moved by an eddy. As some cells are carried upward by eddies others will be hastened in their descent, and the average sinking rate will be the same as if the water were still. Given good luck, a protistan may meet a succession of eddies that will keep it near the surface for a long time, but it must eventually sink to the dark depths if it does not swim. Turbulence alone is not enough to maintain a population in the upper waters. Some other influence is needed, and that turns out to be reproduction.

Protistans generally reproduce by splitting into two halves, each of which becomes a half-sized protistan and then grows. If the cells that are moved upward by eddies reproduce fast enough, they may be able to replace the ones lost by sinking. Turbulence and reproduction together may be able to maintain the population. The more slowly the cells sink and the faster they reproduce, the more likely a population will survive near the surface.

Small objects sink more slowly than large ones of equal density, both in air and in water. For example, raindrops fall fast, but fine droplets of mist sink to the ground very slowly. Pebbles sink rapidly through water, but fine silt particles settle out slowly. The smaller a planktonic organism is, the more slowly it is likely to sink.

Small organisms also breed fast. A population of bacteria may double in a few hours, but even in the most favorable conditions a population of whales (unfortunately) need many years to do the same. Because small organisms sink slowly and breed fast, turbulence may be enough to maintain their populations even if they do not swim.

I have make a rough calculation to find the largest size an organism can be if a population is to remain near the surface without swimming, assuming a typical degree of turbulence in the sea. The maximum seems to be about 2 micrograms, about the mass of a moderately large ciliate. An organism of this mass would have a diameter of about 0.15 millimeter: it would probably sink in still water at rate of about 0.6 millimeter per second; and its generation time would most likely be about 22 hours. Flagellates generally seem to be small enough to survive in the surface waters without swimming, despite being denser than the water, but most of the larger plankton that feed on them must swim.

Some planktonic protistans have no means of swimming and appear to stay high in the water by just these means. Yet although flagellates are small enough to do likewise, they can and do swim. Many of them (*Chlamydomonas* is an example) are bottom-heavy, denser at the hind end and less dense at the front end bearing the flagella. This means that when they swim in still water, gravity will tend to pull them into a bottom-down position, and their swimming action will move them upward. Swimming counteracts the tendency to sink, reducing the losses that would otherwise occur.

Swimming by flagellates gives rises to some curious effects that have been investigated by John Kessles of the University of Arizona and Tim Pedley of Leeds University. When Kessler lectures about these effects, he likes to show a very simple experiment. He sets up a vertical glass tube with a tap at the bottom so that he can allow liquid to flow slowly down through the tube. He takes water containing a high concentration of *Chlamydomonas* (their chlorophyll makes the suspension bright green) and stirs it well, then pours it into the top of the tube. As it flows down, the cells become concentrated in the tube's center, leaving clear water around a green core.

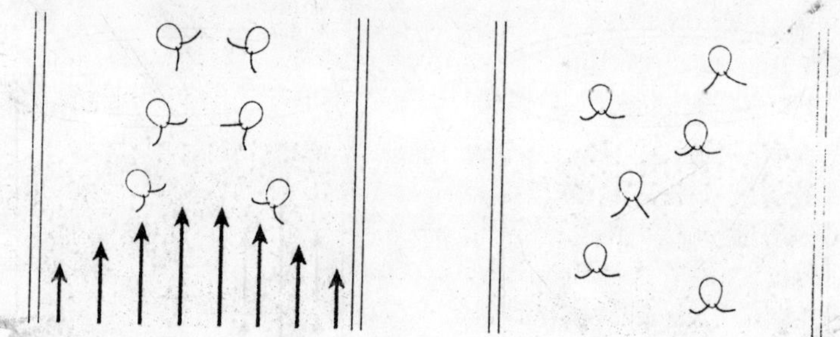

Fig. 10.9 : Chlamydomonas swim vertically upward in still water in a tube (left). However, if the water is made to flow downward (right), it rotates the organsisms, making them swim toward the center.

This phenomenon is called hydrodynamic focusing. It happens because fluid flows relatively slowly near the walls of a tube, where it is slowed down by viscosity, and fastest near the center. The bottom heavy protistans would swim vertically upward if the water were still, but they are swimming against a current that is faster on one side of the cell (toward the center of the tube) and slower on the other (toward the wall). This uneven current swivels the cells, turning them toward the center of the tube so that their swimming moves them toward the center. Hydrodynamic focusing is the result of their swimming, of their being bottom-heavy, and of the faster flow in the center of the tube.

In another simple experiment. John Kessler fills a glass beaker with a suspension of *Chlamydomonas*, stirs it well, and leaves it sitting on a laboratory bench. At first the liquid is uniformly green, but within a few minutes vertical streaks of deeper green become distinct from the paler green of the rest of the suspension. The reason is that the upward swimming of the flagellates concentrates them near the surface. Because they are denser than water, the uppermost layer of the suspension becomes denser than the rest, an unstable situation. Parts of the dense suspension start sinking from the surface, and as they sink, hydrodynamic focusing draws in nearby cells. As more cells are drawn into the sinking regions, the regions become even denser and sink faster. The result is a pattern of dense, deep-green plumes sinking through the much paler bulk of the suspension. The pale parts must rise to make room for the sinking plumes, and the water is set circulating in a pattern very much like the convection currents in a heated saucepan. Indeed, the phenomenon is called bioconvection. In shallow dishes, bioconvection can set up strikingly regular patterns, as an illustration shows.

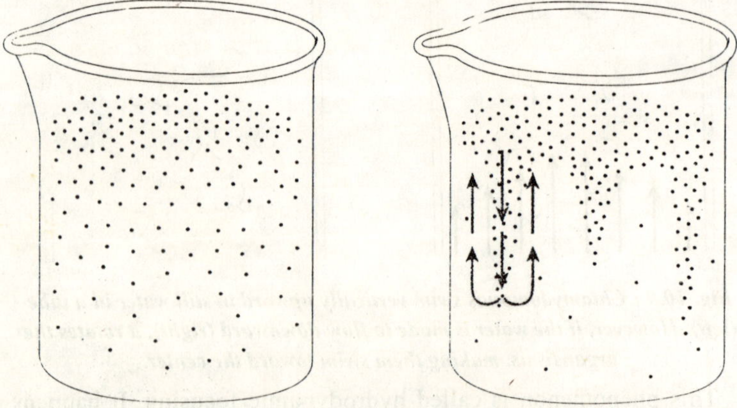

Fig. 10.10 : Bioconvection in a suspension of a flagellate such as Chlamydomonas. The concentration of cells at the surface (left) leads to instability and convection (right).

The tiny organisms that we have been discussing in this chapter move without muscles. Amoebas crawl by forming pseudopods and flowing into them. Flagellate protozoans swin by undulating their long flagella or by moving them like oars. Ciliate protozoans are driven by large numbers of cilia, which are much shorter than flagella but are powered by the same internal mechanism. All these organisms can swim only slowly and so are at the mercy of water currents and turbulence, but some tiny plankton that do not swim depend on turbulence for their very survival : it stirs up the water so that they do not all sink to the dark depths where they could not photosynthesize. Finally, we have seen how the swimming of green flagellates can make them clump together in strange patterns, by the process of bioconvection.